W9-ADZ-305

Growth Factors in Cell and Developmental Biology

Proceedings of the British Society for Cell Biology –
Journal of Cell Science Symposium
Manchester, April 1990

Organized and Edited by

M. D. Waterfield
Ludwig Institute for Cancer Research
Middlesex Hospital/University College London Branch

CARL A. RUDISILL LIBRARY
LENOIR-RHYNE COLLEGE

SUPPLEMENT 13 1990
JOURNAL OF CELL SCIENCE
Published by THE COMPANY OF BIOLOGISTS LIMITED, Cambridge

Typeset, Printed and Published by
THE COMPANY OF BIOLOGISTS LIMITED
Department of Zoology, University of Cambridge, Downing Street,
Cambridge CB2 3EJ

© The Company of Biologists Limited 1990

ISBN: 0 948601 27 2

QP
552
.G76
B7
1990
150864
Feb. 1991

JOURNAL OF CELL SCIENCE SUPPLEMENTS

This volume is the latest in a continuing series on important topics in cell and molecular biology. All supplements are available free to subscribers to *Journal of Cell Science* or may be purchased separately from Portland Press Ltd, PO Box 32, Commerce Way, Colchester CO2 8HP, UK.

1 (1984) Higher Order Structure in the Nucleus (US$23.00)
2 (1985) The Cell Surface in Plant Growth and Development (US$30.00)
3 (1985) Growth Factors: Structure and Function (US$30.00)
4 (1986) Prospects in Cell Biology (US$30.00) **SOLD OUT**
5 (1986) The Cytoskeleton: Cell Function and Organization (US$30.00) **SOLD OUT**
6 (1987) Molecular Biology of DNA Repair (US$70.00)
7 (1987) Virus Replication and Genome Interactions (US$70.00)
8 (1987) Cell Behaviour: Shape, Adhesion and Motility (US$60.00)
9 (1988) Macrophage Plasma Membrane Receptors: Structure and Function (US$50.00)
10 (1988) Stem Cells (US$65.00)
11 (1989) Protein Targeting (US$65.00)
12 (1989) The Cell Cycle (US$60.00)
13 (1990) Growth Factors in Cell and Developmental Biology (US$55.00)

WITHDRAWN
L. R. COLLEGE LIBRARY

CARL A. RUDISILL LIBRARY
LENOIR-RHYNE COLLEGE

Growth Factors in Cell and Developmental Biology

The cover illustration taken from the paper by E. Hafen and K. Basler is a histological section through the compound eye of *Drosophila* showing the pattern of rhodopsin-containing rhabdomeres in the photoreceptor cells.

ISBN: 0 948601 27 2

PREFACE

Major advances are taking place in our ability to elucidate the role of growth regulatory molecules, including those often referred to as growth factors, in the control of the generation of cell diversity, the proliferation of cells and in the specification of the overall patterns which determine embryological development and eventually produce the adult organism. The chapters of this volume illustrate the diversity of systems and the wide range of complementary experimental approaches that are playing a role in the elucidation of growth regulator structure, function and biology.

The introduction into growth factor studies of reporter molecules which can monitor changes in cytosolic ions and messengers in living cells illustrates how the sophisticated organic synthesis of such molecules used in conjunction with optical imaging techniques can reveal the intimate physiological changes in a single cell and identify new details of events – previously only seen as an average response in a population of cells. The use of techniques which are familiar to the biophysicist, but with which the biologist must become conversant, is illustrated by the series of papers on the structural characterisation of growth factors. Knowledge of 3-dimensional features which determine function is essential to understand and manipulate interacting proteins such as factors and receptors and will surely govern future experimental directions – particularly in the design of molecular anatgonists for use in the analysis or treatment of growth and differentiation that occurs in cancer and other diseases.

The use of recombinant DNA techniques to modify growth factors or their receptors is an essential adjunct to the biophysical and biochemical analysis of growth regulation, and the application of such technology to manipulate platelet derived growth factors is described in section II. Also in this section, we see the application of detailed cell biology and biochemical analysis to probe the receptor signal transduction process for the Bombesin system inside the cell. Such studies, which are representative of mitogenic stimulation and response systems, serve to probe the mechanisms which occur following receptor–ligand binding at its external surface.

An enormous number of growth regulatory polypeptides are now known to exist, but as the numbers grow we find that many fall into families which either have a functional (and not necessarily structural) or structural (and not necessarily functional) relationship to each other.

Some of the most important and clinically relevant advances have taken place in our understanding of the role of growth factors in haemopoiesis. Recent progress in this area, which is focused on the myeloid haemopoietic cells, is described in section III. Analysis of the biology of stem cells using embryological systems has provided remarkable overlaps with the study of haemopoiesis. The discovery of leukemia inhibiting factor which is described in section IV provides a striking example of the way different approaches and systems can lead to the charactisation of a multifunctional growth regulator.

The discovery of an ever-growing number of fibroblast growth factor- and transforming growth factor β-related genes, and the characterization of their biological roles in normal and cancer cells, illustrates the way that variants on a basic molecular theme provided by a growth regulator gene can arise and serve as a family of subtly diversified signal molecules.

Section VI provides key examples of the advantages to be gained for the analysis of growth regulation in the *Drosophila* system, which is amendable to molecular genetic as well as biochemical analysis.

Finally the importance of developmental regulators which are not peptides is well illustrated by recent advances that have taken place in our understanding of the control of limb development in the chick. Here in the final section (V) we see described the importance of retinoic acid–receptor interactions that provide a basis for a new definition of the term growth regulatory molecule that must include both polypeptide and non-polypeptide molecules.

M. D. Waterfield
Ludwig Institute for Cancer Research
Middlesex Hospital/University College London Branch

GROWTH FACTORS IN CELL AND DEVELOPMENTAL BIOLOGY

CONTENTS

I. BIOPHYSICAL APPROACHES TO THE STUDY OF GROWTH RECEPTOR STRUCTURE AND FUNCTION

1 HAROOTUNIAN, A. T., KAO, J. P. Y., ADAMS, S. R. and TSIEN, R. Y. Imaging and manipulation of cytosolic ions and messengers during cell activation

5 CAMPBELL, I. D. and COOKE, R. M. Structure function relationships in EGF, TGF-α and IGFI

11 JONES, E. Y., STUART, D. I. and WALKER, N. P. C. The structure of tumour necrosis factor – implications for biological function

19 McDONALD, N. Q., LAPATTO, R., MURRAY-RUST, J. and BLUNDELL, T. L. X-ray crystallographic studies on murine nerve growth factor

II. ANALYSIS OF GROWTH REGULATOR–RECEPTOR INTERACTIONS

31 LAROCHELLE, W. J., GIESE, N., MAY-SIROFF, M., ROBBINS, K. C. and AARONSON, S. A. Chimeric molecules map and dissociate the potent transforming and secretory properties of PDGF A and PDGF B

43 ROZENGURT, E., FABREGAT, I., COFFER, A., GIL, J. and SINNETT-SMITH, J. Mitogenic signalling through the bombesin receptor: role of a guanine nucleotide regulatory protein

III. MYELOID HAEMOPOIETIC GROWTH FACTORS

57 HEYWORTH, C. M., VALLANCE, S. J., WHETTON, A. D. and DEXTER, T. M. The biochemistry and biology of the myeloid haemopoietic cell growth factors

IV. GROWTH AND DIFFERENTIATION FACTORS FOR STEM CELLS

75 HEATH, J. K., SMITH, A. G., HSU, L.-W. and RATHJEN, P. D. Growth and differentiation factors of pluripotential stem cells

87 DICKSON, C., ACLAND, P., SMITH, R., DIXON, M., DEED, R., MACALLAN, D., WALTHER, W., FULLER-PACE, F., KIEFER, P. and PETERS, G. Characterization of *int*-2: a member of the fibroblast growth factor family

97 WESTERMANN, R., GROTHE, C. and UNSICKER, K. Basic fibroblast growth factor (bFGF), a multifunctional growth factor for neuroectodermal cells

119 SLACK, J. M. W. Growth factors as inducing agents in early *Xenopus* development

131 BOYD, F. T., CHEIFETZ, S., ANDRES, J., LAIHO, M. and MASSAGUÉ, J. Transforming growth factor-β receptors and binding proteoglycans

139 WAKEFIELD, L., KIM, S.-J., GLICK, A., WINOKUR, T., COLLETTA, A. and SPORN, M. Regulation of transforming growth factor-β subtypes by members of the steroid hormone superfamily

149 WOZNEY, J. M., ROSEN, V., BYRNE, M., CELESTE, A. J., MOUTSATSOS, I. and WANG, E. A. Growth factors influencing bone development

Continued overleaf

V. DECIPHERING THE CONTROL OF GROWTH REGULATION IN *DROSOPHILA*

157 HAFEN, E. and BASLER, K. Mechanisms of positional signalling in the developing eye of *Drosophila* studied by ectopic expression of *sevenless* and *rough*

169 BRYANT, P. J. and SCHMIDT, O. The genetic control of cell proliferation in *Drosophila* imaginal discs

VI. LIMB DEVELOPMENT AND REGENERATION

191 BROCKES, J. P. Retinoic acid and limb regeneration

199 WOLPERT, L. Signals in limb development: STOP, GO, STAY and POSITION

209 INDEX

J. Cell Sci. Suppl. 13, 1–4 (1990)

Printed in Great Britain © The Company of Biologists Limited 1990

Imaging and manipulation of cytosolic ions and messengers during cell activation

ALEC T. HAROOTUNIAN, JOSEPH P. Y. KAO, STEPHEN R. ADAMS AND ROGER Y. TSIEN

Howard Hughes Medical Institute M-047 and Department of Pharmacology, University of California, San Diego, La Jolla, CA 92093, USA

Summary

Optical methods have recently become available for continuously imaging the free concentrations of important ions and second messengers such as calcium, sodium and hydrogen inside living cells. These ion levels are found to undergo remarkable changes upon stimulation of quiescent cells with growth factors known to stimulate phosphoinositide breakdown. In serum-starved REF-52 fibroblasts, growth factors such as serum, vasopressin, or PDGF (platelet-derived growth factor) cause intracellular $[Na^+]$ to increase from about 4 mM to 8 mM. If mitogen treatment is combined with pharmacological depolarization of the membrane potential, repetitive $[Ca^{2+}]_i$ spikes result in these rat fibroblasts. The mechanism of this oscillation has been investigated by light-flash release of intracellular messengers such as inositol 1,4,5-trisphosphate (Ins(1,4,5)P_3), Ca^{2+}, and diacylglycerol, as well as more traditional biochemical techniques. The key feedback pathway appears to be Ca^{2+}-stimulation of phospholipase C production of Ins(1,4,5)P_3.

Introduction

Recent technical advances now enable cytosolic free Ca^{2+} concentrations ($[Ca^{2+}]_i$), Na^+ concentrations ($[Na^+]_i$), and pH to be continuously imaged inside individual living cells with micrometer spatial resolution and subsecond time resolution (for recent reviews see Tsien, 1988, 1989). This methodology relies on the molecular engineering of indicator dyes whose fluorescence is strong and highly sensitive to those ions (Grynkiewicz *et al.* 1985; Minta and Tsien, 1989; Rink *et al.* 1982). Binding of these ions shifts the fluorescence spectrum of the corresponding indicator. The ratio of excitation or emission amplitudes at two wavelengths measures the free ion concentration while canceling out intensity variations due to non-uniform cell thickness or dye content. A fluorescence microscope equipped to acquire images at two wavelengths and ratio them can thus produce dynamic images of intracellular messenger levels.

Results and discussion

Currently our most fully developed use of this methodology is the analysis of the response of the REF-52 line of rat fibroblasts to mitogenic stimulation. Serum-starved cells typically show $[Na^+]_i$ of only about 4 mM. Within a few minutes after addition of serum vasopressin, or PDGF, $[Na^+]_i$ doubles to about 8 mM

Key words: cytosolic ions, messengers, cell activation.

(Harootunian *et al.* 1989*a*). This surprisingly large increase cannot be blocked by cytoplasmic alkalinization, a finding that suggests that the Na^+ influx is primarily due to pathways other than Na^+/H^+ exchange. When similar mitogenic stimulation using serum, vasopressin, bradykinin, thrombin, bombesin or ATP is combined with membrane potential depolarization by gramicidin, high $[K^+]$, or sodium pump blockage, repetitive spikes of high $[Ca^{2+}]_i$ result every few tens of seconds to minutes (Harootunian *et al.* 1988). Though these oscillations require static depolarization and Ca^{2+} influx, they do not involve cyclic fluctuations of membrane potential as in excitable cells, but rather rhythmic dumping of intracellular stores. Similar oscillations can be elicited without depolarization if GTPγS (guanosine-5′-*O*-(3-thiotriphosphate) or AlF_4^- are administered to activate G proteins (Harootunian *et al.* 1989*b*). Many other cases of $[Ca^{2+}]_i$ oscillations have been reported in other cell types (for reviews see Berridge and Galione, 1988; Berridge and Irvine, 1989; Rink and Hallam, 1989; Rink and Jacob, 1989), but the amenability of REF-52 cells to microinjection and the reproducibility and precise rhythm of their $[Ca^{2+}]_i$ spikes make them an unusually favorable system for analyzing their biochemical mechanism. An essential adjunct to single-cell imaging is the ability suddenly to release Ca^{2+}, inositol 1,4,5-trisphosphate (Ins(1,4,5)P_3), and diacylglycerol (DG) by photolysis of caged precursors. Ca^{2+} is released from the light-sensitive chelator nitr-7, whose high pre-photolysis affinity for Ca^{2+} makes it better for these experiments than the commercially available nitr-5 (Adams *et al.* 1988). Ins(1,4,5)P_3 is generated by photocleavage of its 1-(2-nitrophenyl)ethyl ester (Walker *et al.* 1987); *sn*-1,2-dioctanoylglycerol is released by irradiation of its 2-nitro-4,5-dimethoxybenzyl ether.

Experiments were designed to differentiate between several oscillation mechanisms already proposed in the literature: (1) Cobbold and collaborators (Woods *et al.* 1987; Berridge *et al.* 1988) have suggested that phosphorylations due to protein kinase C are the dominant negative feedback mechanism controlling the interspike interval. (2) Berridge and collaborators (Berridge *et al.* 1988; Berridge and Irvine, 1989; Goldbeter *et al.* 1990) have proposed that Ins(1,4,5)P_3 releases Ca^{2+} from one intracellular pool, causing Ca^{2+} overload and Ca^{2+}-dependent Ca^{2+} release (CICR) from an Ins(1,4,5)P_3-insensitive pool. (3) Payne *et al.* (1988) and Parker and Ivorra, (1990) have noted that high $[Ca^{2+}]_i$ may inhibit the ability of Ins(1,4,5)P_3 to release more Ca^{2+}; this negative feedback could generate oscillations if some kinetic delays were introduced. (4) Meyer and Stryer (1988) have suggested that positive cooperativity of Ins(1,4,5)P_3 in releasing Ca^{2+} from internal stores, together with Ca^{2+} stimulation of phospholipase C to generate more Ins(1,4,5)P_3 (Eberhard and Holz, 1988), would be a potent feedback loop capable of generating oscillations.

Our current data suggest that the Meyer and Stryer (1988) model, with some modifications, is the most likely to describe the $[Ca^{2+}]_i$ oscillations in REF-52 cells. This conclusion is based on the following key observations: (1) although phorbol esters do slow or inhibit oscillations, *cyclical* activation/deactivation of protein kinase C does not seem necessary for the rhythm generation, since flash photolytic generation of diacylglycerol slows the rhythm for several successive cycles, not just the first after the flash. Also, protein kinase C can be thoroughly down-regulated by12–24 h exposure to phorbol esters, after which oscillations can be produced as normal, though of course they can no longer be inhibited by reapplication of phorbol esters. (2) In the Berridge model, Ins(1,4,5)P_3 levels are elevated in a steady non-oscillatory manner upon agonist stimulation. Sudden delivery of additional Ins(1,4,5)P_3 should either have no effect, or, if it releases some Ca^{2+}, should advance the next cycle of CICR. In fact, photorelease of Ins(1,4,5)P_3 elevates $[Ca^{2+}]_i$ at all phases of the oscillation cycle and causes phase-resetting (and sometimes considerable delay) of the subsequent endogenous oscillations. Also, REF-52 cells are quite insensitive to the traditional pharmacological tests for CICR such as caffeine and ryanodine. Cells can be subjected to deliberate moderate to massive $[Ca^{2+}]_i$ elevation, for example by wounding with a micropipet, but oscillations never result from such Ca^{2+} overload alone, and still require addition of the mitogenic hormone, presumably to stimulate PIP_2 breakdown. (3) Photorelease of Ca^{2+} in oscillating cells causes not only an immediate but also a secondary rise in $[Ca^{2+}]_i$ about 10 s later; the second phase is blocked by heparin, which is known to block Ins(1,4,5)P_3 receptors (e.g. see Kobayashi *et al.* 1989). Also, in some microinjected cells, the endogenous oscillations can be seen to proceed as waves which are initiated at localized sites of high Ca^{2+}, due to imperfect resealing from the microinjection. Thus in REF-52 cells, Ca^{2+} exerts mainly *positive* feedback on its own release, and this is probably Ins(1,4,5)P_3 mediated. (4) GTPγS causes $[Ca^{2+}]_i$ oscillations that proceed at lower $[Ca^{2+}]_i$ levels and that are relatively independent of extracellular Ca^{2+}, an effect explicable by the reported ability of GTPγS to shift the Ca^{2+}-activation curve of phospholipase C to lower Ca^{2+} concentrations (e.g. see Taylor and Exton, 1987). (5) We are attempting to synchronize the $[Ca^{2+}]_i$ oscillations sufficiently so that direct assay of Ins(1,4,5)P_3, which necessarily must be done on large populations of cells, can show whether its levels oscillate as well.

Future investigations of intracellular signaling, especially to determine the functional significance of complex single-cell responses such as oscillations and spatial gradients, would be aided if further optical methods could be developed. For example, a light-sensitive Ca^{2+} chelator whose affinity could be reversibly cycled from high to low and back by different wavelengths of illumination (unlike current designs, which use irreversible photochemistry: Adams *et al.* 1988, 1989) would be useful for generating artificial oscillations and spatial gradients. Methods for measuring protein phosphorylation and reporter gene transcription at the single cell level would also be highly desirable to complement the imaging of second messenger signals.

References

Adams, S. R., Kao, J. P. Y., Grynkiewicz, G., Minta, A. and Tsien, R. Y. (1988). Biologically useful chelators that release Ca^{2+} upon illumination. *J. Am. chem. Soc.* **110**, 3212–3220.

Adams, S. R., Kao, J. P. Y. and Tsien, R. Y. (1989). Biologically useful chelators that take up Ca^{2+} upon illumination. *J. Amer. chem. Soc.* **111**, 7957–7968.

Berridge, M. J., Cobbold, P. H. and Cuthbertson, K. S. R. (1988). Spatial and temporal aspects of cell signalling. *Phil. Trans. R. Soc. Lond* B **320**, 325–343.

Berridge, M. J. and Galione, A. (1988). Cytosolic calcium oscillators. *Fedn Proc. Fedn Am. Socs exp. Biol. J.* **2**, 3074–3082.

Berridge, M. J. and Irvine, R. F. (1989). Inositol phosphates and cell signalling. *Nature* **341**, 197–205.

Eberhard, D. A. and Holz, R. W. (1988). Intracellular Ca^{2+} activates phospholipase C. *Trends Neurosci.* **11**, 517–520.

Goldbeter, A., Dupont, G. and Berridge, M. J. (1990). Minimal model for signal-induced Ca^{2+} oscillations and for their frequency encoding through protein phosphorylation. *Proc. natn. Acad. Sci. U.S.A.* **87**, 1461–1465.

Grynkiewicz, G., Poenie, M. and Tsien, R. Y. (1985). A new generation of Ca^{2+} indicators with greatly improved fluorescence properties. *J. biol. Chem.* **260**, 3440–3450.

Harootunian, A. T., Kao, J. P. Y. and Tsien, R. Y. (1988). Agonist-induced calcium oscillations in depolarized fibroblasts and their manipulation by photoreleased Ins(1,4,5)P_3, Ca^{2+} and Ca^{2+} buffer. *Cold Spring Harbor Symposia on Quantitative Biology* **53**, 945–953.

Harootunian, A. T., Kao, J. P. Y., Eckert, B. K. and Tsien, R. Y. (1989*a*). Fluorescence ratio imaging of cytosolic free Na^{+} in individual fibroblasts and lymphocytes. *J. biol. Chem.* **264**, 19 458–19 467.

Harootunian, A. T., Kao, J. P. Y. and Tsien, R. Y. (1989*b*). G-protein activation is sufficient to generate $[Ca^{2+}]_i$ oscillations in REF52 fibroblasts. *J. Cell Biol.* **109**, 99a (abstr.)

Kobayashi, S., Kitazawa, T., Somlyo, A. V. and Somlyo, A. P. (1989). Cytosolic heparin inhibits muscarinic and α-adrenergic Ca^{2+} release in smooth muscle. *J. biol. Chem.* **264**, 17 997–18 004.

Meyer, T. and Stryer, L. (1988). Molecular model for receptor-stimulated calcium spiking. *Proc. natn. Acad. Sci. U.S.A.* **85**, 5051–5055.

Minta, A. and Tsien, R. Y. (1989). Fluorescent indicators for cytosolic sodium. *J. biol. Chem.* **264**, 19 449–19 457.

Parker, I. and Ivorra, I. (1990). Inhibition by Ca^{2+} of inositol trisphosphate-mediated Ca^{2+} liberation: A possible mechanism for oscillatory release of Ca^{2+}. *Proc. natn. Acad. Sci. U.S.A.* **87**, 260–264.

Payne, R., Walz, B., Levy, S. and Fein, A. (1988). The localization of calcium release by inositol trisphosphate in *Limulus* photoreceptors and its control by negative feedback. *Phil. Trans. R. Soc. Lond.* B **320**, 359–379.

Rink, T. J. and Hallam, T. J. (1989). Calcium signalling in non-excitable cells: Notes on oscillations and store refilling. *Cell Calcium* **10**, 385–395.

Rink, T. J. and Jacob, R. (1989). Calcium oscillations in non-excitable cells. *Trends Neurosci.* **12**, 43–46.

Rink, T. J., Tsien, R. Y. and Pozzan, T. (1982). Cytoplasmic pH and free Mg^{2+} in lymphocytes. *J. Cell Biol.* **95**, 189–196.

Taylor, S. J. and Exton, J. H. (1987). Guanine-nucleotide and hormone regulation of polyphosphoinositide phospholipase C activity of rat liver plasma membranes. Bivalent-cation and phospholipid requirements. *Biochem. J.* **248**, 791–799.

Tsien, R. Y. (1988). Fluorescence measurement and photochemical manipulation of cytosolic free calcium. *Trends Neurosci.* **11**, 419–424.

Tsien, R. Y. (1989). Fluorescent probes of cell signaling. *A. Rev. Neurosci.* **12**, 227–253.

Walker, J. W., Somlyo, A. V., Goldman, Y. E., Somlyo, A. P. and Trentham, D. R. (1987). Kinetics of smooth and skeletal muscle activation by laser pulse photolysis of caged inositol 1,4,5-trisphosphate. *Nature* **327**, 249–252.

Woods, N. M., Cuthbertson, K. S. and Cobbold, P. H. (1987). Agonist-induced oscillations in cytoplasmic free calcium concentration in single rat hepatocytes. *Cell Calcium* **8**, 79–100.

J. Cell Sci. Suppl. 13, 5–10 (1990)
Printed in Great Britain *© The Company of Biologists Limited 1990*

Structure function relationships in EGF, TGF-α and IGFI

IAIN D. CAMPBELL AND ROBERT M. COOKE*
Department of Biochemistry, University of Oxford, South Parks Road, Oxford OX1 3QU, UK

Summary

The solution structures of the homologous growth factors hEGF and hTGF-α, have been determined independently from high resolution nuclear magnetic resonance (NMR) data. A model of the insulin-like growth factor structure based on insulin coordinates (Blundell *et al.* (1978) *Proc natn. Acad. Sci. U.S.A.* **75**, 180–184), has also been refined using molecular dynamics simulations with NMR-determined restraints. Knowledge of these structures, together with known sequences of other homologous proteins and experiments with site-specific residue changes, allows predictions to be made about growth factor residues which might be involved in the receptor–ligand interfaces.

Introduction

There is considerable pharmaceutical interest in mapping the parts of growth factors which interact with their receptors; this, in principle, could lead to the rational design of growth factor agonists and antagonists. The ultimate aim would be a complete structure of the growth factor–receptor complex but this is not likely to be available for some time. An alternative is to combine knowledge of growth factor structure with experiments on the receptor-binding properties of growth factor analogues, *e.g.* peptide fragments and various site-specific mutations. Often, however, there is no knowledge about the structure of the variant protein produced, although it is clearly important to know whether any observed change in biological activity is brought about by changes in local or global protein structure.

We have been studying the structures of three growth factors which have not proved amenable to X-ray crystallography because they do not crystallize readily. We have used high resolution nuclear magnetic resonance (NMR) to study recombinant human epidermal growth factor (hEGF), its homologue, transforming growth factor-alpha (hTGF-α) and insulin-like growth factor (IGFI). We are also beginning to investigate the effect of site-specific changes on the conformation of the proteins. Some of our recent work on hEGF, hTGF-α and IGFI will be briefly described and the implications for predictions and experiments to identify receptor–ligand interfaces will be outlined.

* Present address: Glaxo Group Research, Greenford Road, Greenford, Middlesex UB6 OHE, UK.

Key words: EGF, TGF-α, IGF, NMR, solution structure, site specific mutagenesis, receptor binding.

Structures

a) hEGF and hTGF-α

EGF and TGF-α are members of a family of homologous polypeptides with three disulphide bonds, which bind to the EGF receptor (Burgess, 1989). Many sequences of this family are known, with an overall amino-acid sequence homology of about 30%. In addition, there are many modules or domains of extracellular proteins which have sequences homologous to EGF (for recent surveys of the EGF family see, for example, Campbell *et al.* 1989, 1990; Burgess, 1989; Shoyab *et al.* 1989).

The structures of hEGF and hTGF-α have been determined independently in this laboratory (Cooke *et al.* 1987; Tappin *et al.* 1989; Campbell *et al.* 1989, 1990) using high resolution ^{1}H NMR and computer-based methods which have become established in recent years (Wüthrich, 1989; Cooke and Campbell, 1988). The molecule can be considered to consist of two domains, an N-terminal domain (1–32) and a C-terminal domain (32–53). The dominant motif is a double stranded β-sheet formed between residues 18 to 33. Three disulphide bonds radiate (up) from one face of this platform. The N-terminal strand is weakly associated to the main sheet. There is also a short anti-parallel β-sheet in the C-terminal domain. There are a number of loops and turns and intimate contacts between the N- and C-terminal domains, especially between the loop around positions 13–16 and the turn around positions 40–43.

There is good agreement between the various NMR studies which have been carried out on human, mouse and rat EGF and TGF-α (Cooke *et al.* 1987; Tappin *et al.* 1989; Campbell *et al.* 1989, 1990; Montelione *et al.* 1987; Mayo *et al.* 1989; Kohda *et al.* 1988). There is some evidence that the structures of the EGF family are relatively mobile compared to inhibitor proteins like bovine trypsin inhibitor. In both hEGF and TGF-α, few nuclear Overhauser effects (NOEs) are observed for the N- and C-terminal residues. This lack of information leads to a wider variation within the families of calculated structures in these regions. Much of this variation is probably a reflection of protein mobility, and this is supported by the observation of relatively narrow resonances from these parts of the molecule. These results, together with relatively fast amide exchange rates, suggest that the EGF family are relatively flexible molecules which are susceptible to conformational interconversions on a millisecond time scale.

b) IGFI

IGFI, which is highly homologous to insulin and IGFII, has 70 residues. The high degree of homology with insulin led to the production of a model of IGFI based on the X-ray structure of insulin (Blundell *et al.* 1978).

In a recent study we have assigned the NMR spectrum of IGFI and 344 distant restraints were obtained from NOE data. The NOE data were incorporated into restrained molecular dynamics simulations of the IGFI structure. This resulted in a refined structure which, although largely similar to the model, has some significant differences especially in the side chain orientations. Like the model, residues 3–6 adopt a β-strand-like conformation, the first helix is between residues 8–17, residues 23–26 are defined but extended, the second helix encompasses residues 44–49 and the third helix runs from residues 54–59. The protein has three disulphide bridges, Cys6–Cys48 is on the surface, while Cys18–Cys61 and Cys47–52 are buried in a hydrophobic protein core.

Some parts of the protein are well defined by the NMR restraints although other parts are not. The average root mean square deviation (RMSD) for the backbone atoms of residues 3–19, 22–26 and 43–61 was 0.17 nm in the various molecular dynamics simulations; if all parts of the protein are considered, this figure rises to 0.51 nm. As with the EGF family there is quite strong evidence, from the NMR spectra, that the ill-defined regions undergo considerable conformational flexibility.

The receptor binding surfaces

Now that structural information about these growth factors is available, it is useful to combine this with information from sequences and receptor-binding studies on analogues to predict the parts of the growth factors which might be in contact with their receptors.

a) EGF and TGF-α

By comparing the sequences of different polypeptides which are known to bind to the EGF receptor with those which form the growth factor-like modules in a wide variety of mosaic proteins, we have previously predicted residues on the EGF molecule which might bind to the receptor (Campbell *et al.* 1989, 1990). This was done on the assumption that the structures of all these members of the EGF family are similar and that the EGF modules do not bind to the receptor. These assumptions have been borne out to some extent by our experiments, since although EGF and TGF-α only share 40 % of residues their NMR structures are very similar. In addition, recent structural and calcium-binding studies on an EGF module from factor IX produced by recombinant techniques (Handford *et al.* 1990) show that while the structures of EGF and the EGF module are similar, the receptor binding activity was less than 1000 times that observed for EGF itself.

Residues 6, 14, (16), 18, 20, 31, 33, 36, 37, 39, 42 and (43) were predicted to be important for the integrity of the EGF structure while residues 13, 15, 41 and 47 were at the EGF/EGF receptor interface (the brackets indicate uncertainty about structural or functional roles) (Campbell *et al.* 1989, 1990). Although Y13, L15 and R41 are in different loops they are close to each other in space at the domain interface. L47, on the other hand, is significantly separated from the other three.

These predictions can be checked since a large number of studies have been carried out on the binding and mitogenic activity of variants of the EGF and TGF-α structures (Heath and Merrifield, 1986; Engler *et al.* 1988; Defeo-Jones *et al.* 1988). These studies are largely, although not entirely, consistent with the identification of residues 13, 15, 41 and 47 as interface residues. One difficulty is that in some cases a mutation might affect the overall structure of the molecules rather than a local change. One way to check this is to use high resolution NMR. We chose to do this with hEGF and have been producing the 1–52 wild type molecule with several mutations at the 'structural' and 'interface' sites defined above. One of the best studied residues in the EGF family is L47. Several studies have shown that deletion or change at this site seriously affects receptor binding. We have recently carried out receptor binding and NMR studies of four L47 mutants (V, A, D and E) (Dudgeon *et al.* 1990). In receptor binding assays, comparisons with wild type hEGF showed that L47 V bound approximately seven times more weakly while L47A, L47E and L47D bound approximately 50 times more weakly. These data are consistent with those of other groups (e.g. Defeo-Jones *et al.* 1988; Engler *et al.* 1988; Moy *et al.* 1989). We also carried out detailed 1D and 2D NMR studies and observed no major changes in structure, although minor effects were observed in the C-terminal region of the molecule. These results confirm the notion that L47 is a receptor interface residue and that it is not very important for the overall structure of the EGF molecule. In other cases, however, e.g. R41H, significant structural as well as receptor binding changes have been observed (Hommel, U., Cooke, R. M., Dudgeon, T. and Campbell, I. D., unpublished data).

b) IGFI

The IGFs are peculiar in that they bind to more than one receptor. Two distinct IGF receptors, type 1 and type 2, are found in many cell lines. In addition IGFI binds to the insulin receptor and a number of serum-binding proteins (Czech, 1989). Structural information about IGFI is sparse although much is known about its homologue insulin (Baker *et al.* 1988).

A number of modified IGF molecules have been investigated. The most extensive recent studies have been carried out by Cascieri and Bayne (1990). They investigated the binding of IGFI, and many variants, to the type 1 IGF receptor (from human placenta) the type 2 receptor (from rat liver), the insulin receptor (from human placenta) and human serum-binding proteins. They found analogues which retained binding to some of these receptors but lost activity in others. This implies that the structure of these proteins is not significantly changed, at least in some regions, by the amino acid substitutions.

It is of interest to identify, on our refined structure, the regions of IGFI which these studies suggest to be involved in binding to the various receptors. In summary these are: residues 1–3 and 49–51 for the binding proteins; 1, 2, 8, 9, 12, 49–51 for the type 2 receptor; 21, 23–25, 42–44, 46, 60, 62 for the insulin receptor (this list is also based on insulin results, Baker *et al.* 1988). There is overlap between the type 2 receptor and the binding protein surface and considerable overlap between the insulin and type 1 receptors. These two regions of overlap are on opposite sides of the IGFI molecule. Now that the NMR spectrum of IGFI has been assigned we are in a position to investigate, in more detail, some of the structural properties of the variant IGFs that are available. This should allow these regions to be defined more precisely.

Conclusions

The NMR method for determining protein structure has been shown to be viable for the growth factors EGF, TGF-α and IGFI. The structures of all three seem to be relatively flexible compared, for example, to a trypsin inhibitor, and sensitive to solution conditions such as pH and temperature. The growth factor–receptor interfaces also seem to involve relatively large patches on the surface of the growth factors with residues from widely different parts of the sequence. These observations imply that it may be rather difficult to make a stable small agonist or antagonist for these growth factors.

This is a contribution from the Oxford Centre for Molecular Science which is supported by SERC and MRC. We also thank ICI Pharmaceuticals, British Biotechnology and Monsanto for their financial and technical support. The work owes much to many colleagues in Oxford, including Martin Baron, Jonathan Boyd, Tim Harvey, Tim Dudgeon, Uli Hommel and Mike Tappin.

References

BAKER, E. N., BLUNDELL, T. L., CUTFIELD, J. F., CUTFIELD, S. M., DODSON, E. J., DODSON, G. G., CROWFOOT-HODGKIN, D. M., HUBBARD, R. E., ISAACS, N. W., RENOLDS, C. D., SAKABE, K., SAKABE, N. AND VIJAYAN, N. M. (1988). The structure of 2Zn pig insulin crystals at 1.5Å resolution. *Phil. Trans. R. Soc. Lond.* **319**, 369–456.

BLUNDELL, T. L., BEDARKER, S., RINDERKNICHT, E. AND HUMBEL, R. E. (1978). Insulin-like growth factor: a model for tertiary structure accounting for immunoreactivity and receptor binding. *Proc. natn. Acad. Sci. U.S.A.* **75**, 180–184.

BURGESS, A. W. (1989). Epidermal growth factor and transforming growth factor-α. *Br. Med. Bull.* **45**, 401–424.

CAMPBELL, I. D., COOKE, R. M., BARON, M., HARVEY, T. S. AND TAPPIN, M. J. (1989). The solution structures of EGF and TGF-α. *Prog. in Growth Factor Research* **1**, 13–22.

CAMPBELL, I. D., BARON, M., COOKE, R. M., DUDGEON, T. J., FALLON, A., HARVEY, T. S. AND TAPPIN, M. J. (1990). Structure function relationships in EGF and TGF-α. *Biochem. Pharmac.* in press.

CASCIERI, M. A. AND BAYNE, M. L. (1989). Identification of the domains of IGFI which interact with the IGF receptors and binding proteins. In *Molecular and Cellular Biology and Insulin-like Growth Factors and their Receptors* (ed. leRoith, D. and Raizada, M. K.) pp. 285–296, Plenum Press, New York.

COOKE, R. M. AND CAMPBELL, I. D. (1988). Protein structure determination by NMR. *Bioessays* **8**, 52–56.

COOKE, R. M., WILKINSON, A. J., BARON, M., PASTORE, A., TAPPIN, M. J., CAMPBELL, I. D., GREGORY, H. AND SHEARD, B. (1987). The solution structure of human epidermal growth factor. *Nature* **327**, 339–341.

CZECH, M. P. (1989). Signal transmission by the insulin-like growth factors. *Cell* **59**, 235–238.

DEFEO-JONES, D., TAI, J. Y., WEGRZYN, R. J., VUOCOLO, G. A., BAKER, A. E., PAYNE, L. S., GARSKY, V. M., OLIFF, A. AND RIEMEN, M. W. (1988). Structure-function analysis of synthetic and recombinant derivatives of TGF-α. *Molec. cell Biol.* **8**, 2999–3007.

DUDGEON, T. J., BARON, M., COOKE, R. M., CAMPBELL, I. D., EDWARDS, R. M. AND FALLON, A. (1990). Structure and function of hEGF: receptor binding and NMR. *FEBS Lett.* **261**, 392–396.

ENGLER, D. A., MATSUNAMI, R. K., CAMPION, S. R., STRINGER, C. D., STEVENS, A. AND NIYOGI, S.

K. (1988). Cloning of authentic human epidermal growth factor as a bacterial secretory protein and its initial structure-function analysis by site directed mutagenesis. *J. biol. Chem.* **263**, 12 384–12 390.

Handford, P., Baron, M., Mayhew, M., Willis, A., Beesley, T., Brownlee, G. G. and Campbell, I. D. (1990). The first EGF domain of Factor IX has a high affinity calcium binding site. *EMBO J.* **9**, 475–480.

Heath, W. F. and Merrifield, R. B. (1986). A synthetic approach to structure-function relationships in the murine EGF molecule. *Proc. natn. Acad. Sci. U.S.A.* **83**, 6367–6371.

Kohda, D., Shimida, I., Miyake, T., Fuwa, T. and Inagaki, F. (1988). Polypeptide chain fold of hTGF-α analogous to those of mouse and human epidermal growth factors as studied by 2D ^{1}H NMR. *Biochemistry* **28**, 953–958.

Mayo, K. H., Caballi, R. C., Peters, A. R., Boelens, R. and Kaptein, R. (1989). Sequence specific ^{1}H NMR assignments and peptide backbone conformation in rat EGF. *Biochem. J.* **257**, 197–205.

Montelione, G. T., Wüthrich, K., Nice, E. C., Burgess, A. W. and Sheraga, H. A. (1987). Solution structure of murine EGF; determination of the polypeptide backbone chain-fold by NMR and distance geometry. *Proc. natn. Acad. Sci. U.S.A.* **84**, 5226–5230.

Moy, F. L., Sheraga, H. A., Liu, J.-F., Wu, R. and Montelione, G. T. (1989). Conformational characterization of a single-site mutant of murine epidermal growth factor (EGF) by ^{1}H NMR provides evidence that leucine-47 is involved in the interactions with the EGF receptor. *Proc. natn. Acad. Sci. U.S.A.* **86**, 9836–9840.

Shoyab, M., Plowman, G. D., McDonald, V. L., Bradley, J. G. and Todaro, G. J. (1989). Structure and function of human amphiregulin: a member of the EGF family. *Science* **243**, 1074–1076.

Tappin, M. J., Cooke, R. M., Fitton, J. and Campbell, I. D. (1989). A high resolution H NMR study of hTGF-α: structure and pH dependent conformational interconversion. *Eur. J. Biochem.* **179**, 629–637.

Wüthrich, K. (1989). Protein structure determination in solution by NMR. *Science* **243**, 45–50.

J. Cell Sci. Suppl. 13, 11–18 (1990)
Printed in Great Britain © The Company of Biologists Limited 1990

The structure of tumour necrosis factor – implications for biological function

E. Y. JONES, D. I. STUART
Laboratory of Molecular Biophysics, University of Oxford, Oxford, OX1 3QU, UK

AND N. P. C. WALKER
BASF AG, Ludwigshafen, FRG

Summary

The three-dimensional structure of TNF has been determined at 0.29 nm using the technique of X-ray crystallography. Published data on site-directed mutagenesis and antibody binding may now be assessed in the light of the structure, thus the links between structure and function for TNF may be addressed.

TNF is a compact trimer composed of three identical subunits of 157 amino acids. The mainchain topology for a single subunit is essentially a β-sandwich structure formed by two antiparallel β-pleated sheets. This mainchain fold corresponds to the 'jelly roll' motif observed in viral coat proteins such as VP1, VP2 and VP3 of rhinovirus, or the hemagglutinin molecule of influenza. TNF is the first non-viral protein to contain this motif. The subunits associate tightly about a threefold axis interacting through a simple edge-to-face packing of the β-sandwich to form the solid, conical shaped trimer. A large number of the residues conserved between the amino acid sequences of TNF and lymphotoxin lie within the β-sandwich or at the threefold axis of the trimer. This implies the presence of the same β-sandwich motif in the lymphotoxin monomer and preservation of the edge-to-face mode of trimeric association.

The detailed three dimensional structure for TNF explains a wide range of observations, including data on antibody binding and site directed mutagenesis. The currently available evidence points to a region of biological importance situated at the interface between two subunits on the lower half of the trimer.

Introduction

The biological action of the cytokine tumour necrosis factor (TNF) is dependent on its interactions with other protein molecules such as cell surface receptors, inhibitors etc. These interactions are governed by the precise arrangement adopted within three dimensions by the amino acids of the linear polypeptide sequence (primary structure) on formation of the correctly folded tertiary structure. Thus in order to understand how the TNF molecule performs its biological function at the level of amino acid interactions, one must not only know the amino acid sequence but also the three dimensional structure.

The three dimensional structure of human TNF has been determined to a resolution of 0.29 nm by X-ray crystallography (Jones *et al.* 1989) and has been

Key words: tumour necrosis factor, X-ray crystallography, structure/function, receptor binding.

refined to a current R factor of 20.4% (on all data from 0.6 to 0.29 nm with no model for the solvent structure). We present here a brief summary of the general topology and structural characteristics of the TNF molecule followed by a discussion, in the light of the structure, of data published by other authors on site directed mutagenesis and antibody binding.

The structure

The overall shape of a single 157 amino acid subunit of the TNF trimer is wedge-like with height 5.5 nm and maximum breadth 3.5 nm near the base. The main-chain topology is illustrated in Fig. 1A; it is essentially a β-sandwich structure formed by two antiparallel β-pleated sheets. The mainchain fold conforms to that of the classic jelly roll motif (Richardson, 1981, Fig. 1B) common in viral capsid proteins. The nomenclature adopted in Fig. 1 for the labels of the secondary structural units follows the established convention for the viral structures (Rossmann *et al.* 1983).

The crystallographic data yield a measure of the relative flexibility of the various parts of the structure. The β-strands form a fairly inflexible scaffold; in particular the back β-sheet is situated at the core of the trimer and consequently is particularly rigid, and the C terminus is embedded in the base of this secondary structural unit. As would be expected it is the loops which adorn the outer, solvent-accessible surface of the molecule which exhibit high levels of flexibility/mobility. The N terminus is highly flexible and as far as residue 10 is rather independent of the rest of the molecule. Overall there is a general decrease in rigidity as the core becomes more loosely packed in the upper half of the molecule.

Three TNF monomeric subunits associate non-covalently to form a compact, conical trimer of length 5.5 nm and maximum breadth 5.0 nm. The β-strands of the three individual β-sandwiches lie approximately parallel to the threefold axis. The interaction between subunits related by the threefold axis is through a simple edge-to-face packing of the β-sandwich; the edge of the β-sandwich consisting of strands F and G from one subunit lies across the back β-sheet (GDIB) of a threefold-related subunit (see Fig. 2). This mode of packing produces an extremely tight association between the subunits. Thus the core of the trimer is completely inaccessible to solvent. Loss of solvent-accessible surface area may be equated to hydrophobic free energy gain: each 0.01 nm^2 of accessible surface area removed from contact with water gives a free energy gain of approximately 25 cal mol^{-1} (1 cal=4.184 Joules) (Chothia, 1974). The solvent accessible surface area buried for a single TNF subunit on formation of the trimer is over 20 nm^2, the energetic equivalent of about 30 hydrogen bonds. The stability of the TNF trimer is placed in perspective if one considers that for a typical antibody antigen complex the epitope comprises a solvent-accessible area of some 7 to 8 nm^2 (Tulip *et al.* 1989).

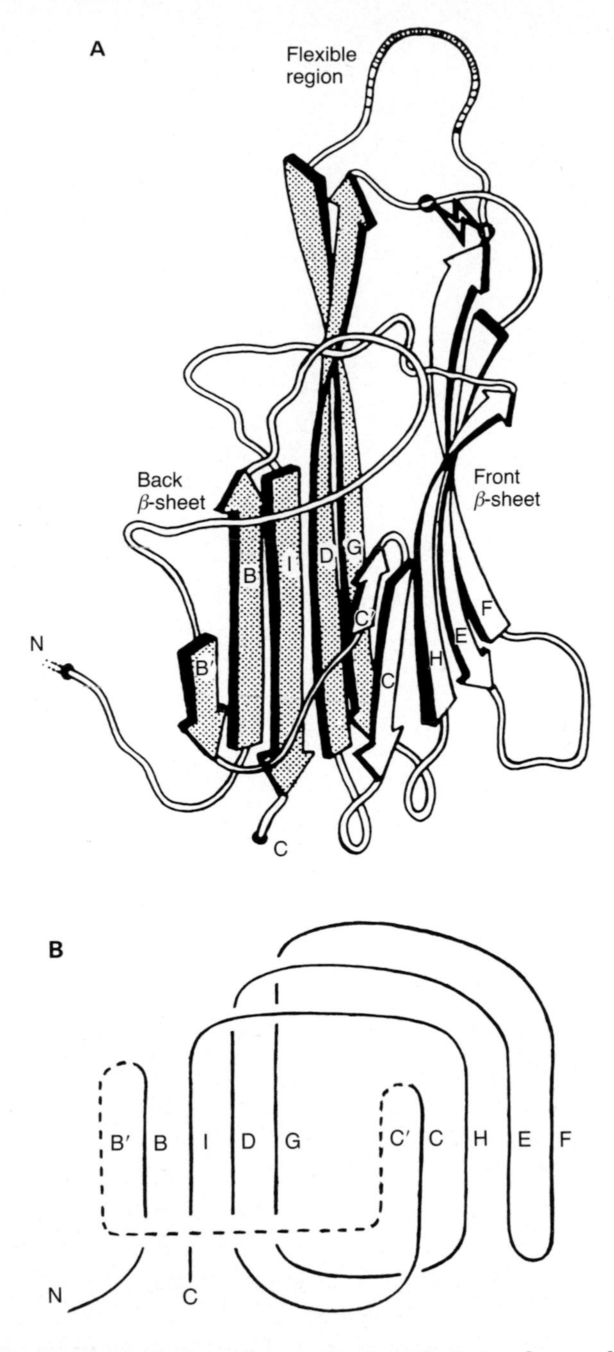

Fig. 1. A. Diagrammatic sketch of the subunit fold. β-strands are shown as thick arrows in the amino to carboxy direction and connecting loops are depicted as thin lines. The disulphide bridge is denoted by a lightening flash and a region of high flexibility is cross-hatched. The trimer threefold axis would be vertical for this orientation. B. The jelly-roll motif. The insertion between β-strands B and C is shown in dashed lines, normally the connection between B and C would run straight across at the top of the molecule.

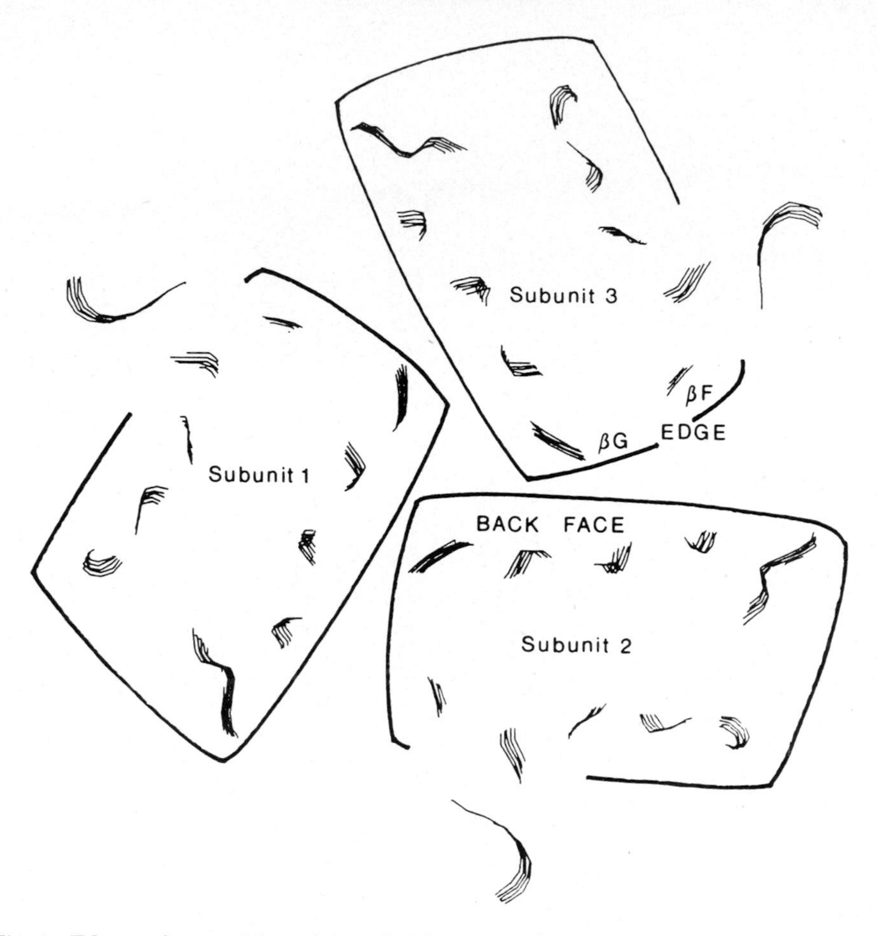

Fig. 2. Edge-to-face packing of β-sandwiches in the TNF trimer. The view, down the threefold axis, shows a narrow slab of the trimer with β-strands represented by ribbons running into and out of the page.

The overall distribution of residue types in the three dimensional structure of TNF echoes the general rule for proteins; namely the hydrophobic residues cluster in the core of the molecule whilst charged residues decorate the surface. Thus the core of the TNF β-sandwich has the expected filling of tightly intercalating, large, apolar residues. The detailed distribution of residue types becomes particularly noteworthy in the region near the trimeric threefold axis. At the top of the trimer the interaction between subunits involves charged sidechains, towards the center it is mediated by polar sidechains and at the base by a patch of large apolars. The threefold interaction at the heart of the trimer is produced by the edge-to-face packing of the aromatic rings from a nest of tyrosine residues (59, 119 and 151) centred on Tyr119. There is 28% sequence identity between sequences of TNF and lymphotoxin (Pennica *et al.* 1984). The majority of the residues conserved between the amino acid sequences of TNF and

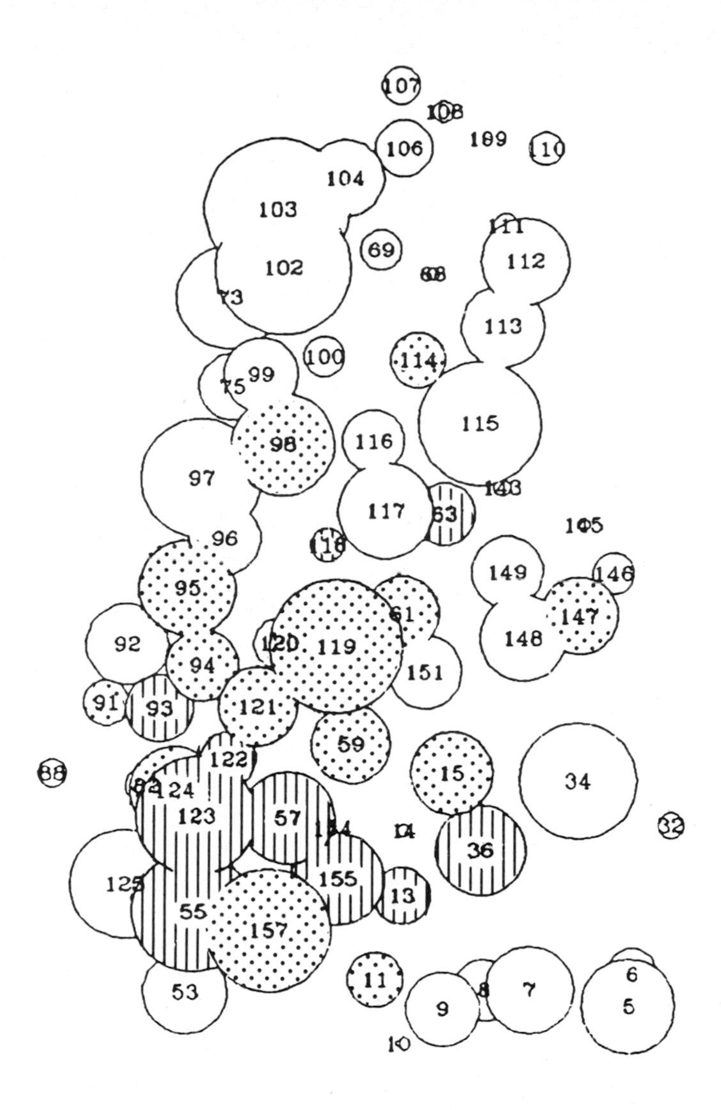

Fig. 3. Residues buried at the trimeric interface. The residues for a TNF monomer are represented by spheres, with area proportional to their loss of solvent-accessible area on the association of three such monomers to form a trimer. The view is onto the back β-sheet. Residues which are absolutely conserved in all TNF and lymphotoxin sequences are stippled. Residues which maintain a particular class identity (*e.g.* large apolars) are striped. Variable residues remain blank.

lymphotoxin may be classified, on the basis of the TNF structure, as internal to the β-sandwich or at the trimeric interface; the conservation of the residues which form the surface buried in trimer formation is illustrated in Fig. 3. This conservation in lymphotoxin of residues which play a key structural role in TNF strongly implies the presence of the same β-sandwich motif in the lymphotoxin monomer and preservation of the edge-to-face mode of trimeric association (Jones *et al.* 1990*a*).

Structure/function relationship

For the three-dimensional structure of TNF a set of prominent, surface residues, mainly from the loops connecting the β-strands of the β-sandwich, form the antibody-accessible surface of the trimer (Jones *et al.* 1990*b*). It has been observed that antibodies raised against TNF from one species (*e.g.* human) do not crossreact with TNF from another species (*e.g.* mouse), despite a sequence identity in excess of 80 % and the ability of TNF to bind to the TNF receptors of other species (Fiers *et al.* 1987). However, it is immediately obvious from the structure that the most sequence-variable regions of the molecule correspond to the antibody-accessible surface loops. Thus the epitope for an antibody against TNF will always contain some residues which will vary between species thus abolishing antibody binding. This implies that the characteristics of the interaction between TNF and its receptors must somehow differ from those required for binding of an antibody to TNF.

Neutralising antibodies have so far been mapped in two regions of TNF, amino acids 1–15 (Socher *et al.* 1987) and an epitope involving Arg131 (Fiers *et al.* 1990). However, antibodies are rather blunt instruments when employed to probe the properties of the TNF molecule; the relative sizes of the TNF trimer (3×157 residues) and a single antibody Fab fragment (about 400 residues) are comparable. A rather more specific probe of the structure/function relationship is provided by site-directed mutagenesis. Fiers *et al.* (1990) report more than a thousand fold loss of biological activity on mutating serine 86 to phenylalanine. A double mutation of the two external arginines 31 and 32 to asparagine and threonine by Tsujimoto *et al.* (1987) also resulted in a dramatic loss of biological activity. Yamamoto *et al.* (1989) have shown that site-directed mutagenesis of histidine 15 may abolish biological activity; this residue is involved in stabilising the threefold interaction in a region close to residues 31–32 and, in terms of the trimer, adjacent to serine 86. In contrast, mutation of tryptophan residues 28 and 114 to phenylalanines does not markedly affect biological activity (Van Ostade *et al.* 1988) nor does changing histides 73 and 78 to glutamines (Yamamoto *et al.* 1989). Finally Narachi *et al.* (1987) have disrupted the disulphide bridge between cysteines 69 and 101 without deleterious effects. This set of functionally tolerated mutations generally lies in the upper portion of the molecule.

In Fig. 4 the various pieces of evidence are combined in a schematic diagram of the trimer structure. From the results of site-directed mutagenesis a region of functional importance appears to be located at the interface between two subunits on the lower half of the trimer. Both neutralising antibodies could serve to block receptor binding throughout this general region by simple steric hindrance. In addition, Creasey *et al.* (1987) have reported that truncation of the N terminus by four or seven amino acids actually increases biological activity; this would be

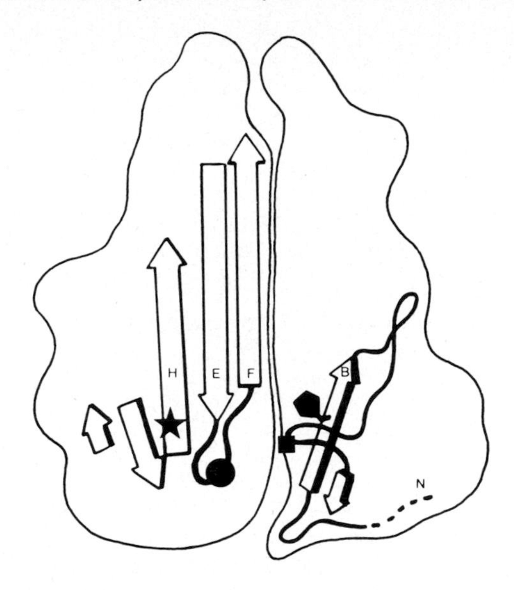

Fig. 4. Putative area of receptor binding. This schematic diagram highlights the relative positions of Ser 86 (circle) and Arg 131 (star) on one subunit, also His 15 (pentagon) and residues Arg 31 and 32 (square) on the neighbouring subunit in the trimer.

consistent with the extremely flexible/mobile N terminus causing a slight steric hindrance effect at a receptor site in the region of residues 15, 31–32 and 86. Thus current circumstantial evidence points to a receptor-binding site which involves the interface between two subunits in this region near the base of the TNF trimer.

References

Chothia, C. (1974). Hydrophobic bonding and accessible surface area in proteins. *Nature* **248**, 338–339.

Creasey, A. A., Doyle, L. V., Reynolds, M. T., Jung, T., Lin, L. S. and Vitt, C. R. (1987). Biological effects of recombinant human tumor necrosis factor and its novel muteins on tumor and normal cell lines. *Cancer Res.* **47**, 145–149.

Fiers, W., Beyaert, R., Brouckaert, P., Everaerdt, B., Haegeman, G., Libert, C., Suffys, P., Takahashi, N., Tavernier, J., Van Bladel, S., Vanhaesebroeck, B., Van Ostade, X. and Van Roy, F. (1990). *In vitro* and *in vivo* action of tumor necrosis factor. In *Tumor Necrosis Factor: Structure, Mechanism of Action, Role in Disease and Therapy* (ed. Bonavida, B. and Granger, G.), pp. 31–37. Karger, Basle.

Fiers, W., Brouckaert, P., Goldberg, A. L., Kettelhut, I., Suffys, P., Tavernier, J., Vanhaesebroeck, B. and Van Roy, F. (1987). Structure-function relationship of tumour necrosis factor and its mechanism of action. In *Tumour necrosis factor and related cytotoxins*. (Ciba Foundation Symposium **131**) pp. 109–123. Wiley, Chichester.

Jones, E. Y., Stuart, D. I. and Walker, N. P. C. (1989). Structure of Tumour Necrosis Factor. *Nature* **338**, 225–228.

Jones, E. Y., Stuart, D. I. and Walker, N. P. C. (1990*a*). The three-dimensional structure of TNF at 2.9 Å resolution. In *Tumor Necrosis Factor: Structure, Mechanism of Action, Role in Disease and Therapy* (ed. Bonavida, B. and Granger, G.), pp. 31–37. Karger, Basle.

Jones, E. Y., Stuart, D. I. and Walker, N. P. C. (1990*b*). The three-dimensional structure of tumour necrosis factor. In *Molecular and Cellular Biology of IL-1, TNF and Lipocortins in*

Inflammation and Differentiation. (Proceedings of the Sclavo International Conference, ed. L. Parente), pp. 321–327. Alan R. Liss: New York.

Narachi, M. A., Davis, J. M., Hsu, Y.-R. and Arakawa, T. (1987). Role of single disulfide in recombinant human tumor necrosis factor-α. *J. biol. Chem.* **262**, 13 107–13 110.

Pennica, D., Nedwin, G. E., Hayflick, J. S., Seeburg, P. S., Derynck, R., Palladino, M. A., Kohr, W. J., Aggarwal, B. B. and Goeddel, D. V. (1984). Human tumour necrosis factor: precursor structure, expression and homology to lymphotoxin. *Nature* **312**, 724–729.

Richardson, J. S. (1981). Protein anatomy. *Adv. Protein Chem.* **34**, 167–339.

Rossmann, M. G., Abad-Zapatero, C., Murthy, M. R. N., Liljas, L., Jones, T. A. and Strandberg, B. (1983). Structural comparisons of some small spherical plant viruses. *J. molec. Biol.* **165**, 711–736.

Socher, S. H., Riemen, M. W., Martinez, D., Friedman, A., Tai, J., Quintero, J. C., Garsky, V. and Oliff, A. (1987). Antibodies against amino acids 1–15 of tumor necrosis factor block its binding to cell-surface receptor. *Proc. natn. Acad. Sci. U.S.A.* **84**, 8829–8833.

Tsujimoto, M., Tanaka, S., Sakuragawa, Y., Tsuruoka, N., Funakoshi, K., Butsugan, T., Nakazato, H., Nishihara, T., Noguchi, T. and Vilcek, J. (1987). Comparative studies of the biological activities of human tumor necrosis factor and its derivatives. *J. Biochem., Tokyo.* **101**, 919–925.

Tulip, W. R., Varghese, J. N., Webster, R. G., Air, G. M., Laver, W. G. and Colman, P. M. (1989). *Cold Spring Harbor Symposium on Immune Recognition*. In press.

Van Ostade, X., Tavernier, J. and Fiers, W. (1988). Two conserved tryptophan residues of tumor necrosis factor and lymphotoxin are not involved in the biological activity. *FEBS Lett.* **238**, 347–352.

Yamamoto, R., Wang, A., Vitt, C. R. and Lin, L. S. (1989). Histidine-15: an important role in the cytotoxic activity of human tumor necrosis factor. *Protein Engineering* **2**, 553–558.

J. Cell Sci. Suppl. 13, 19–30 (1990)
Printed in Great Britain *© The Company of Biologists Limited 1990*

X-ray crystallographic studies on murine nerve growth factor

N. Q. McDONALD, R. LAPATTO, J. MURRAY-RUST AND T. L. BLUNDELL

Laboratory of Molecular Biology and ICRF Structural Molecular Biology Unit, Dept. of Crystallography, Birkbeck College, Malet St., London, WC1, UK

Summary

The largest and best characterised family of neurotrophic growth factors is that of nerve growth factor (NGF) and its relatives. In order to understand the relation of structure and function, we have undertaken X-ray analyses of murine NGF. The active component β-NGF crystallises as hexagonal bipyramids that give good X-ray diffraction data using a synchrotron to 2.3Å resolution. We have prepared several heavy atom derivatives that are being used in the method of multiple isomorphous replacement to solve the phase problem and determine the three-dimensional structure. We have also prepared crystals of the precursor, 7S NGF, which is a complex of three different subunits of composition $\alpha_2\beta_2\gamma_2$. We have collected X-ray data to 3Å resolution on two crystal forms with related cell dimensions and orthorhombic spacegroups. Detailed analyses of the structures of NGF in these crystal forms, taken together with data on sequence and biological activity, should give clues concerning the role of the precursor complex in storage and assist the identification of the surface region involved in receptor binding.

Introduction

During the development of the vertebrate nervous system, many populations of neurons, including the sympathetic and sensory neurons, depend for their survival on their interactions with target cells. These neuron–target cell interactions are controlled by specific proteins or neurotrophic factors, which are released by target cells in both the peripheral and central nervous system (Thoenen *et al.* 1985 and Levi-Montalcini, 1987). The best characterised neurotrophic factor is nerve growth factor (NGF), first purified from murine submaxillary gland by Cohen (1959). Three decades of intense study on the chemistry and biology of NGF have provided evidence for the trophic role that NGF plays both in the development and throughout the lifetime of the neuron.

When isolated from murine submaxillary gland, the active neurotrophic molecule, β-NGF, is found incorporated in a complex with three different subunits of composition $\alpha_2\beta_2\gamma_2$ (Darling, 1983). This is referred to as the 7S NGF complex, named after its sedimentation coefficient. γ-NGF is an esteropeptidase with a specificity for the C-terminal Arg-Arg sequence of β-NGF and it processes the β-NGF precursor (Edwards *et al.* 1988). The α-subunit is 80 % homologous to the γ-

Abbreviations used: The heavy atom derivatives for NGF are called $NaAuCl_4$ for sodium aurichloride; EMP, for ethyl mercury phosphate; $K_2Pt(NO_2)_4$, for platinum tetranitrate; PCMBS, for p-chloromercuricbenzenesulphonic acid.

Key words: X-ray diffraction, nerve growth factor, structure, neurotrophic, crystallisation.

NGF but lacks esteropeptidase activity. It is clear that both these subunits are related to the kallikrein family of serine proteases (Bothwell *et al.* 1979; Isackson *et al.* 1987).

The β-NGF protomer has a molecular weight of 12 650 M_r, comprises 118 amino acids and associates as a non-covalently bonded dimer (Server and Shooter, 1977). A single asparagine-linked complex oligosaccharide is present in about 2 % of murine NGF protomers, and contributes 10 % of the total mass of such a protomer (Murphy *et al.* 1989). Each protomer contains three disulphide bridges that give a strongly crosslinked, stable structure analogous to other growth factors (James and Bradshaw, 1978). The biological activity of β-NGF is strongly conformation dependent, since the reduction of the disulphide bonds abolishes NGF activity (Frazier *et al.* 1973). An analysis of the sequences of the NGF family of neurotrophic factors shows five clearly separated and highly conserved regions that may be involved in signal transduction (Fig. 1). The high similarity of sequences indicates similar three-dimensional structures. This is supported by the biological activities and immunological properties of β-NGF from different species (Harper and Theonen, 1980).

β-NGF is a highly basic protein with a pI of 9.3. Chemical modification and site specific mutagenesis have implicated several arginine residues in receptor binding (Bradshaw *et al.* 1976; Ibanez *et al.* 1990). A complementary surface on the NGF receptor must exist and it is likely that multiple ionic interactions are responsible for the high affinity constant observed for β-NGF (K_d=10^{-11} M). Indeed the recent sequencing of the rat and human NGF receptors (Radeke *et al.* 1987; Johnson *et al.* 1986) showed an overall negative charge distribution of amino acids in their extracellular domains. Other residues such as valine 21

```
                  ----V1---                                    -----V2----       --
Mouse  NGF SSTHPVFHMGEFSVCDSVSVWV**GDKTTATDIKGKEVTVLAEVNINNSVFRQYFFETKCRA
Human  NGF SSSHPIFHRGEFSVCDSVSVWV**GDKTTATDIKGKEVMVLGEVNINNSVFKQYFFETKCRD
Bovine NGF SSSHPVFHRGEFSVCDSISVWV**GDKTTATDIKGKEVMVLGEVNINNSVFKQYFFETKCRD
Guinea NGF SSTHPVFHMGEFSVCDSVSVWV**ADKTTATDIKGKEVTVLAEVNVNNNVFKQYFFETKCRD
Chick  NGF  TAHPVLHRGEFSVCDSVSMWV**GDKTTATDIKGKEVTVLGEVNINNNVFKQYFFETKCRD
Snake  NGF  EDHPVHNLGEHSVCDSVSAWV***TKTTATDIKGNTVTVMENVNLDNKVYKQYFFETKCKN
Pig   BDNF   HSDPARRGELSVCDSISEWVTAADKKTAVDMSGGTVTVLEKVPVSKGQLKQYFYETKCNP
Mouse NT-3  YAEHKSHRGEYSVCDSESLWVT**DKSSAIDIRGHQVTVLGEIKTGNSPVKQYFYETRCKE

Prediction     TTTTTTTBBBBTT       TTT          BBBBBB   TTTTTBBBB       TT

           -V3---                      ---V4---
Mouse  NGF SNPVESGCRGIDSKHWNSYCTTTHTFVKALTTDEKQ*AAWRFIRIDTACVCVLSRKATRRG
Human  NGF PNPVDSGCRGIDSKHWNSYCTTTHTFVKALTMDGKQ*AAWRFIRIDTACVCVLSRKAVRRA
Bovine NGF PSPVESGCRGIDSKHWNSYCTTTHTFVKALTTDNKQ*AAWRFIRIDTACVCVLNRKAARRG
Guinea NGF PNPVDSGCRGIDAKHWNSYCTTTHTFVKALTMDGKQ*AAWRFIRIDTACVCVLSRKTGQRA
Chick  NGF PRPVSSGCRGIDAKHWNSYCTTTHTFVKALTMEGKQ*AAWRFIRIDTACVCVLSRKSGRP
Snake  NGF PNPEPSGCRGIDSSHWNSYCTETDTFIKALTMEGNQ*ASWRFIRIETACVCVITKKKGN
Pig   BDNF MGYTKEGCRGIDKRHWNSQCRTTQSYVRALTMDSKKRIGWRFIRIDTSCVCTLTIKRGR
Mouse NT-3 ARPVKNGCRGIDDKHWNSQCKTSQTYVRALTSENNKLVGWRWIRIDTSCVCALSRKIGRT

Prediction TTT  TTTTT      TTTTBBBBBBBBBBB         BBBBBBTTTBBBBBBB
```

Fig. 1. Sequence alignment of NGF from different species and related neurotrophic factors. A cumulative secondary structure prediction is shown below using the Leeds Prediction Package (Eliopoulos, 1989).

(Ibanez *et al.* 1990) and the tryptophan side chains have also been implicated in receptor binding (Cohen *et al.* 1980). However, recent mutagenesis experiments suggest that the tryptophans are structurally but not functionally important, since their replacement alters but does not abolish NGF activity or receptor binding (Ibanez *et al.* 1990).

A detailed three-dimensional structure of this molecule has so far been elusive despite the availability of X-ray quality crystals for many years (Wlodawer *et al.* 1975). Here we report on recent progress towards defining the active conformation of NGF. We also report on the preliminary characterisation of crystals of the high molecular weight form of NGF in the submaxillary gland, the 7S NGF protein complex.

Purification and crystallisation of β-NGF

With the advent of recombinant DNA techniques, large quantities of relatively scarce proteins such as growth factors can be produced for structural analysis. However, milligram quantities of NGF and EGF (epidermal growth factor) are available from certain tissue sources (Walker, 1982). The adult male submaxillary gland is one such tissue source, and the 7S NGF comprises about 0.2% by weight of this gland (Server and Shooter, 1977).

Purification of the 7S NGF complex to homogeneity by the method of Stach *et al.* (1977) helps protect the β-NGF from proteolytic modification. Proteolytic enzymes in the submaxillary gland can modify the native β-NGF molecule at both the amino and carboxyl termini to give multiple forms of NGF heterodimers (Moore *et al.* 1974). In an effort to reduce the proportion of heterodimers, we have adapted the purification method of Mobley *et al.* (1976) to include an FPLC ion exchange step in place of conventional ion exchangers. After purifying the complex, it is then dissociated by acid pH into its constituent subunits. A Mono-S column gives a crude fractionation of the different β-NGF forms (Fig. 2A) and a higher yield of the final product (typically 3–4 mg from 40 g of glands). The β-NGF is further purified by gel filtration in 2 M acetic acid to remove contaminating γ-NGF and mouse IgG proteins (Fig. 2B).

β-NGF produced in this manner crystallises reproducibly to give well formed hexagonal bipyramids (Fig. 3A) as previously described by Wlodawer *et al.* (1975). This is in contrast to earlier problems with disordered 'bulges' where the two pyramids join (Gunning *et al.* 1982). The crystals diffract to greater than 2.3 Å resolution (1Å = 0.1 nm). A *h0l* precession photograph is shown in Fig. 3B. The crystals belong to the space group $P6_122$ (or its enantiomorph $P6_522$) and have cell dimensions a=b=56.5 Å, c=182.4 Å. A high resolution native dataset has been collected for β-NGF to 2.3 Å using the X-ray synchrotron radiation facilities at the EMBL station at DESY, Hamburg (Table 1).

Multiple isomorphous replacement of β-NGF

After producing well ordered crystals, the remaining obstacle to a crystal

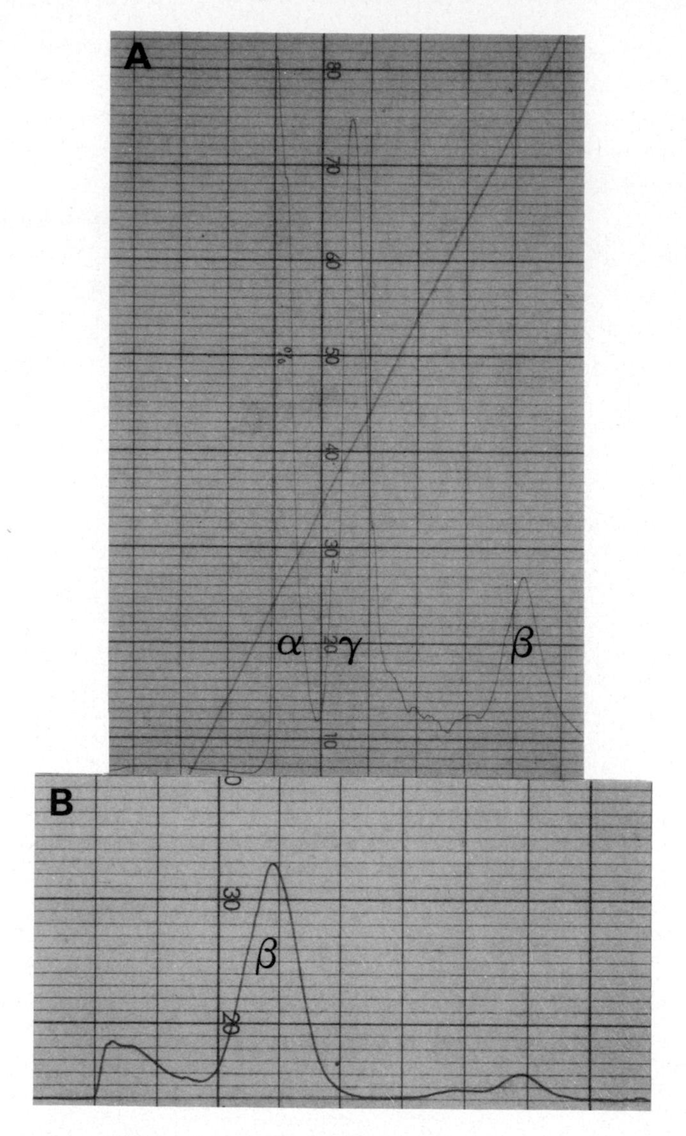

Fig. 2. (A) A 1 ml FPLC Mono S chromatograph of individual subunits from dissociated 7S NGF complex. Buffer conditions are pH 4.0, 50 mM sodium acetate. A flow rate of 1 ml min^{-1}, and a salt gradient of 0–1 M NaCl in 50 ml was used. (B) Gel filtration of β-NGF in 2 M acetic acid using G75 Sephadex.

structure is the solution of the phase problem. Since only the amplitude of each diffracting X-ray wave can be measured, the phase relationship is lost. Therefore a way to evaluate each phase is required. Traditionally new macromolecular structure analyses have used isomorphous replacement with heavy atoms. This is performed by diffusing reactive heavy atom compounds into the native protein crystals. This utilises the large solvent channels present between packed protein molecules in the crystal lattice.

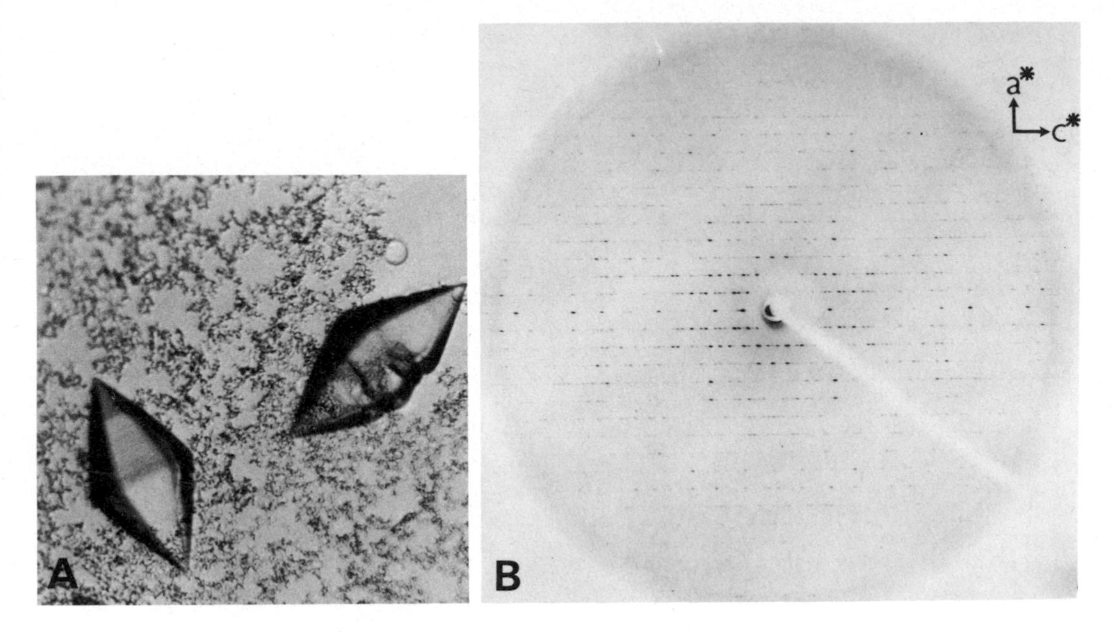

Fig. 3. (A) Hexagonal bipyramid crystals of β-NGF (dimensions 0.5–0.75 mm) grown at pH 6.8 according to Wlodawer *et al.* (1975). (B) A *h0l* precession photograph of a β-NGF crystal, showing systematic absences characteristic of a 6_1 (or 6_5) screw axis along the c axis.

The amino acid sequence of β-NGF indicates three possible types of side chains for heavy atom binding (Angeletti *et al.* 1971). A single methionine and three histidine side chains are potential sites for soft, polarisable heavy atom compounds that form covalent bonds. Also the disulphide bonds are potential interaction sites for mercury and platinum compounds, though these interactions can often give conformational changes (Blundell and Johnson, 1976). As predicted, the best quality heavy atom derivatives have been produced using mercurial and platinum compounds (see Table 1). The harder, less polarisable metals such as the lanthanides often disordered the crystals or changed the cell dimensions (Gunning, 1983).

X-ray data for several heavy atom derivatives have been measured for the β-NGF crystals and are described in Table 1. The quality of the data can be assessed by the internal agreement within the dataset and by the differences found when compared to the native dataset. Previous work at Birkbeck College highlighted the problems of collecting isomorphous heavy atom derivative datasets for β-NGF (Gunning, 1983). Two particular problems were encountered. First, the variation of native crystal cell parameters required that all X-ray data be measured from crystals produced from the same preparation of NGF. Second, even small cell dimension changes during the preparation of heavy atom derivatives resulted in noisy, uninterpretable difference Patterson maps used for locating the heavy

Table 1. *Beta nerve growth factor data processing statistics*

Dataset	Cell		Concentration (mM)	R_{deriv} (%)	R_{merge} (%)	Resolution (Å)	% Data
Native	56.5	182.4	–	–	6.6	2.3	90.3
$SmCl_3$	56.6	182.4	1	14	5.5	3.1	96.5
PCMBS	56.09	183.2	5	19.9	13	3.1	90.8
EMP	56.10	180.32	1	16.4	12	3.3	90.4
$K_2Pt(NO_2)_4$	56.19	181.36	1	16.2	12	3.1	88.2
$NaAuCl_4$	56.10	181.1	5	28.2	11.8	3.3	93.3

$R_{merge}=\sum(|I-\langle I\rangle|)/\sum I$
$R_{deriv}=\sum(|F_{ph}-F_p|)/\sum F_{ph}$.
1 Å=0.1 nm.

atom positions. So a careful control of soaking conditions for heavy atom reagents was necessary.

Single sites in each of the Pt and Hg derivatives have allowed calculation of phases for these derivatives. These phases can be used to locate heavy atom positions for other derivatives, using cross-phased difference Fourier maps (see Blundell and Johnson, 1976) and to check the agreement with the difference Patterson Harker sections. One example is shown in Fig. 4A and 4B, where the phases from the Hg derivative were used with the Pt structure factor amplitudes. A clear single site is evident. The three corresponding Harker vectors from this single site can be seen in the Harker section of the Pt difference Patterson map (Fig. 4B). We have successfully interpreted four of the derivatives listed in Table 1, though the PCMBS and EMP derivatives have sites in common. Therefore the PCMBS dataset was not used. Also the $NaAuCl_4$ dataset showed two heavy atom sites and was included in the phase calculations.

On the basis of the heavy atom positions in Pt, Sm, Au and Hg derivatives described, an electron density map has been calculated to 3.5 Å resolution. It is clear that part of the NGF structure is observable from this map. However, further heavy atom derivatives are currently being screened to improve the MIR phases produced from just three derivatives. We expect these data to allow the solution of the NGF structure in the near future.

X-ray analysis of the 7S NGF protein complex

Varon *et al.* (1967) first reported that NGF exists as a high molecular weight complex. The molecular weight of 7S NGF is 130 000 as estimated from a variety of sedimentation techniques. From the known molecular mass of each subunit and also the relative amounts of each subunit obtained on dissociation, the subunit composition of the complex is two α-NGF subunits of 26 500, one β-NGF dimer of 26 000, two γ-NGF subunits of 26 500 and one or two zinc ions (Pattison and Dunn, 1975).

Three crystal forms of this complex suitable for X-ray analysis have been crystallised using polyethylene glycol as a precipitant (Fig. 5 and McDonald and

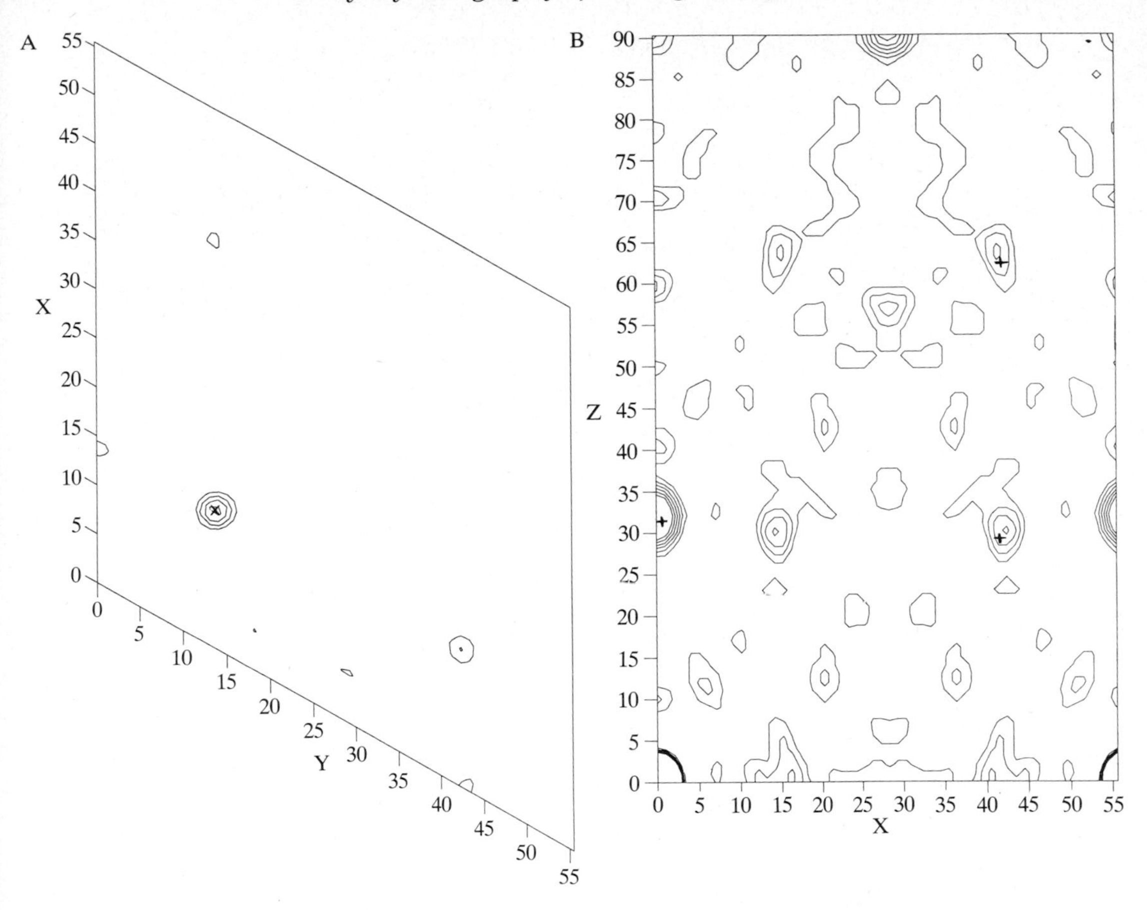

Fig. 4. (A) Cross-phased difference Fourier map section using the phases calculated from the Hg derivative with the Pt isomorphous differences. A single site is located at X and agrees with (B) the three Harker vectors marked X from the difference Patterson map (y=0 section) for the Pt derivative.

Blundell, 1990). Table 2 indicates the cell dimensions and space group for each form. Since the crystal form A_1 is most commonly observed, and has the more convenient cell dimensions for data collection, our X-ray studies have concentrated on this form. The cell dimensions for all forms were estimated from precession photographs, and the space group was derived from the systematic absences from these photos (Fig. 6). The cell parameters shown for form A_1 are the refined cell dimensions used during data collection. A native dataset for form A_1 has been collected to 3.0 Å resolution using synchrotron radiation. Although several high resolution structures are known for the serine protease family (Bode *et al.* 1983), the complexity of the oligomer implies that use of these as search models for a molecular replacement solution of 7S NGF (Rossmann, 1972) may not succeed. Thus, heavy atom trials are in progress so that multiple isomorphous replacement can be used to solve the phase problem.

Electron micrographs using negative staining indicate two features of the

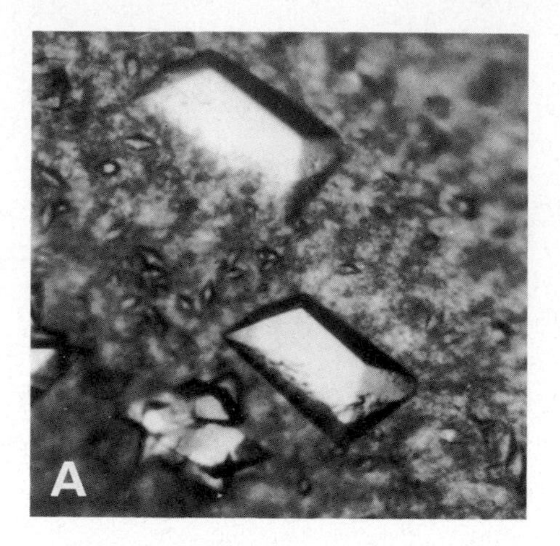

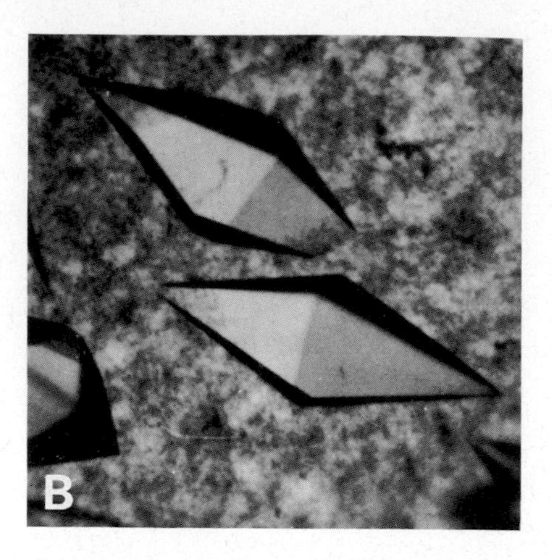

Fig. 5. Crystal forms A_1 (A) and B_1 (B) of 7S NGF grown at pH 4.5 in the presence of zinc according to McDonald and Blundell (1990).

Table 2. *Crystal forms of the 7S nerve growth factor protein complex, crystallized using polyethylene glycol as the precipitant*

Crystal Form	Cell Å			Space Group	V_m(Å^3 Da^{-1})	Z	Resln
Orthorhombic A_2	86.3	91.4	148.5	$P2_12_12_1$	2.3	1 7S complex	3 Å
Orthorhombic A_1	95.4	96.7	146.5	$P2_12_12_1$	2.6	1 7S complex	3 Å
Orthorhombic B_1	93.4	97.8	308.0	$P22_12_1$	2.5	2 7S complexes	3.6 Å

1 Å=0.1 nm.
V_m = volume of asymmetric unit/molecular mass.
Z=number of molecules per asymmetric unit.

complex (Fig. 7). First, a negatively stained 'hole', surrounded by several unstained protein subunits, can be seen in most of the complex molecules, indicating the subunits are arranged in a symmetric fashion with a closed point group symmetry. Two possibilities exist for the arrangement of the two α and two γ-NGF subunits (which are twice the size of the β-NGF subunits), either with a cyclic or a dihedral point group symmetry. Unfortunately these cannot be distinguished by standard negative stain electron microscopy. Second, the micrographs indicate the globular shape of the complex with an approximate diameter of 80 Å (which comprises two subunit diameters). This is in agreement with approximate dimensions of 35 Å by 30 Å observed for a single serine protease molecule. The size of the 7S complex imposes certain restrictions on the packing of 7S molecules, since four or eight molecules must pack within the unit cell. However, the packing arrangement for the different forms and the relationship between the three crystal forms is not clear at present.

The α and γ-NGF contain primarily β-strand secondary structure motifs, by analogy with other serine protease structures (Thomas *et al.* 1981). The β-NGF is also predicted to be principally a β type protein from spectroscopic evidence (Williams and Gaber, 1982) and from secondary structure prediction using seven different prediction algorithms (see Fig. 1). In the light of these predictions, the circular dichroism spectra for the whole complex confirm the presence of a large β-sheet contribution in the far UV peptide region (Fig. 8) (see Curtis Johnson, 1988 for review). In addition aromatic side chains from different subunits can become

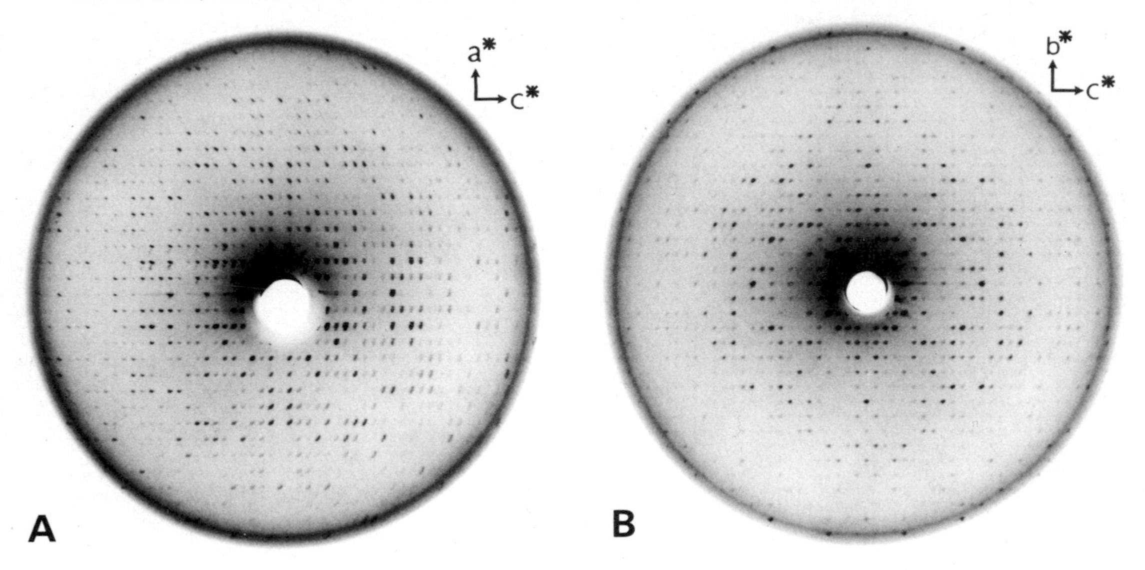

Fig. 6. (A) A *h0l* precession photograph of crystal forms A_2 and (B) an ϕkl photograph of A_1 of 7S NGF. 2_1 screw axes along a and c axes are evident from the systematic absences in these photos.

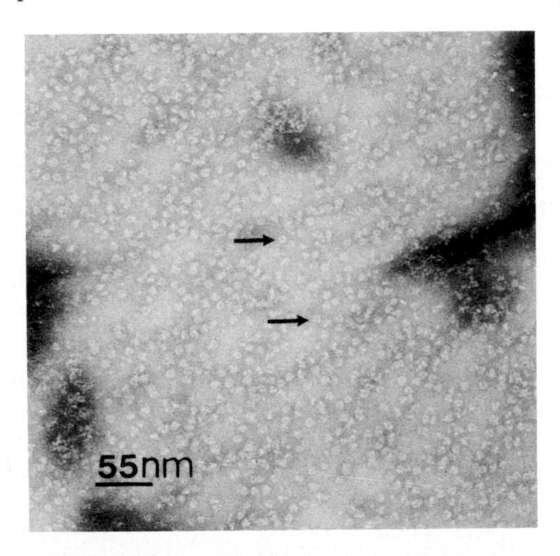

Fig. 7. Electron micrograph of the 7S NGF complex negatively stained with uranyl acetate. The two arrows indicate clear images with a 'hole' and an arrangement of subunits around this 'hole'. Magnification is ×126 500; the bar represents 550 Å (55 nm).

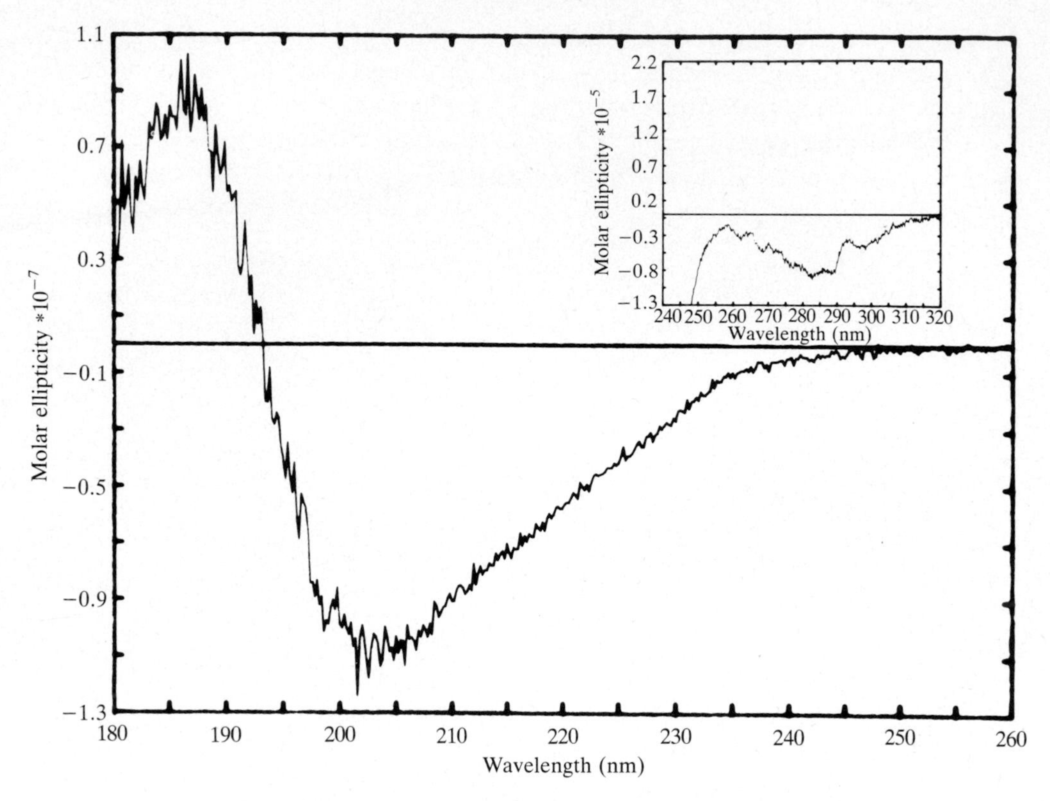

Fig. 8. Far and near (inset) UV circular dichroism spectra of 7S NGF, measured at pH 6.8 at 1 mg ml^{-1} concentration in the presence of zinc, showing a β-sheet contribution in the peptide region.

trapped on association in protein complexes (Wood *et al.* 1975). The small CD signals seen in the near UV (inset Fig. 8) are removed on dissociation of the 7S NGF complex (McDonald, unpublished results). This feature of the near UV CD spectra of the associated complex may be a direct consequence of buried aromatic residues between the NGF subunits.

In addition to its high affinity ligand–receptor interaction which is responsible in part for its signalling properties, β-NGF undergoes another important protein–protein interaction with its processing enzmyes α and γ-NGF. The exact physiological role of this complex is not fully understood, though a role in NGF biosynthesis and protection is likely. The serine protease subunits can together competitively inhibit NGF binding to its cell surface receptor and thus inhibit the NGF neurotrophic activity (Woodruff and Neet, 1986; Stach and Shooter, 1980). The three-dimensional structure of such a growth factor-processing enzyme complex should provide information on the receptor binding region of NGF, as well as the manner in which it inhibits the γ-NGF serine protease activity. Thus in parallel with studies on β-NGF, we are pursuing the structure of the 7S NGF complex.

Thanks to Dr Helen Saibil for the electron micrograph of 7S NGF, and to Professor Axel Wollmer for use of the AVIV spectrometer. We also acknowledge the use of SRS, Daresbury and the EMBL station at DESY, Hamburg in these studies. Thanks to Alex Wlodawer, Julia Goodfellow and Jenny Gunning for making their X-ray data available.

References

ANDRES, R. Y., JENG, I. AND BRADSHAW, R. A. (1977). Nerve growth factor receptors: Identification of distinct classes in plasma membranes and nuclei of embryonic dorsal root neurons. *Proc. natn. Acad. Sci. U.S.A.* **74**, 2785–2789.

ANGELETTI, R. H. AND BRADSHAW, R. A. (1971). Nerve growth factor from mouse submaxillary gland: amino acid sequence. *Proc. natn. Acad. Sci. U.S.A.* **68**, 2417–2420.

BLUNDELL, T. L. AND JOHNSON, L. N. (1976). *Protein Crystallography*. London: Academic Press.

BODE, W., CHEN, Z., BARTELS, K., KUTZBACH, C., SCHMIDT-KASTNER, G. AND BARTUNIK, H. (1983). Refined 2Å X-ray structure of porcine pancreatic kallikrein A, a specific trypsin-like serine proteinase. *J. molec. Biol.* **164**, 237–282.

BOTHWELL, M. A., WILSON, W. H. AND SHOOTER, E. M. (1979). The relationship between glandular kallikrein and growth factor-processing proteases of mouse submaxillary gland. *J. biol. Chem.* **254**, 7287–7294.

BRADSHAW, R. A., JENG, I., ANDRES, R. Y., PULLIAM, M. W., SILVERMAN, R. E., RUBIN, J. AND JACOBS, J. W. (1976). The structure and function of nerve growth factor. In *Endocrinology* vol. 2 (ed. H. V. T. James), pp. 206–212. Excerpta Medica: Amsterdam.

BRADSHAW, R. A. (1978). Nerve growth factor. *A. Rev. Biochem.* **47**, 191–216.

COHEN, P., SUTTER, A., LANDRETH, G., ZIMMERMAN, A. AND SHOOTER, E. M. (1980). Oxidation of tryptophan-21 alters the biological activity and receptor binding characteristics of mouse nerve growth factor. *J. biol. Chem.* **255**, 2949–2954.

COHEN, S. (1959). Purification and metabolic effects of a nerve growth-promoting protein from snake venom. *J. biol. Chem.* **234**, 1129–1137.

CURTIS JOHNSON JR., W. (1988). Secondary structure of proteins through circular dichroism spectroscopy. *A. Rev. Biophys. Chem.* **17**, 145–166.

DARLING, T. L., PETRIDES, P. E., BEGUIN, P., FREY, P., SHOOTER, E. M., SELBY, M. J. AND RUTTER, W. J. (1983). The biosynthesis and processing of proteins in the mouse 7S nerve growth factor complex. *Cold Spring Harbour Symp.* Part 1, 427–434.

EDWARDS, R. H., SELBY, M. J., GARCIA, P. D. AND RUTTER, W. J. (1988). Processing of the native nerve growth factor precursor to form biologically active nerve growth factor. *J. biol. Chem.* **263**, 6810–6815.

ELIOPOULOS, E. E. (1989). Documentation for the Leeds Prediction Package, Dept. of Biophysics, University of Leeds, U.K.

FRAZIER, W. A., HOGUE-ANGELETTI, R. A., SHERMAN, R. AND BRADSHAW, R. A. (1973). Topography of mouse 2.5S nerve growth factor. Reactivity of tyrosine and tryptophan. *Biochemistry* **12**, 3281–3293.

GUNNING, J., BEDARKAR, S., TAYLOR, G. L., GOODFELLOW, J. M., BLUNDELL, T. L., WLODAWER, A., HODGSON, K. O., SHOOTER, E. M., FOURME, R., GABER, B. AND WILLIAMS, R. (1982). Conformational studies of polypeptide growth factors: IGF and NGF. In *Cell Function and Differentiation* Part A, pp. 221–230. New York: Alan R. Liss Inc.

GUNNING, J. (1983). Ph.D. Thesis, University of London.

HARPER, G. E. AND THOENEN, H. (1980). Nerve growth factor: Biological significance, measurement and distribution. *J. Neurochem.* **34**, 5–16.

IBÁÑEZ, C. F., HALLBÖÖK, F., EBENDAL, T. AND PERSSON, H. (1990). Structure-function studies of nerve growth factor: functional importance of highly conserved amino acid residues. *EMBO J.* **9**, 1477–1483.

ISACKSON, P. J., DUNBAR, J. C. AND BRADSHAW, R. A. (1987). Role of glandular kallikreins as growth factor processing enzymes: structural and evolutionary considerations. *J. Cell. Biochem.* **33**, 65–75.

JAMES, R. AND BRADSHAW, R. A. (1984). Polypeptide growth factors. *A. Rev. Biochem.* **53**, 259–92.

JOHNSON, D., LANAHAN, A., BUCK, C. R., SEHGAL, A., MORGAN, C., MERCER, E., BOTHWELL, M. AND CHAO, M. (1986). Expression and structure of the human NGF receptor. *Cell* **47**, 545–554.

LEVI-MONTALCINI, R. (1987). The nerve growth factor: thirty-five years later. *EMBO J.* **6**, 1145–1154.
MCDONALD, N. Q. AND BLUNDELL, T. L. (1990). Manuscript in preparation.
MOBLEY, W. C., SCHENKER, A. AND SHOOTER, E. M. (1976). Characterization and isolation of proteolytically modified nerve growth factor. *Biochemistry* **15**, 5543–5552.
MOORE, J. B., MOBLEY, W. C. AND SHOOTER, E. M. (1974). Proteolytic modification of the β nerve growth factor protein. *Biochemistry* **13**, 833–840.
MURPHY, R. A., CHLUMECKY, V., SMILLIE, L. B., CARPETER, M., NATRISS, M., ANDERSON, J. K., RHODES, J. A., BARKER, P. A., SIMINOSKI, K., CAMPENOT, R. G. AND HASKINS, J. (1989). Isolation and characterization of a glycosylated form of β nerve growth factor in mouse submandibular glands. *J. biol. Chem.* **264**, 12 502–12 509.
PATTISON, S. E. AND DUNN, M. F. (1975). On the relationship of zinc ion to the structure and function of the 7S nerve growth factor protein. *Biochemistry* **14**, 2733–2739.
ROSSMANN, M. (1972). *The Molecular Replacement Method.* Gordon and Breach: New York.
SERVER, A. C. AND SHOOTER, E. M. (1977). Nerve growth factor. *Adv. Protein Chem.* **31**, 339–409.
STACH, R. W., WAGNER, B. J. AND STACH, B. M. (1977). A more rapid method for the isolation of the 7S nerve growth factor complex. *Analyt. Biochem.* **83**, 26–32.
STACH, R. W. AND SHOOTER, E. M. (1980). Cross-linked 7S nerve growth factor is biologically inactive. *J. Neurosci.* **34**, 1499–1505.
THOENEN, H., KORSCHING, S., HEUMANN, R. AND ACHESON, A. (1985). Nerve growth factor. *CIBA Foundation Symposium* **116**, pp. 113–128. Pitman: London.
THOMAS, K. A. AND BRADSHAW, R. A. (1981). *Methods in Enzymology* **80**, pp. 609–620. Academic Press.
VARON, S., NOMURA, J. AND SHOOTER, E. M. (1967). The isolation of the mouse nerve growth factor protein in a high molecular weight form. *Biochemistry* **6**, 2202–2209.
WALKER, P. (1982). The mouse submaxillary gland: a model for the study of hormonally dependent growth factors. *J. Endocrin. Invest.* **5**, 183–196.
WILLIAMS, R. AND GABER, B. (1982). Raman spectroscopic determination of the secondary structure of crystalline nerve growth factor. *J. biol. Chem.* **257**, 13 321–13 323.
WLODAWER, A., HODGSON, K. O. AND SHOOTER, E. M. (1975). Crystallization of nerve growth factor from mouse submaxillary glands. *Proc. natn. Acad. Sci. U.S.A.* **72**, 777–779.
WOOD, S. P., BLUNDELL, T. L., WOLLMER, A., LAZARUS, N. R. AND NEVILLE, R. W. J. (1975). The relation of conformation and association of insulin to receptor binding: X-ray and circular dichroism studies on bovine and hystricomorph insulins. *Eur. J. Biochem.* **55**, 531–542.
WOODRUFF, N. R. AND NEET, K. E. (1986). Inhibition of β nerve growth factor binding to PC12 cells by α nerve growth factor and γ nerve growth factor. *Biochemistry* **25**, 7967–7974.

J. Cell Sci. Suppl. 13, 31–42 (1990)
Printed in Great Britain © *The Company of Biologists Limited 1990*

Chimeric molecules map and dissociate the potent transforming and secretory properties of PDGF A and PDGF B

WILLIAM J. LAROCHELLE, NEILL GIESE, MARY MAY-SIROFF, KEITH C. ROBBINS* AND STUART A. AARONSON

Laboratory of Cellular and Molecular Biology, Building 37, Room 1E24, National Cancer Institute, Bethesda, MD 20892, USA

Summary

Human platelet-derived growth factor (PDGF) is a connective tissue cell mitogen comprising two related chains encoded by distinct genes. The B chain is the homolog of the v-*sis* oncogene product. Properties that distinguish these ligands include greater transforming potency of the B chain and more efficient secretion of the A chain. By a strategy involving the generation of PDGF A and B chimeras, these properties were mapped to distinct domains of the respective molecules. Increased transforming efficiency segregated with the ability to activate both alpha and beta PDGF receptors. These findings genetically map PDGF B residues 105 to 144 as responsible for conformational alterations critical to beta PDGF receptor interaction, and provide a mechanistic basis for the greater transforming potency of the PDGF B chain.

Introduction

Human platelet-derived growth factor (PDGF) is a major mitogen for cells of connective tissue origin that is involved in development and wound healing (Ross *et al.* 1986). Abnormal expression of this growth factor has also been implicated in a variety of pathologic states including cancer (Eva *et al.* 1982; Gazit *et al.* 1984). PDGF is a disulfide linked dimer consisting of two related polypeptide chains, designated A and B, that are products of different genes. The gene encoding the human PDGF B chain is the normal counterpart of the v-*sis* oncogene (Waterfield *et al.* 1983; Doolittle, 1983; Devare *et al.* 1983). PDGF A and B chains are approximately 40% related (Betsholtz *et al.* 1986) and contain eight conserved cysteine residues (Giese *et al.* 1987; Sauer and Donoghue, 1988). PDGF A and B chains can form homodimers as well as the AB heterodimer, and there is evidence for the natural occurrence of all three isoforms (Johnsson *et al.* 1982).

Although homodimers of either PDGF A or B are mitogenic as well as chemotactic for cells possessing the appropriate PDGF receptor (Matsui *et al.* 1989*b*), major differences in their biological properties have been observed. The PDGF B chain gene exhibits 10 to 100 fold greater transforming efficiency in the NIH/3T3 transfection assay (Beckman *et al.* 1988). Moreover, its product remains

* Laboratory of Cellular Development and Oncology, National Institute of Dental Research, National Institutes of Health, Bethesda, MD 20892, USA.

Key words: platelet-derived growth factor, growth factors, protein structure/function, chimeras, PDGF receptors.

tightly cell associated (Robbins *et al.* 1985), whereas the PDGF A chain is efficiently secreted (Beckman *et al.* 1988). In addition, the two molecules differentially bind and activate the products of two distinct genes, encoding respectively the alpha and beta PDGF receptor (Matsui *et al.* 1989*a*; Yarden *et al.* 1986; Claesson-Welsh *et al.* 1989). While PDGF B interacts with either receptor, PDGF A binds and triggers only the alpha PDGF receptor (Matsui *et al.* 1989*a*,*b*; Hart *et al.* 1988; Heldin *et al.* 1988). In the present study, we constructed chimeras of PDGF A and B chains in an effort to map domains of each that potentially influence their normal functions and role in pathologic processes.

Results

Strategy for construction of PDGF chimeric molecules

We initially constructed ten chimeric PDGF molecules to investigate structural regions of PDGF A or PDGF B associated with specific PDGF functions (LaRochelle *et al.* 1990). The chimeric constructs were developed utilizing pre-existing, or engineered, common restriction endonuclease sites within the PDGF A- or PDGF B-coding sequences. All PDGF A- or PDGF B-coding sequences altered for this purpose by oligonucleotide-mediated, site-directed mutagenesis (Kunkel, 1985) were first shown to possess biological activities indistinguishable from those of their respective parental cDNAs. Each chimera was designated on the basis of the codon at which the recombination was performed. Four of the chimeric constructs, $A^{97}B^{99}$, $B^{98}A^{98}$, $A^{177}B^{179}$, and $B^{178}A^{178}$, were designed to maintain the functional integrity of the PDGF B minimal transforming domain (King *et al.* 1985; Hannink *et al.* 1986) or the analogous region of PDGF A (Fig. 1). An additional six constructs further dissected the minimal transforming domain. The $A^{143}B^{145}$ and $B^{144}A^{144}$ chimeras divided the transforming region roughly in half, while $A^{104}B^{106}$ and $B^{105}A^{105}$ chimeras as well as $A^{153}B^{155}$ and $B^{154}A^{154}$ chimeras further subdivided the minimal transforming domains (Fig. 1).

Mapping of a PDGF B sub-domain responsible for its potent transforming activity

All wild type parental and recombinant PDGF constructs (LaRochelle *et al.* 1990) were transferred into a vector containing the metallothionein promoter (MMTneo) and analyzed for transforming activity by transfection of NIH/3T3 cells. Since the MMTneo vector also contained a dominant, selectable neomycin marker gene, it was possible to score neomycin-resistant colony formation for each plasmid as well. Thus, we were able to compare precisely the specific transforming efficiencies of each construct.

As shown in Fig. 1, the PDGF B expression vector showed around twenty-fold higher transforming efficiency than that of PDGF A, as previously reported (Beckman *et al.* 1988). $A^{97}B^{99}$ and $B^{178}A^{178}$ chimeric constructs, which both contained the minimal transforming domain of the PDGF B gene product, demonstrated high transforming efficiency, indistinguishable from that of the

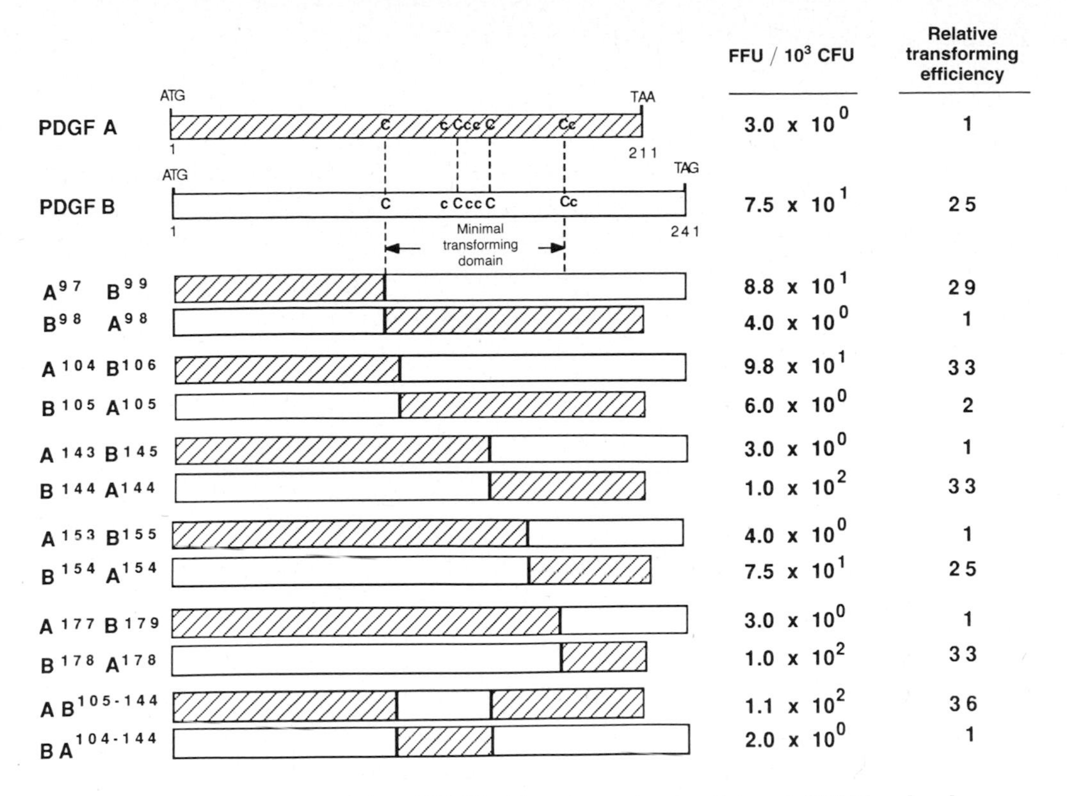

Fig. 1. Transforming activity of PDGF chimeric constructs. Chimeric PDGF molecules were constructed by recombination of the PDGF A (hatched box) and B (open box) genes at common pre-existing or engineered (Kunkel, 1985) restriction endonuclease sites. The structure of each recombinant was verified by a combination of restriction endonuclease mapping and nucleotide sequence determination. NIH/3T3 cells were transfected with the recombinant plasmid DNA and 40 μg of carrier calf thymus DNA by the calcium phosphate precipitation technique (Wigler *et al.* 1977). Transfected cultures were either scored for colony formation in the presence of G418 (Southern and Berg, 1982) or for focus-forming activity as described (Giese *et al.* 1987; Beckman *et al.* 1988). Data shown represent the mean values of three experiments. Relative transforming efficiency was calculated by dividing the number of foci (FFU) by the number of colonies (CFU) per ng of DNA relative to that of PDGF A. Cysteine residues essential (C) or non-essential (c) for PDGF B transformation are shown.

wild type PDGF B construct. In contrast, $B^{98}A^{98}$ and $A^{177}B^{179}$ chimeric constructs, which possessed the analogous domain of PDGF A, showed ten to twenty fold lower specific transforming efficiency, equivalent to that of PDGF A (Fig. 1). Chimeras $A^{104}B^{106}$, $B^{144}A^{144}$ and $B^{154}A^{154}$ also possessed high specific transforming efficiency, whereas the reciprocal chimeras $B^{105}A^{105}$, $A^{143}B^{145}$ and $A^{153}B^{155}$, respectively, were only weakly transforming. All of these findings suggested that amino acid residues 105–144 of PDGF B were responsible for its more potent transforming activity (Fig. 1).

To test this hypothesis, we substituted only the minimal regions mapped above that were responsible for differences in transforming activities of the native

PDGF A and B molecules. As shown in Fig. 1, the $AB^{105-144}$ chimera possessed high specific transforming activity, indistinguishable from that of PDGF B. Conversely, substitution of PDGF A codons 104–144 for those of PDGF B reduced transforming activity of the resulting chimera to that of the PDGF A molecule. All of these results conclusively demonstrated that the subdomain encompassed by PDGF B amino acid residues 105–144 was responsible for its more potent transforming properties.

Immunochemical characterization of PDGF chimeric proteins

We next sought to verify the chimeric constructs by analysis of their translational products. Thus, each NIH/3T3 transfectant was metabolically labeled and subjected to immunoprecipitation analysis using antisera specific to PDGF A or PDGF B amino or carboxy termini, respectively. As shown in Fig. 2A, the primary PDGF A translational product, p42, as well as its 38×10^3 and $32\times10^3\,M_r$ processed forms, were detected in lysates of PDGF A transfected cells. Similarly, the primary PDGF B translational product, p54, was processed at amino and carboxy termini to $40\times10^3\,M_r$, $34\times10^3\,M_r$ and $24\times10^3\,M_r$ species (Fig. 2B).

A representative chimera, $A^{143}B^{145}$, encoded a p45 primary translational product, and was processed at amino and carboxy termini to 40 and $38\times10^3\,M_r$ forms, as well as a major $26\times10^3\,M_r$ species (Fig. 2C). The reciprocal chimeric construct directed synthesis of a p46 translational product, which was amino terminally processed to a major p32 species (Fig. 2D). Thus, each of these chimeras possessed the correct amino and carboxy terminal epitopes of their respective parental molecules and was of an intermediate molecular weight relative to PDGF A or PDGF B homodimers. Immunoprecipitation analysis of the remaining transfectants revealed the expected PDGF A or PDGF B antigenic determinants and predicted intermediate sizes (data not shown).

Mapping of a domain responsible for differences in PDGF A and B secretion

Previous studies of the compartmentalization of PDGF A and PDGF B in transformed NIH/3T3 fibroblasts have shown that PDGF B remains tightly membrane-associated, whereas PDGF A is efficiently secreted into culture fluids (Beckman *et al.* 1988). No obvious structural motif that might cause retention of PDGF B, such as hydrophobic stretches, has been observed. Thus, we sought to identify the domain(s) responsible for differences in secretion of the two molecules and whether such a domain could account for their different transforming potencies.

Following metabolic radiolabeling of cultures for 4 h, medium conditioned by each transfectant, as well as the crude cell membrane fraction of each, were subjected to immunoprecipitation analysis with a panel of PDGF antibodies. Crude membrane preparations of each transfectant showed roughly comparable

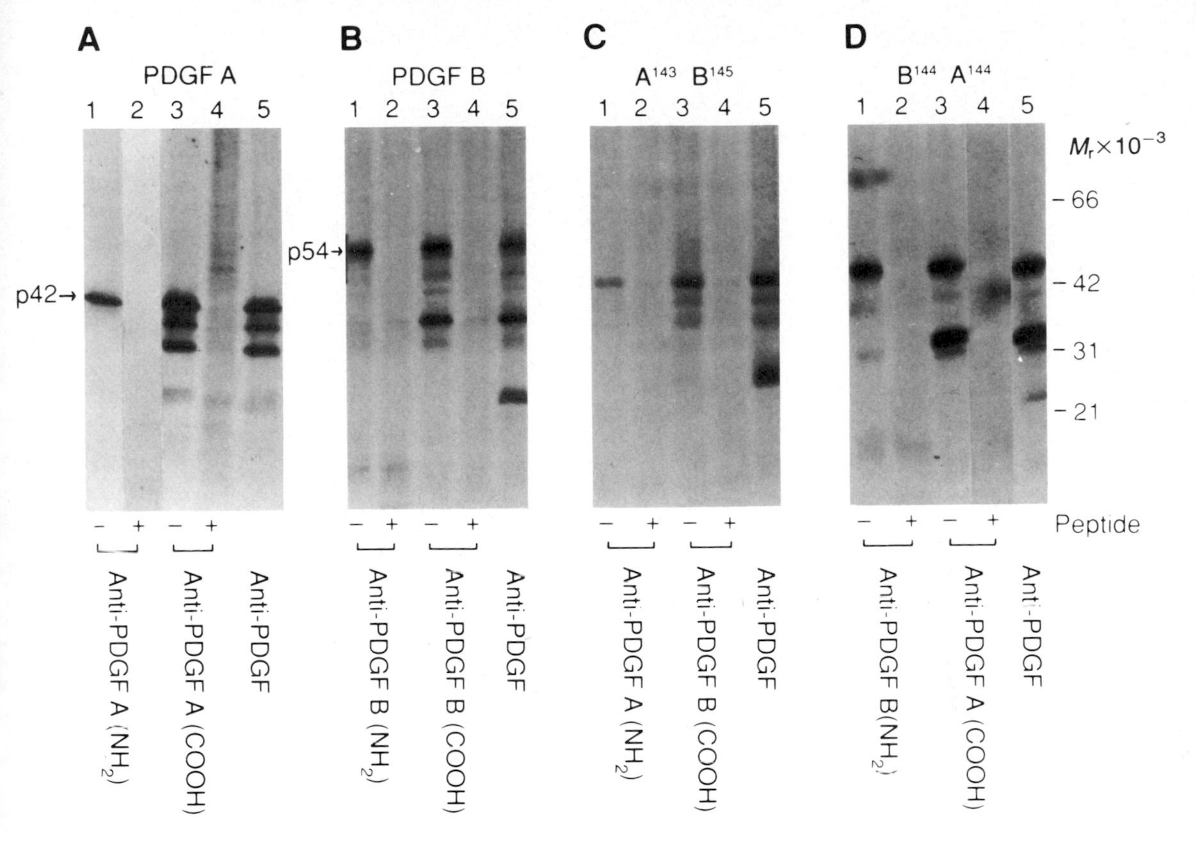

Fig. 2. Immunochemical characterization of PDGF chimeric proteins. Mass populations of 10^7 cells transfected with the MMTneo vector containing either PDGF A (panel A), PDGF B (Panel B), PDGF $A^{143}B^{145}$ (panel C), or PDGF $B^{144}A^{144}$ (panel D), were preincubated overnight in DMEM containing 10 % calf serum and 25 μM $ZnCl_2$. The medium was then replaced for three hours with cysteine- and methionine-free DMEM containing [^{35}S]methionine and [^{35}S]cysteine at 125 μCi ml^{-1} and 25 μM $ZnCl_2$. Crude membrane fractions were immunoprecipitated with PDGF A amino terminal antibody, PDGF A carboxy terminal antibody, PDGF B amino terminal antibody, PDGF B carboxy terminal antibody, or PDGF antibody. In some cases (lanes 2 and 4), antibodies were preincubated with excess homologous peptide. Immune complexes were analyzed by SDS–PAGE under non-reducing conditions and results were visualized by fluorography for ten days.

levels of PDGF immunoreactive protein. However, only those chimeras which contained PDGF A carboxy terminal amino acid residues 178–211, namely $B^{98}A^{98}$, $B^{105}A^{105}$, $B^{144}A^{144}$, $B^{154}A^{154}$ and $B^{178}A^{178}$, were found to be efficiently secreted. Fig. 3 shows that the $B^{178}A^{178}$ chimera, which contained only PDGF A amino acid residues 178–211, was efficiently released, whereas the reciprocal chimera, $A^{177}B^{179}$, remained more than 90 % membrane associated.

To confirm our immunological findings, we analyzed mitogenic activities associated with culture fluids and crude membrane preparations of transfectants containing parental or chimeric PDGF constructs. Comparable mitogenic activity

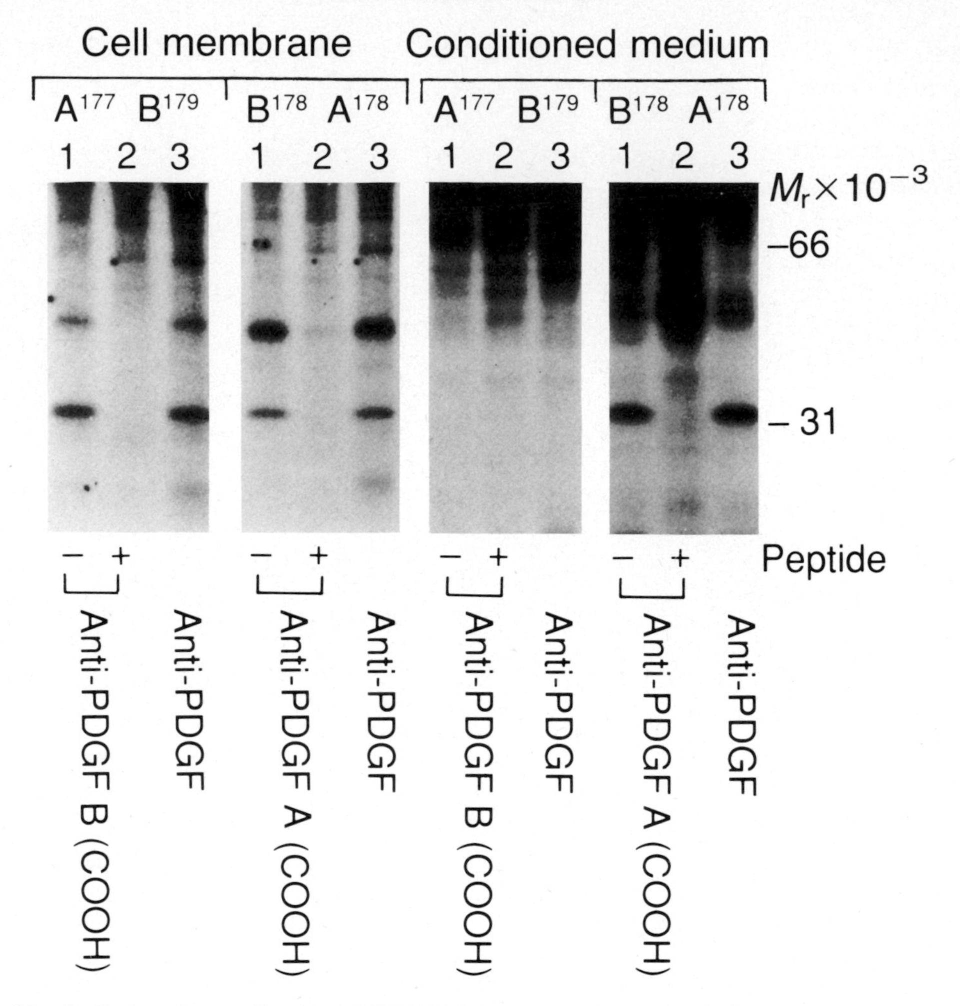

Fig. 3. Compartmentalization of PDGF chimeric constructs. NIH/3T3 cells transfected with $A^{177}B^{179}$ or $B^{178}A^{178}$ were metabolically labeled as described in the text. Crude cellular membrane or conditioned media were examined by immunoprecipitation using the antibodies indicated. In some cases, antibodies were preincubated with excess homologous peptide (lane 2). Immune complexes were analyzed as indicated above and the results were visualized by fluorography for twelve days (Cell membrane) or 36 h (Conditioned medium).

was detected in each crude membrane fraction. However, only in the case of PDGF A and those chimeras containing at least the carboxy terminal thirty-four amino acid residues of PDGF A, was mitogenic activity detectable in culture fluids (data not shown). In each case, the mitogenic activity was specifically inhibited by neutralizing PDGF antibody, establishing the PDGF-related nature of the secreted mitogen. As shown above, potent transforming activity mapped to PDGF B amino acid residues 105–144 (Fig. 1). Thus, localization of the domain responsible for differences in PDGF A and B secretory properties to their carboxy terminal regions excluded this property from being responsible for their different transforming activities.

Activation of alpha and beta PDGF receptors in NIH/3T3 cells expressing PDGF chimeras

We next investigated whether PDGF receptor binding and/or activation might be responsible for the differences in oncogenic potency of PDGF A and B. Thus, we examined the steady state level of tyrosine phosphorylation of alpha and beta receptors expressed in NIH/3T3 transfectants containing either wild type PDGF A or B constructs, as well as each of the chimeras. To do so, cell lysates were enriched for each receptor by immunoprecipitation with alpha or beta PDGF receptor-specific peptide antisera, followed by immunoblotting with anti-phosphotyrosine antibody. As shown in Fig. 4, NIH/3T3 cells showed no

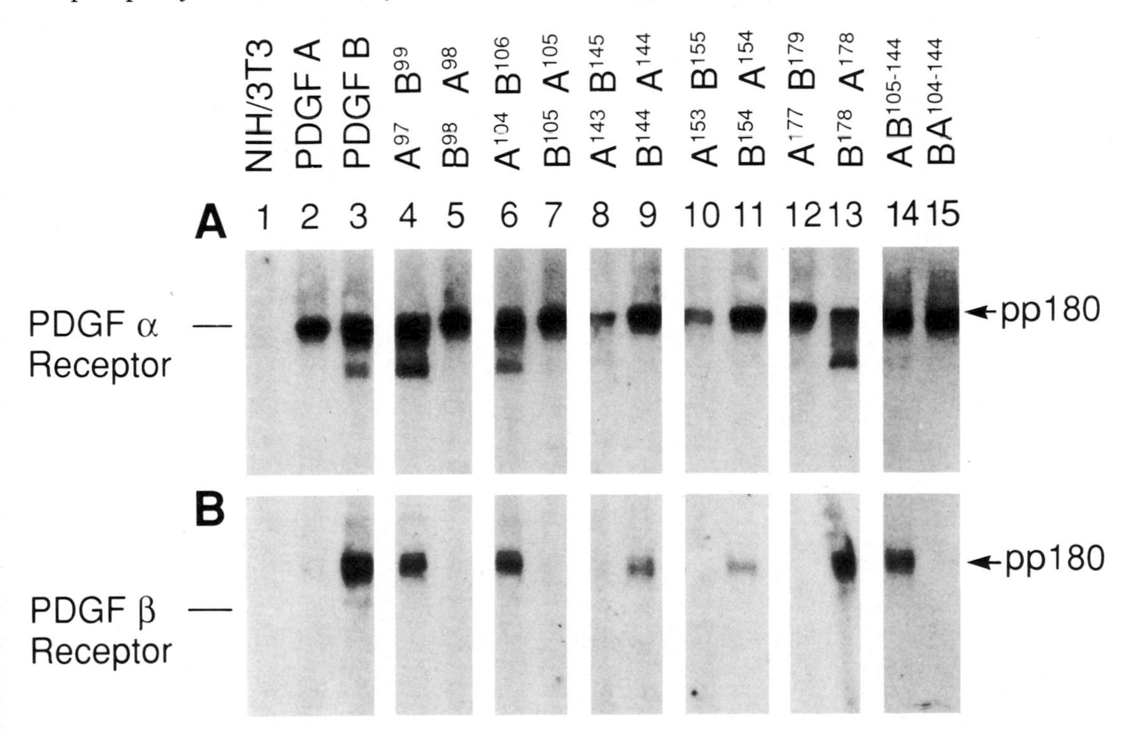

Fig. 4. Tyrosine phosphorylation of alpha and beta PDGF receptors in NIH/3T3 cells expressing PDGF chimeras. NIH/3T3 cells (lane 1) or transfectants expressing PDGF A (lane 2), PDGF B (lane 3), or the chimeric PDGF constructs (lane 4–15) were incubated overnight in DMEM, 25 μM $ZnCl_2$. After 16 h, the cells were washed with PBS/1.0 mM sodium orthovanadate and lysed as described by Matsui *et al.* (1989*a*,*b*). Protein extracts were immunoprecipitated with anti-peptide antibodies specific for the alpha (panel A) or beta (panel B) PDGF receptors (Matsui *et al.* 1989*a*,*b*). Immunoprecipitated proteins were blotted to Immobilon-P and probed with anti-phosphotyrosine specific antibody. Filters were treated with ^{125}I-labeled protein A and subjected to autoradiography. The electrophoretic mobility of the pp180 alpha or beta PDGF receptors are shown. In some cases immature forms of PDGF receptors were recognized as well.

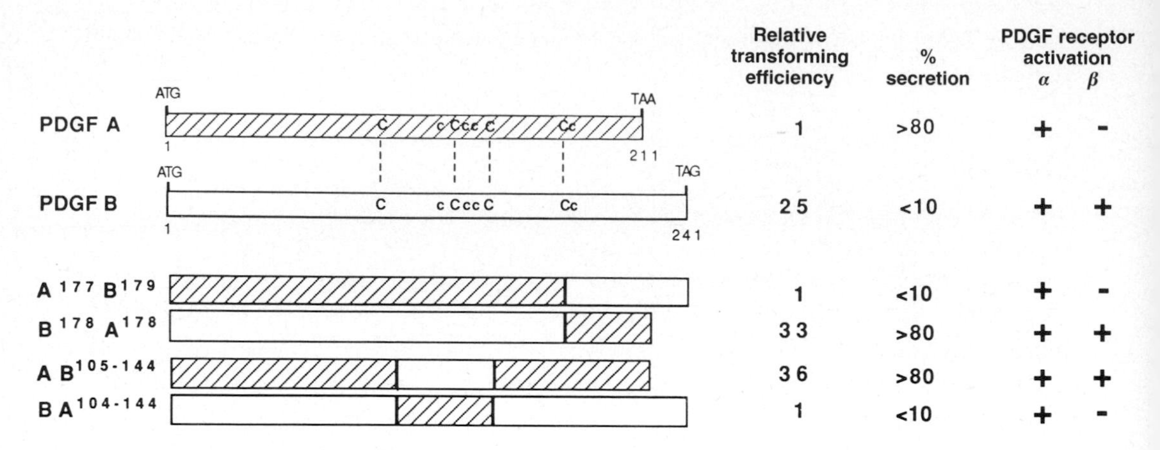

Fig. 5. Summary of biological properties mapped by PDGF chimeras. PDGF A (hatched box) and PDGF B (open box) cDNAs are represented. Cysteine residues essential (C) or nonessential (c) for PDGF B transformation are shown. Percent secretion was determined by comparing the quantity of PDGF protein immunologically detected in cell membrane preparations with that secreted into the culture fluid. Autophosphorylation was utilized as a measure of receptor activation. Relative transforming efficiency is defined in Fig. 1.

detectable alpha or beta PDGF receptor tyrosine phosphorylation. As expected from known receptor binding properties of each ligand (Matsui *et al.* 1989*a*,*b*; Hart *et al.* 1988; Heldin *et al.* 1988), NIH/3T3 cells expressing PDGF A demonstrated tyrosine phosphorylated $180\times10^3\,M_r$ alpha but not beta receptor species. In contrast, both $180\times10^3\,M_r$ alpha and beta receptor species were tyrosine phosphorylated in PDGF B-producing cells (Fig. 4). The specificity of the antibody was demonstrated by the ability of phosphotyrosine to compete for immunodetection of these proteins (data not shown).

When the steady state level of PDGF receptor tyrosine phosphorylation was examined in transfectants containing the PDGF chimeras, readily detectable levels of the activated $180\times10^3\,M_r$ alpha PDGF receptor were observed in each case (Fig. 4). However, there was chronic activation of the $180\times10^3\,M_r$ beta PDGF receptor species of cells expressing $A^{97}B^{99}$, $A^{104}B^{106}$, $B^{144}A^{144}$ and $B^{154}A^{154}$, as well as $B^{178}A^{178}$ chimeras, all of which contained at least PDGF B amino acid residues 105–144. As shown in Fig. 4, the $AB^{105-144}$ chimera which substituted only PDGF B amino acid residues 105–144 into the analogous region of PDGF A, demonstrated the same pattern of receptor tyrosine phosphorylation as observed with PDGF B. Conversely, the switch of analogous PDGF A amino acid residues into PDGF B led to a pattern of receptor tyrosine phosphorylation indistinguishable from that of PDGF A. These results demonstrate that amino acid residues 105–144 are responsible for the ability of PDGF B to bind and activate preferentially the beta PDGF receptor. Since increased transforming activity of PDGF B molecules mapped to these same residues, we conclude that the ability to activate beta receptors in addition to alpha receptors provides the basis for the greater transforming potency of the PDGF B chain gene for NIH/3T3 cells (LaRochelle *et al.* 1990).

Discussion

The minimal transforming domain of PDGF B has been mapped and encompasses 84 amino acids (King *et al.* 1985; Hannink *et al.* 1986). Biochemical studies of cysteine residues involved in disulfide linkages (Giese *et al.* 1987; Vogel and Hoppe, 1989) suggest that intrachain disulfide bonds, essential to the transforming function of PDGF B, fold this minimal transforming domain into two major loops, according to all models which have been proposed (Giese *et al.* 1987; Vogel and Hoppe, 1989). PDGF A shares functional and structural similarities with PDGF B, including the ability to activate the PDGF alpha receptor (Matsui *et al.* 1989*a*,*b*; Hart *et al.* 1988; Heldin *et al.* 1988) as well as conservation of all cysteine residues and spacing between these residues. Less is known about the structure and disulfide linkages of PDGF A. In the present studies, chimera PDGF molecules were generated between PDGF A and B. The analysis of the biological activity of each of the twelve reciprocal chimeras generated argues strongly that very similar disulfide linkages must be formed by the PDGF A chain to allow for such interchangeability of portions of these two related molecules.

Analysis of transforming potency of PDGF B single codon deletion mutants (Giese *et al.* 1990) has localized a domain which is much more sensitive to mutagenesis than other domains. Moreover, monoclonal antibody *sis* 1 neutralization of PDGF B (LaRochelle *et al.* 1989) complimented by epitope mapping (Giese *et al.* 1990; LaRochelle *et al.* 1989), suggests that this same region contains a surface epitope that is critical for PDGF receptor binding and activation. Analysis of these PDGF chimeras directly demonstrated that amino acid residues 105–144 of PDGF B, within this same critical domain, are responsible for the more potent transforming properties of the PDGF B chains. PDGF A activated only alpha PDGF receptors whereas PDGF B triggered both alpha and beta PDGF receptors in an autocrine fashion, and the differences in transforming potency of these molecules directly correlated with those constructs containing PDGF B codons 105–144 and their activation of beta PDGF receptors. Thus, this domain is not only critical for ligand–receptor interaction but is the major determinant of the subtle amino acid changes that must determine differences in the ability of the PDGF molecule to interact with other receptors. Moreover, the direct correlation between the ability to activate beta as well as alpha PDGF receptors with greater transforming potency of the chimeras containing PDGF B codons 105–144, provides a mechanistic basis for greater transforming efficiency of the PDGF B chain.

The quantitative differences in transforming activities of PDGF A and B chain genes for NIH/3T3 cells *in vitro* correlate with *in vivo* findings that a PDGF B encoding retrovirus induces fibrosarcomas in nude mice (Pech *et al.* 1989), while an analogous retrovirus encoding PDGF A has as yet not produced detectable tumors (P. Arnstein and S. A. Aaronson, unpublished observations). Nonetheless, we have shown that the alpha PDGF receptors can couple with mitogenic signalling pathways as efficiently as the beta PDGF receptor when either is independently expressed in a null hematopoietic cell (Matsui *et al.* 1989*a*,*b*). Thus, it is possible that in fibroblasts which express both alpha and beta receptors, the alpha receptor may be somewhat less efficient than the beta receptor in stimulating this pathway. Alternatively, non-saturating ligand concentrations expressed in autocrine transformation of NIH/3T3 must be quantitatively more effective in mitogenic signalling when both receptors are stimulated, as opposed to the alpha receptor alone.

We also mapped an additional domain of the two PDGF molecules that was responsible for the differences in their PDGF secretory properties. This domain localized to the carboxy terminal amino acid residues of PDGF, and was demonstrated not to be a determinant of the differences in transforming potency of the two molecules. These differences in the secondary properties could be explained either by a tendency of the PDGF B carboxy terminus to cause association with the cell membrane or an effect of the analogous domain of the PDGF A to promote secretion. Further studies will be necessary to resolve this question.

There is increasing evidence that functional domains of protein correspond to individual exons (Gilbert, 1978; 1985). In the case of the PDGF A molecule, the signal peptides and proteolytic processing sites correspond to specific exons (Heldin and Westermark, 1989). Amino acid residues 105–144 which specify beta PDGF receptor activation are included exclusively within exon four of PDGF B. In contrast, the secretory properties of the long form of PDGF A or the membrane retention properties of PDGF B correspond to amino acids predominantly encoded by exon 6 of either molecule. Thus, the functional differences which we mapped correspond to individual exons specific for either of the PDGFs.

Comparison of the PDGF B 105–144 region with PDGF A reveals two subdomains (residues 106–114 and 135–143) which contain little amino acid homology, separated by an almost entirely conserved stretch of twenty amino acids (residues 115–134). Thus, combinations of PDGF B amino acid substitutions within these two subdomains of PDGF A will determine which PDGF B amino acids bind and activate the beta PDGF receptor. Amino acids of PDGF B directly involved in beta PDGF receptor binding and activation may then be altered to develop competitive antagonists.

W.J.L. is supported by an American Cancer Society Postdoctoral Fellowship (PF no. 3030). We thank C. Betsholtz, B. Westermark and C.-H. Heldin for providing the PDGF A cDNA.

References

Beckman, M. P., Betsholtz, C., Heldin, C.-H., Westermark, B., Dimarco, E., Difiore, P. P., Robbins, K. C. and Aaronson, S. A. (1988). Comparison of the biological properties and transforming potential of human PDGF-A and PDGF-B. *Science* **241**, 1346–1348.

Betsholtz, C., Johnsson, A., Heldin, C.-H., Westermark, B., Lind, P., Urdea, M. S., Eddy, R.,

Shows, T. B., Philipott, K., Mellor, A. L., Knott, T. J. and Scott, J. (1986). cDNA sequence and chromosomal localization of human platelet-derived growth factor A-chain and its expression in tumor cell lines. *Nature* **320**, 695–699.

Claesson-Welsh, L., Eriksson, A., Westermark, B. and Heldin, C.-H. (1989). cDNA cloning and expression of the A-type platelet-derived growth factor (PDGF) receptor establishes structural similarity to the B-type PDGF receptor. *Proc. natn. Acad. Sci. U.S.A.* **86**, 4917–4921.

Devare, S. G., Reddy, E. P., Law, D. J., Robbins, K. C. and Aaronson, S. A. (1983). Nucleotide sequence of the simian sarcoma virus genome: demonstration that its acquired cellular sequences encode the transforming gene product $p28^{sis}$. *Proc. natn. Acad. Sci. U.S.A.* **80**, 731–735.

Doolittle, R. F., Hunkapiller, M. W., Hood, L. E., Devare, S. G., Robbins, K. C., Aaronson, S. A. and Antoniades, H. N. (1983). Simian sarcoma virus *onc* gene, v-*sis*, is derived from the gene (or genes) encoding a platelet-derived growth factor. *Science* **221**, 275–277.

Eva, A., Robbins, K. C., Andersen, P. R., Srinivasan, A., Tronick, S. R., Reddy, E. P., Ellmore, N. W., Galen, A. T., Lautenberger, J. A., Papas, T. S., Westin, E. H., Wong-Staal, F., Gallo, R. C. and Aaronson, S. A. (1982). Cellular genes analogous to retroviral *onc* genes are transcribed in human tumour cells. *Nature* **295**, 116–119.

Gazit, A., Igarashi, H., Chiu, I.-M., Srivivasan, A., Yaniv, A., Tronick, S. R., Robbins, K. C. and Aaronson, S. A. (1984). Expression of the normal human *sis*/PDGF-2 coding sequence induces cellular transformation. *Cell* **39**, 89–97.

Giese, N., Larochelle, W. J., May-Siroff, M., Robbins, K. C. and Aaronson, S. A. (1990). A small v-*sis*/PDGF B protein domain in which subtle conformational changes abrogate PDGF receptor interaction and transforming activity. *Molec. cell. Biol.* **10**, in press.

Giese, N., Robbins, K. C. and Aaronson, S. A. (1987). The role of individual cysteine residues in the structure and function of the v-*sis* gene product. *Science* **236**, 1315–1318.

Gilbert, W. (1978). Why genes in pieces? *Nature* **271**, 501.

Gilbert, W. (1985). Genes in pieces revisited. *Science* **228**, 823–824.

Hannink, M., Sauer, M. K. and Donoghue, D. J. (1986). Deletions in the C-terminal coding region of the v-*sis* gene: dimerization is required for transformation. *Molec. cell. Biol.* **6**, 1304–1314.

Hart, C. E., Forstrom, J. W., Kelly, J. D., Seifert, R. A., Smith, R. A., Ross, R., Murray, M. J. and Bowen-Pope, D. F. (1988). Binding of different dimeric forms of PDGF to human fibroblasts: evidence for two separate receptor types. *Science* **240**, 1529–1531.

Heldin, C.-H., Backstrom, G., Ostman, A., Hammacher, A., Ronnstrand, L., Rubin, K., Nister, M. and Westermark, B. (1988). Binding of different dimeric forms of PDGF to human fibroblasts: evidence for two separate receptor types. *EMBO J.* **7**, 1387–1394.

Heldin, C.-H. and Westermark, B. (1989). Platelet-derived growth factor: three isoforms and two receptor types. *Trends in Genet.* **5**, 108–111.

Johnsson, A., Heldin, C.-H., Westermark, B. and Wasteson, A. (1982). Platelet-derived growth factor: identification of constituent polypeptide chains. *Biochem. biophys. Res. Commun.* **104**, 66–74.

King, C. R., Giese, N., Robbins, K. C. and Aaronson, S. A. (1985). *In vitro* mutagenesis of the v-*sis* transforming gene defines functional domains of its growth factor-related product. *Proc. natn. Acad. Sci. U.S.A.* **82**, 5295–5299.

Kunkel, T. A. (1985). Rapid and efficient site-specific mutagenesis without phenotypic selection. *Proc. natn. Acad. Sci. U.S.A.* **82**, 488–492.

Larochelle, W. J., Giese, N., May-Siroff, M., Robbins, K. C. and Aaronson, S. A. (1990). Molecular localization of the transforming and secretory properties of PDGF A and PDGF B. *Science* **248**, 1541–1544.

Larochelle, W. J., Robbins, K. C. and Aaronson, S. A. (1989). Immunochemical localization of the epitope for a monoclonal antibody that neutralizes human platelet-derived growth factor mitogenic activity. *Molec. cell. Biol.* **9**, 3538–3542.

Matsui, T., Heidaran, M., Miki, T., Popescu, N., Larochelle, W. J., Kraus, M., Pierce, J. and Aaronson, S. A. (1989*a*). Isolation of a novel receptor cDNA establishes the existence of two PDGF receptor genes. *Science* **243**, 800–804.

Matsui, T., Pierce, J., Fleming, T. P., Greenberger, J. S., Larochelle, W. J., Ruggiero, M.

AND AARONSON, S. A. (1986*b*). Independent expression of human alpha or beta platelet-derived growth factor receptor cDNAs in a naive hematopoietic cell leads to functional coupling with mitogenic and chemotactic signaling pathways. *Proc. natn. Acad. Sci. U.S.A.* **86**, 8314–8318.

PECH, M., GAZIT, A., ARNSTEIN, P. AND AARONSON, S. A. (1989). Generation of fibrosarcoma *in vivo* by a retrovirus that expresses the normal B chain of platelet-derived growth factor and mimics the alternative splice pattern of the v-*sis* oncogene. *Proc. natn. Acad. Sci. U.S.A.* **86**, 2693–2697.

ROBBINS, K. C., LEAL, F., PIERCE, J. H. AND AARONSON, S. A. (1985). The v-*sis*/PDGF-2 transforming gene product localizes to cell membranes but is not a secretory protein. *EMBO J.* **4**, 1783–1792.

ROSS, R., RAINES, E. W. AND BOWEN-POPE, D. F. (1986). The biology of platelet-derived growth factor. *Cell* **46**, 155–169.

SAUER, M. K. AND DONOGHUE, D. J. (1988). Identification of the nonessential disulfide bonds and altered conformations in the v-*sis* protein, a homolog of the B chain of platelet-derived growth factor. *Molec. cell. Biol.* **8**, 1011–1018.

SOUTHERN, P. J. AND BERG, P. (1982). Transformation of mammalian cells to antibiotic resistance with a bacterial gene under control of the SV40 early promoter. *J. Molec. Appl. Genet.* **1**, 327–341.

VOGEL, S. AND HOPPE, J. (1989). Binding domains and epitopes in platelet-derived growth factor. *Biochemistry* **28**, 2961–2966.

WATERFIELD, M. D., SCRACE, G. T., WHITTLE, N., STROOBANT, P., JOHNSSON, A., WASTESON, A., WESTERMARK, B., HELDIN, C.-H., HUANG, J. S. AND DEUEL, T. F. (1983). Platelet-derived growth factor is structurally related to the putative transforming protein $p28^{sis}$ of simian sarcoma virus. *Nature* **304**, 35–39.

WIGLER, M., SILVERSTEIN, S., LEE, L.-S., PELLICER, A., CHENG, Y.-C. AND AXEL, R. (1977). Transfer of purified herpes virus thymidine kinase gene to cultured mouse cells. *Cell* **11**, 223–232.

YARDEN, Y., ESCOBEDO, J. A., KUANG, W.-J., YANG-FENG, T. L., DANIEL, T. O., TREMBLE, P. M., CHEN, E. Y., ANDO, M. E., HARKINS, R. N., FRANCKE, U., FRIED, V. A., ULLRICH, A. AND WILLIAMS, L. T. (1986). Structure of the receptor for platelet-derived growth factor helps define a family of closely related growth factor receptors. *Nature* **323**, 226–232.

J. Cell Sci. Suppl. 13, 43–56 (1990)
Printed in Great Britain © *The Company of Biologists Limited 1990*

Mitogenic signalling through the bombesin receptor: role of a guanine nucleotide regulatory protein

ENRIQUE ROZENGURT*, ISABEL FABREGAT, ARNOLD COFFER, JOAN GIL AND JAMES SINNETT-SMITH

Imperial Cancer Research Fund, PO Box 123, Lincoln's Inn Fields, London WC2A 3PX, UK

Summary

Bombesin and structurally related peptides including gastrin releasing peptide (GRP) are potent mitogens for Swiss 3T3 cells. The early cellular and molecular responses elicited by bombesin and structurally related peptides have been elucidated in detail. Further understanding of the molecular basis of the potent mitogenic response initiated by bombesin is required in order to elucidate the mechanism by which the occupied receptor communicates with effector molecules in the cell. Transmembrane signalling mechanisms involving either a tyrosine kinase or a guanine nucleotide-binding regulatory protein (G protein) have been proposed. Here we summarize our experimental evidence indicating that a G protein(s) is involved in the coupling of the bombesin receptor to the generation of intracellular signals related to mitogenesis. Evidence for the role of G proteins in bombesin signal transduction pathways has been obtained by assessing the effects of guanine nucleotide analogues on both receptor-mediated responses in permeabilized cells and ligand binding in membrane preparations. We found that [^{125}I]GRP–receptor complexes were solubilized from Swiss 3T3 cell membranes by using the detergents taurodeoxycholate or deoxycholate. Addition of guanosine 5-[γ-thio]triphosphate (GTPγS) to ligand–receptor complexes isolated by gel filtration enhanced the rate of ligand dissociation in a concentration-dependent and nucleotide-specific manner. These results demonstrate the successful solubilization of [^{125}I]GRP–receptor complexes from Swiss 3T3 cell membranes and provide evidence for the physical association between the ligand–receptor complex and a guanine nucleotide binding protein(s).

Introduction

The cells of many tissues and organs *in vivo* are maintained in a non-proliferating state (G_0/G_1). However, such cells remain viable and can be induced to resume DNA synthesis and cell division when exposed to external stimuli such as hormones, antigens or growth factors. In this manner the growth of individual cells is regulated according to the requirements of the whole organism. The elucidation of the molecular mechanism by which these mitogens regulate growth and differentiation at the cellular level may prove crucial to understanding both the normal proliferative response and the unrestrained growth of cancer cells.

Many studies of growth factors have used cultured fibroblasts, such as murine 3T3 cells, as a model system. These cells cease to proliferate when they deplete the medium of its growth-promoting activity, and can be stimulated to re-initiate

* Author for correspondence

Key words: neuropeptides, receptor solubilization, signal transduction, growth control.

DNA synthesis and cell division either by replenishing the medium with fresh serum, or by the addition of purified growth factors or pharmacological agents in serum-free medium. Studies performed using such quiescent cells and defined combinations of growth factors have revealed the existence of potent and specific synergistically acting signal transduction pathways, initiated almost immediately following mitogen addition (Rozengurt, 1986, 1989).

It is increasingly recognized that small regulatory peptides act as molecular messengers in a complex network of information-processing by cells throughout the body. They may act on post-ganglionic receptors (neurotransmitters), nearby cells (paracine hormones) or distant target organs (endocrine hormones). The classical role of these peptides as fast-acting neurohumoral signallers has recently been challenged by the discovery that they also stimulate slow-acting mitogenesis (reviewed in Rozengurt, 1986; Zachary *et al.* 1987*a*; Woll and Rozengurt, 1989*a*). In particular, bombesin (Rozengurt and Sinnett-Smith, 1983), vasopressin (Rozengurt *et al.* 1979, 1981), bradykinin (Woll and Rozengurt, 1988), vasoactive intestinal peptide (Zurier *et al.* 1988), endothelin (Takuwa *et al.* 1989) and vasoactive intestinal contractor (Fabregat and Rozengurt, 1990) can act as growth factors for cultured 3T3 cells. Collectively, these findings demonstrate that the mitogenic actions of regulatory peptides are mediated by multiple signalling pathways, and imply that the participation of regulatory peptides in the control of cell proliferation may be broader and more fundamental than previously thought.

The peptides of the bombesin family, including gastrin-releasing peptide (GRP), are of particular significance. These peptides are potent mitogens for Swiss 3T3 cells in the absence of other growth-promoting factors (Rozengurt and Sinnett-Smith, 1983) and may act as autocrine growth factors for small cell lung cancer (SCLC) (Woll and Rozengurt, 1989*a*). Indeed, the autocrine growth loop of bombesin-like peptides may be only a part of an extensive network of autocrine and paracine interactions involving a variety of neuropeptides in SCLC (Woll and Rozengurt, 1989*b*). Thus, murine Swiss 3T3 cells which express receptors for several mitogenic neuropeptides and are capable of multiple neuropeptide regulation, provide a model system relevant to SCLC. A detailed understanding of the signal transduction pathways in this model system may identify novel targets for therapeutic intervention.

Signal transduction

The early cellular and molecular responses elicited by bombesin and structurally related peptides (listed in Table 1) have been elucidated in detail. The cause–effect relationships and temporal organization of these early signals and molecular events have been reviewed elsewhere (Rozengurt, 1988, 1989; Rozengurt *et al.* 1988). They provide a paradigm for the study of other growth factors and mitogenic neuropeptides and illustrate the activation and interaction of a variety of signalling pathways.

Table 1. *Events in the action of bombesin in Swiss 3T3 cells*

Event	References
Binding to specific receptors	Zachary and Rozengurt (1985*a*)
Cross-linking to M_r 75 000–85 000 glycoprotein	Zachary and Rozengurt (1987*a*); Kris *et al.* (1987); Sinnett-Smith *et al.* (1988)
Ligand internalization and degradation	Zachary and Rozengurt (1987*b*); Brown *et al.* (1988)
Activation of PKC (intact cells)	Zachary *et al.* (1986); Isacke *et al.* (1986); Rodriguez-Pena *et al.* (1986)
Activation of PKC (permeabilized cells)	Erusalimsky *et al.* (1988)
Elevation of DAG levels	Muir and Murray (1987); Takuwa *et al.* (1987)
Ins(1,4,5)P_3 production	Heslop *et al.* (1986); Takuwa *et al.* (1987); Lopez-Rivas *et al.* (1987)
Ca^{2+} mobilization	Mendoza *et al.* (1986); Takuwa *et al.* (1987); Lopez-Rivas *et al.* (1987)
Ins(1,4,5)P_3 and Ca^{2+} (kinetics)	Nånberg and Rozengurt (1988)
Na^+ influx and Na^+/K^+ pump	Mendoza *et al.* (1986)
Transmodulation of EGF receptor	Zachary and Rozengurt (1985*b*); Zachary *et al.* (1986)
Arachidonic acid release and prostaglandin synthesis	Millar and Rozengurt (1988)
Enhancement of cyclicAMP accumulation	Millar and Rozengurt (1988)
Increase in c-*fos* and c-*myc* mRNA levels	Rozengurt and Sinnett-Smith (1987, 1988); Mehmet *et al.* (1989)
Elevation of c-*fos* protein	Mehmet *et al.* (1989)
Stimulation of DNA synthesis	Rozengurt and Sinnett-Smith (1983)

A central problem in understanding the molecular basis of the potent mitogenic response initiated by bombesin is to elucidate how the occupied receptor communicates with effector molecules in the cell. Transmembrane signalling mechanisms involving either a tyrosine kinase or a guanine nucleotide-binding regulatory protein (G protein) have been proposed to couple growth factor receptors to intracellular effectors (e.g. see Rozengurt, 1986). Bombesin stimulation of tyrosine phosphorylation has been reported (Cirillo *et al.* 1986; Gaudino *et al.* 1988) and the possibility that bombesin receptor signalling is associated with this kinase activity was raised. However, bombesin associated tyrosine kinase activity was not detected by another laboratory (Isacke *et al.* 1986) and the molecular weight of the tyrosine phosphorylated band does not coincide with that of the putative receptor identified by affinity cross-linking in intact cells or membrane preparations (see below).

It was suggested that a pertussis toxin-sensitive G protein could couple the bombesin receptor to the enzymes which hydrolyze polyphosphoinositides (Letterio *et al.* 1986). However, the demonstration that pertussis toxin did not inhibit bombesin stimulation of inositol phosphate formation, Ca^{2+} mobilization or activation of protein kinase C in Swiss 3T3 cells (Zachary *et al.* 1987*b*), did not support this hypothesis. These experiments did not rule out, however, the possibility that another class of G protein, which was insensitive to pertussis toxin, might be involved in the pathway of polyphosphoinositide breakdown and activation of protein kinase C.

Evidence for the role of G proteins in signal transduction pathways can be obtained by assessing the effects of guanine nucleotide analogues on both receptor-mediated responses in permeabilized cells and ligand binding in membrane preparations. In what follows we summarize our experimental evidence indicating that a G protein(s) is involved in the coupling of the bombesin receptor to the generation of intracellular signals related to mitogenesis.

Protein phosphorylation in permeabilized cells

One of the most striking events induced by bombesin and structurally related peptides is a rapid increase in the phosphorylation of an acidic cellular protein of apparent M_r=80 000 (termed 80 K) which is a prominent substrate of protein kinase C in Swiss 3T3 cells (Rozengurt *et al.* 1983; Zachary *et al.* 1986; Isacke *et al.* 1986; Erusalimsky *et al.* 1988). The phosphorylation of this protein is also enhanced by biologically active phorbol esters (Rozengurt *et al.* 1983; Rodriguez-Pena and Rozengurt, 1985, 1986; Blackshear *et al.* 1985, 1986), diacylglycerols (Rozengurt *et al.* 1984; Issandou and Rozengurt, 1989) and other growth factors and neuropeptides (Rozengurt *et al.* 1983; Rodriguez-Pena and Rozengurt, 1985, 1986; Blackshear *et al.* 1985). Furthermore, the same 80 K protein is phosphorylated in cell-free systems either by activation of protein kinase C or by addition of the purified enzyme (Blackshear *et al.* 1986; Rodriguez-Pena and Rozengurt, 1986). A protein closely related to fibroblast 80 K has been purified from rat brain (Morris and Rozengurt, 1988) and partially sequenced (Erusalimsky *et al.* 1989). This protein, in turn, is related to a protein kinase C substrate purified from bovine brain (Albert *et al.* 1987) and recently cloned (Stumpo *et al.* 1989) but does not show significant homology to any other known protein. It is well established that the increase in the phosphorylation of 80 K provides a specific marker for protein kinase C activation.

Recent work from this laboratory has characterized the phosphorylation of 80 K in digitonin permeabilized Swiss 3T3 cells, and employed this technique to study the mechanism of bombesin-induced activation of protein kinase C (Erusalimsky *et al.* 1988; Erpsalimsky and Rozengurt, 1989). A salient feature of the results is that the GDP analogue GDPβS inhibited the stimulation of 80 K phosphorylation by bombesin in a selective manner. GDPβS is known to prevent the activation of G proteins by inhibiting the binding of GTP. The fact that GTP can reverse the inhibitory effect of GDPβS is consistent with this notion. The findings indicate that guanine nucleotides modulate the transduction of the signal from the bombesin receptor and imply that a G protein links the bombesin receptor to the generation of an intracellular signal, which in turn activates protein kinase C in

Swiss 3T3 cells. Further experimental work, using membrane or receptor preparations, was necessary to elucidate whether the putative G protein(s) is associated with the receptor and to clarify the role of tyrosine phosphorylation in the transduction of the bombesin signal.

Bombesin receptor in membranes from Swiss 3T3 cells

Binding measurements and chemical cross-linking experiments using [^{125}I]GRP show that bombesin-like peptides interact with specific, high-affinity receptors located on the cell surface. [^{125}I]GRP-binding is inhibited by various bombesin-like peptides in proportion to their ability to stimulate DNA synthesis, but not by structurally unrelated mitogens (Zachary and Rozengurt, 1985*a*,*b*). Two potent bombesin antagonists, [D-Arg^{1},D-Phe^{5},D-Trp^{7,9},Leu^{11}]substance P and [Leu^{13}-ψ(CH_2NH)-Leu^{14}] bombesin, inhibit both GRP-binding and bombesin/GRP-stimulated mitogenesis (Woll and Rozengurt, 1988; Woll *et al.* 1988). Thus, bombesin and related peptides interact with receptors that are distinct from those for other mitogens in Swiss 3T3 cells.

Recently, the properties of the mitogenic bombesin receptor have been examined in membrane preparations from Swiss 3T3 cells. It is noteworthy that addition of Mg^{2+} (5–10 mM) during the homogenization of these cells is crucial in order to stabilize the bombesin receptor in the resulting membrane preparation (Sinnett-Smith *et al.* 1990). [^{125}I]GRP-binding to such membranes is specific, saturable and reversible. Scatchard analysis indicates the present of a single class of high-affinity binding sites. The K_d obtained from such analysis (2×10^{-10} M) is in excellent agreement with the equilibrium constant derived from rate constants (Rozengurt and Sinnett-Smith, 1990). These results are consistent with the existence of a homogeneous population of bombesin/GRP binding sites in membranes of Swiss 3T3 cells.

The physical properties of the bombesin/GRP receptor have been investigated using an affinity-labelling method. Analysis of extracts of cells which have been preincubated with [^{125}I]GRP and then treated with disuccinimidyl cross-linking agents reveals the presence of a major band migrating with apparent M_r 75 000–85 000 (Zachary and Rozengurt, 1987*a*; Kris *et al.* 1987; Sinnett-Smith *et al.* 1988). Furthermore, [^{125}I]GRP can be cross-linked to an M_r 75 000–85 000 glycoprotein of Swiss 3T3 membranes but not to membranes from cells lacking bombesin receptors (Sinnett-Smith *et al.* 1990). The radio-labelled M_r 75 000–85 000 protein binds to wheat germ lectin–sepharose columns and can be eluted with *N*-acetyl-D-glucosamine, suggesting that it is a glycoprotein (Sinnett-Smith *et al.* 1988; and unpublished results). In addition, treatment with endo-β-*N*-acetylglycosaminidase F reduces the apparent molecular weight of the affinity-labelled band from 75 000–85 000 to 43 000, indicating the presence of *N*-linked oligosaccharide groups (Kris *et al.* 1987; Sinnett-Smith *et al.* 1988). Thus, the bombesin/GRP receptor appears to be a glycoprotein of apparent M_r

75 000–85 000 with *N*-linked carbohydrate side-chains and a polypeptide core of M_r 43 000.

The availability of membrane preparations that retain specific bombesin receptors is useful in the identification of the signal-transduction mechanism(s) that couples the receptor for these neuropeptides with the generation of intracellular events. A decrease in ligand affinity for receptors produced by added guanine nucleotides is characteristic of a receptor–G protein interaction (Bourne *et al.* 1988). Sinnett-Smith *et al.* (1990) demonstrated that the non-hydrolyzable GTP analogue GTPγS caused a specific and concentration-dependent inhibition of [^{125}I]GRP-binding and cross-linking to 3T3 cell membranes. The effect is due primarily to an increase in the equilibrium dissociation constant (K_d) rather than to a decrease in the number of receptors. A typical experiment is shown in Fig. 1A. This modulation of ligand affinity by guanine nucleotides provides further evidence that a G protein couples the mitogenic bombesin receptor with intracellular effector systems.

In view of the preceding results, it was important to determine whether bombesin stimulates tyrosine phosphorylation in membrane preparations in which the bombesin receptor remains functional, as judged by binding activity and modulation by guanine nucleotides. Membranes from Swiss 3T3 cells, prepared as described by Sinnett-Smith *et al.* (1990), were incubated with [γ-^{32}P]ATP in the absence or presence of bombesin and then immunoprecipitated with a monoclonal P-Tyr antibody. Platelet-derived growth factor (PDGF) was also tested for comparison. Fig. 1B shows that bombesin failed to stimulate any discernible tyrosine phosphorylation, whereas PDGF, under identical conditions, caused a marked increase in the phosphorylation of its receptor. Thus, in contrast to the results of Cirillo *et al.* (1986), these results show that the bombesin receptor does not possess a closely associated tyrosine kinase activity. This conclusion is in line with the interpretation that the bombesin receptor signals *via* a G protein but does not rule out that the peptide could stimulate tyrosine phosphorylation in intact cells, in an indirect manner. This possibility warrants further experimental work.

Solubilization of the bombesin receptor from Swiss 3T3 cells membranes: coupling to a G protein

The molecular and regulatory characterization of plasma membrane receptors requires a procedure for their solubilization in a functional state. It is relevant that the binding of various hormonal peptides to their corresponding membrane receptors has been shown to stabilize the receptor molecules, as well as to induce tight association between the receptor and their respective G proteins (Dickey *et al.* 1987; Polakis *et al.* 1988; Marie *et al.* 1989). Consequently, attempts were made to solubilize bombesin/GRP receptors under conditions in which the ligand was pre-bound to the receptor prior to detergent extraction.

Coffer *et al.* (1990) found that [^{125}I]GRP–receptor complexes were solubilized

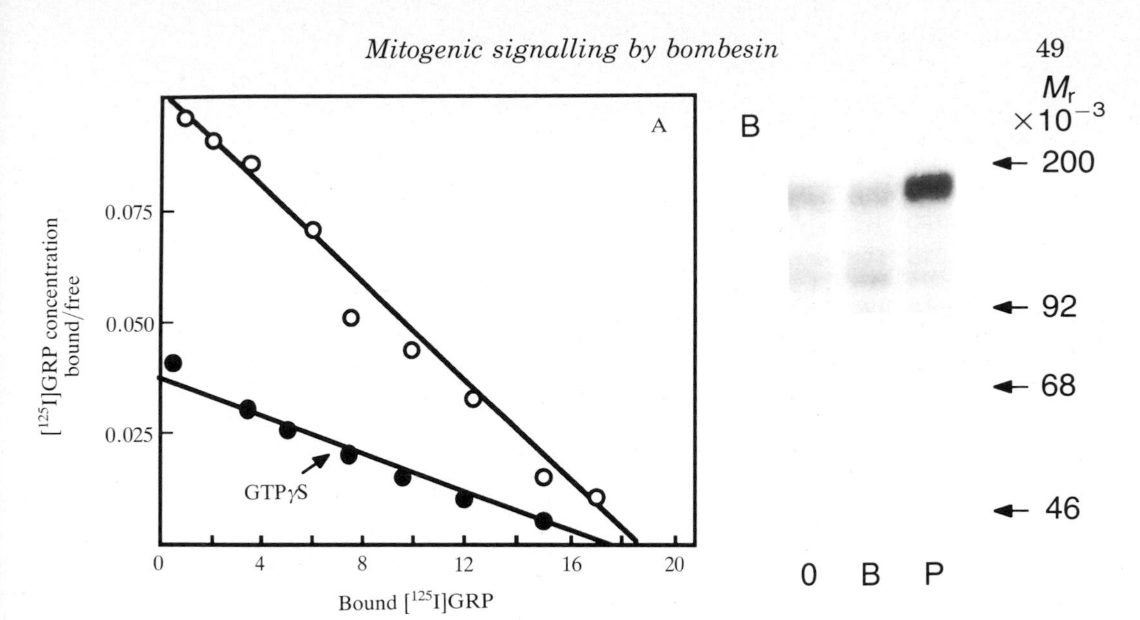

Fig. 1. (A) GTPγS reduces the affinity of the bombesin receptor for [^{125}I]GRP as determined by Scatchard analysis. Membranes in 100 μl of binding medium were incubated in the presence of various concentrations of [^{125}I]GRP at 37 °C either in the absence (○) or in the presence of GTPγS (10 μM), (●). Specific binding was determined after 10 min of incubation. Non-specific binding was measured by the addition of at least a 1000-fold excess of unlabelled bombesin, or 1 μM bombesin for concentrations of [^{125}I]GRP below 1 nM. Scatchard analysis of the data is shown: bound [^{125}I]GRP is expressed as fmol 25 μg^{-1} of membrane protein; the free [^{125}I]GRP concentration is expressed in pM. All other experimental details were as described by Sinnett-Smith *et al.* (1990). (B) Swiss 3T3 membranes show no detectable tyrosine phosphorylation by bombesin. Membranes (200 μg) were incubated in the absence (○) or presence of either bombesin (20 nM) (B) or PDGF (30 nM) (P) at 37 °C for 7.5 min in 250 μl of 50 mM Hepes, 5 mM $MgCl_2$, 0.5 mM $MnCl_2$, 100 μM $NaVO_4$, 20 μM $ZnCl_2$ (buffer A) containing [^{32}P]ATP (15 μCi). The reaction was terminated by the addition of 250 ml of buffer A containing Triton X-100 (2 %). The membranes were solubilized for 20 min at 4 °C, and non-extractable material removed by centrifugation at 16 000 ***g*** for 10 min at 4 °C. The supernatant was then incubated with BSA–agarose (100 μl) for 1 h at 4 °C. Following centrifugation at 16 000 ***g*** for 10 s to remove the BSA–agarose, the supernatant was incubated with monoclonal antiphosphotyrosine antibody coupled to agarose at 4 °C for 2 h. Following 3 washes with buffer A, the phosphotyrosinyl proteins were eluted from the matrix with SDS–sample buffer (0.1 M Tris–HCl, pH 6.8, 10 % (w/v) glycerol, 2 % (w/v) SDS, 0.1 M DTT) and analyzed by PAGE. The gel was then soaked in 1 M KOH at 55 °C for 1 h, dried and autoradiographed.

from Swiss 3T3 cell membranes by using the detergents taurodeoxycholate or deoxycholate. These detergents promoted ligand–receptor solubilization in a dose-dependent manner. In contrast, a variety of other detergents including Triton X-100, octylglycoside, CHAPS, digitonin, cholic acid and *n*-dodecyl-β-D maltoside, were much less effective. A typical experiment is shown in Fig. 2. Membrane preparations from Swiss 3T3 cells were incubated with [^{125}I]GRP and then freed of unbound ligand by centrifugation. The membrane pellet was resuspended in a solution containing 0.5 % deoxycholate and incubated for 30 min at 4 °C. The solubilized material was separated from the non-extractable

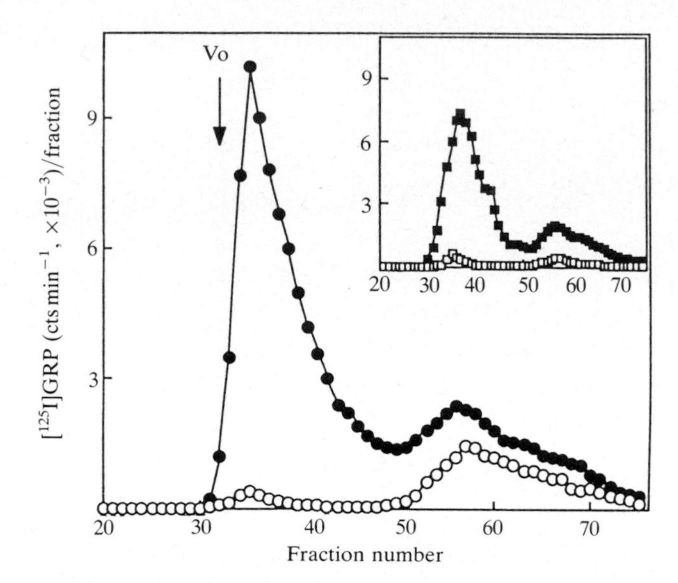

Fig. 2. Gel filtration profiles of solubilized $[^{125}I]$GRP–receptor complexes. Swiss 3T3 membranes (0.9 mg) were incubated in 30 mM Hepes, 5 mM $MgCl_2$, 0.25 M sucrose, 10 $\mu g\,ml^{-1}$ aprotonin, 1 $mg\,ml^{-1}$ bacitracin (binding medium) for 10 min at 37 °C with $[^{125}I]$GRP (0.5 nM) in the absence (●) and presence (○) of 10 μM bombesin. After centrifugation, the membranes were solubilized with 0.5 % taurodeoxycholate in binding medium at a final protein concentration of 4 $mg\,ml^{-1}$. The supernatant (200 μl) obtained was analyzed by gel filtration on a G–200 column (40 cm×0.9 cm) eluted with 30 mM Hepes, 5 mM $MgCl_2$ and 0.1 % taurodeoxycholate. Other experimental conditions were as described by Coffer *et al.* (1990). V_o, void volume of column. Inset: analysis of cross-linked complex by gel filtration. Membranes (100 μg) were chemically cross-linked to $[^{125}I]$GRP (0.5 nM) using EGS as described previously (Sinnett-Smith *et al.* 1990). Following solubilization of the membranes with taurodeoxycholate (0.5 %) as described above, the supernatant was analyzed on a G–200 column eluted with 0.25 % taurodeoxycholate. Fraction of 0.4 ml were collected at a flow rate of 8 $ml\,h^{-1}$ in both cases. The columns were calibrated using Blue Dextran (void volume of 12.8 ml), amylase (200×10^3 M_r), bovine serum albumin (68×10^3 M_r) and carbonic anhydrase (29×10^3 M_r).

material by centrifugation for 60 min at 100 000 ***g*** and chromatographed on a Sephadex G-200 column at 4 °C. Fig. 2A shows that a sharp peak of radioactivity was eluted with an apparent molecular weight of 190 000, while the remaining radioactivity co-eluted with free $[^{125}I]$GRP. The peak of radioactivity eluting near the void volume of the G–200 column was abolished by adding an excess of unlabelled bombesin together with $[^{125}I]$GRP during labelling of the membrane (Fig. 2). This experiment indicates that only a partial dissociation of the solubilized ligand–receptor complex occurred during the chromatographic separation and suggests that the ligand–receptor complex is physically associated to other proteins. This possibility was tested more stringently by using a higher detergent concentration (0.25 % instead of 0.1 %) during the chromatography. To circumvent the extensive dissociation of $[^{125}I]$GRP that occurs at this detergent concentration, the ligand was cross-linked to the receptor prior to solubilization. Fig. 2 (inset) shows that even in the presence of 0.25% deoxycholate, the cross-linked ligand–receptor complex remains associated to other proteins.

Several hormone receptors known to be functionally coupled to G proteins remain physically associated with them after detergent solubilization. To determine whether the solubilized [^{125}I]GRP–receptor complex is functionally coupled to a G protein(s), we tested whether the ligand–receptor complex isolated by gel filtration retains the ability to be regulated by guanine nucleotides. Swiss 3T3 membranes were incubated with [^{125}I]GRP and then the detergent-solubilized extract was chromatographed on a G-100 column. The ligand–receptor complex eluting in the void volume was pooled and incubated in the absence or in the presence of increasing concentrations of GTPγS for 30 min at 37°C. Then, bound [^{125}I]GRP was separated from dissociated ligand by chromatography (Coffer *et al.* 1990). As shown in Fig. 3, GTPγS caused a dose-dependent decrease in the level of [^{125}I]GRP–receptor complex. In contrast, addition of GMP, ATP and ATPγS at 100 μM did not have any detectable effect on the stability of the solubilized [^{125}I]GRP–receptor complex. The results of Coffer *et al.* (1990) and those shown in Figs 2 and 3 demonstrate for the first time the successful

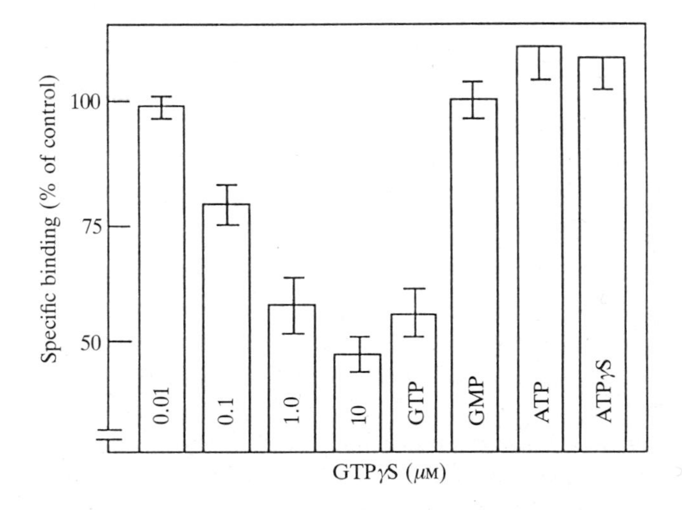

Fig. 3. GTPγS promotes dissociation of [^{125}I]GRP from [^{125}I]GRP–receptor complexes in a concentration-dependent and specific manner. Swiss 3T3 membranes (2×1.0 mg) were labelled with [^{125}I]GRP (0.5 nM) in binding medium at 37°C for 10 min, solubilized with 0.5% deoxycholate and chromatographed on two identical Sephadex G–100 columns (20×0.9 cm) as described in Fig. 2. Fractions containing the solubilized [^{125}I]GRP–receptor were then pooled. Samples (50–100 μl) were incubated in either the absence or the presence of GTPγS at the indicated concentrations, or the specified nucleotides as indicated, all at 100 μM. Following incubation of the reaction mixture at 37°C for 30 min, the dissociated, free [^{125}I]GRP was separated from the [^{125}I]GRP–receptor complex by spun column chromatography as described by Coffer *et al.* (1990). The results are expressed as the percentage of [^{125}I]GRP which remains bound to the receptor complex with respect to the control. The data represent the means±S.E.M.; n=6.

solubilization of [^{125}I]GRP–receptor complexes from Swiss 3T3 cell membranes, and provide functional evidence for a bombesin/GRP receptor–G protein interaction. The solubilization of the bombesin receptor in an active form may prove an important step for attempting its purification and reconstitution into phospholipid vesicles.

Conclusions

Bombesin initiates a complex cascade of signalling events that culminates in the stimulation of DNA synthesis and cell division (summarized in Table 1). PDGF stimulates a similar set of early signals (Rozengurt, 1986; Williams, 1989) and, like bombesin, promotes mitogenesis in the absence of other growth factors. These conclusions were recently substantiated using homodimers of the A chain of PDGF, which binds preferentially to one class of PDGF receptor rather than to all classes of receptors expressed in Swiss 3T3 cells (Mehmet *et al.* 1990). It is now well established that the tyrosine kinase activity of the PDGF receptor plays a central role in transduction of the mitogenic signal by PDGF. In contrast, the experiments summarized here provide strong evidence that the bombesin receptor is directly coupled to a G protein signal transduction pathway. Thus, it is likely that the bombesin receptor belongs to the superfamily of G protein-linked receptors which are glycoproteins of core M_r 40 000–50 000, thought to traverse the cytoplasmic membrane seven times (Lefkowitz and Caron, 1988). Other receptors for the regulatory peptides, substance P (Yokota *et al.* 1989; Hershey and Krause, 1990), substance K (Masu *et al.* 1987) and angiotensin (Jackson *et al.* 1988), belong to this superfamily. As yet, however, the elucidation of the structure of these receptors does not provide information concerning the class (i.e. G_s, G_i, G_o etc.) or number of G proteins to which they are coupled.

The nature of the G protein(s) that couple the bombesin receptor to the activation of phospholipase C remains obscure. Indeed, none of the pertussis toxin-insensitive G proteins involved in regulating phospholipase C have been identified. The isolation and purification of other members of the G protein family (like G_s and G_i) has been greatly assisted by the use of bacterial toxins such as cholera toxin and pertussis toxin, which ADP-ribosylate their target proteins. A similar toxin directed to the G protein(s) that couples various neuropeptide receptors to phospholipase C has not been discovered.

Recently, *Pasteurella multocida* toxin (PMT) has been identified as an extremely potent mitogen for Swiss 3T3 cells and other cultured cells (Rozengurt *et al.* 1990). This toxin causes a dramatic increase in the formation of inositol phosphates (Rozengurt *et al.* 1990) and stimulates protein kinase C activity (Staddon *et al.* 1990). In contrast, PMT does not promote accumulation of cyclic AMP. While several possibilities remain open, it is intriguing to consider the possibility that PMT facilitates signal transduction by altering the properties of a G protein, such as that described here to be coupled to the bombesin receptor.

References

ALBERT, K. A., NAIRN, A. C. AND GREENGARD, P. (1987). The 87-kDa protein, a major specific substrate for protein kinase C: purification from bovine brain and characterization. *Proc. natn. Acad. Sci. U.S.A.* **84**, 7046–7050.

BLACKSHEAR, P. J., WEN, L., GLYNN, B. P. AND WITTERS, L. A. (1986). Protein kinase C-stimulated phosphorylation *in vitro* of a M_r 80 000 protein phosphorylated in response to phorbol esters and growth factors in intact fibroblasts. *J. biol. Chem.* **261**, 1459–1469.

BLACKSHEAR, P. J., WITTERS, L. A., GIRARD, P. R., KUO, J. F. AND QUAMO, S. N. (1985). Growth factor-stimulated protein phosphorylation in 3T3-L1 cells. *J. biol. Chem.* **260**, 13 304–13 315.

BOURNE, H. R., MASTERS, S. B., MILLER, R. T., SULLIVAN, K. A. AND HEIDEMAN, W. (1988). Mutations probe structure and function of G-protein alpha chains. *Cold Spring Harb. Symp. quant. Biol.* **51**, 221–228.

BROWN, K. D., LAURIE, M. S., LITTLEWOOD, C. J., BLAKELEY, D. M. AND CORPS, A. N. (1988). Characterization of the high-affinity receptors on Swiss 3T3 cells which mediate the binding, internalization and degradation of the mitogenic peptide bombesin. *Biochem J.* **252**, 227–235.

CIRILLO, D. M., GAUDINO, G., NALDINI, L. AND COMOGLIO, P. M. (1986). Receptor for bombesin with associated tyrosine kinase activity. *Molec. cell. Biol.* **6**, 4641–4649.

COFFER, A., FABREGAT, I., SINNETT-SMITH, J. AND ROZENGURT, E. (1990). Solubilization of the bombesin receptor from Swiss 3T3 cells membranes: functional association to a guanine nucleotide regulatory protein. *FEBS Lett.* **263**, 80–84.

DICKEY, B. F., FISHMAN, J. B., FINE, R. E. AND NAVARRO, J. (1987). Reconstitution of the rat liver vasopressin receptor coupled to guanine nucleotide-binding proteins. *J. biol. Chem.* **262**, 8738–8742.

ERUSALIMSKY, J. D., FRIEDBERG, I. AND ROZENGURT, E. (1988). Bombesin, diacylglycerols, and phorbol esters rapidly stimulate the phosphorylation of an M_r=80 000 protein kinase C substrate in permeabilized 3T3 cells. *J. biol. Chem.* **263**, 19 188–19 194.

ERUSALIMSKY, J. D., MORRIS, C., PERKS, K., BROWN, R., BROOKS, S. AND ROZENGURT, E. (1989). Internal amino acid sequence analysis of the 80 kDa protein kinase C substrate from rat brain: relationship to the 87 kDa substrate from bovine brain. *FEBS Lett.* **255**, 149–153.

ERUSALIMSKY, J. D. AND ROZENGURT, E. (1989). Vasopressin rapidly stimulates protein kinase C in digitonin-permeabilized Swiss 3T3 cells: involvement of a pertussis toxin-insensitive guanine nucleotide binding protein. *J. cell. Physiol.* **141**, 253–261.

FABREGAT, I. AND ROZENGURT, E. (1990). Vasoactive Intestinal Contractor, a novel peptide, shares a common receptor with endothelin-1 and stimulates Ca^{2+} mobilization and DNA synthesis in Swiss 3T3 cells. *Biochem. biophys. Res. Commun.* **167**, 161–167.

GAUDINO, G., CIRILLO, D., NALDINI, L., ROSSINO, P. AND COMOGLIO, P. M. (1988). Activation of the protein-tyrosine kinase associated with the bombesin receptor complex in small cell lung carcinomas. *Proc. natn. Acad. Sci. U.S.A.* **85**, 2166–2170.

HERSHEY, A. D. AND KRAUSE, J. E. (1990). Molecular characterization of a functional cDNA encoding the rat substance P receptor. *Science* **247**, 958–962.

HESLOP, J. P., BLAKELEY, D. M., BROWN, K. D., IRVINE, R. F. AND BERRIDGE, M. J. (1986). Effects of bombesin and insulin on inositol (1,4,5)trisphosphate and inositol (1,3,4)trisphosphate formation in Swiss 3T3 cells. *Cell* **47**, 703–709.

ISACKE, C. M., MEISENHELDER, J., BROWN, K. D., GOULD, K. L., GOULD, S. J. AND HUNTER, T. (1986). Early phosphorylation events following the treatment of Swiss 3T3 cells with bombesin and the mammalian bombesin-related peptide, gastrin-releasing peptide. *EMBO J.* **5**, 2889–2898.

ISSANDOU, M. AND ROZENGURT, E. (1989). Diacylglycerols, unlike phorbol esters, do not induce homologous desensitization or down-regulation of protein kinase C in Swiss 3T3 cells. *Biochem. biophys. Res. Commun.* **163**, 201–208.

JACKSON, T. R., BLAIR, L. A. C., MARSHALL, J., GOEDERT, M. AND HANLEY, M. R. (1988). The *mas* oncogene encodes an angiotensin receptor. *Nature* **335**, 437–438.

KRIS, R. M., HAZAN, R., VILLINES, J., MOODY, T. W. AND SCHLESSINGER, J. (1987). Identification of the bombesin receptor on murine and human cells by cross-linking experiments. *J. biol. Chem.* **262**, 11 215–11 220.

LEFKOWITZ, R. J. AND CARON, M. G. (1988). Adrenergic receptors. Models for the study of receptors coupled to guanine nucleotide regulatory proteins. *J. biol. Chem.* **263**, 4993–4996.

LETTERIO, J. J., COUGHLIN, S. R. AND WILLIAMS, L. T. (1986). Pertussis toxin-sensitive pathway in the stimulation of c-*myc* expression and DNA synthesis by bombesin. *Science* **234**, 1117–1119.

LOPEZ-RIVAS, A., MENDOZA, S. A., NANBERG, E., SINNETT-SMITH, J. AND ROZENGURT, E. (1987). The Ca^{2+}-mobilizing actions of platelet-derived growth factor differ from those of bombesin and vasopressin in Swiss 3T3 cells. *Proc. natn. Acad. Sci. U.S.A.* **84**, 5768–5772.

MARIE, J.-C., COTRONEO, P., DE CHASSEVAL, R. AND ROSSELIN, G. (1989). Solubilization of somatostatin receptors in hamster pancreatic beta cells. Characterization as a glycoprotein interacting with a GTP-binding protein. *Eur. J. Biochem.* **186**, 181–188.

MASU, Y., NAKAYAMA, K., TAMAKI, H., HARADA, Y., KUNO, M. AND NAKANISHI, S. (1987). cDNA cloning of bovine substance-K receptor through oocyte expression system. *Nature* **329**, 836–837.

MEHMET, H., MOORE, J. P., SINNETT-SMITH, J. W., EVAN, G. I. AND ROZENGURT, E. (1989). Dissociation of c-*fos* induction from protein kinase C-independent mitogenesis in Swiss 3T3 cells. *Oncogene Res.* **1**, 215–222.

MEHMET, H., NANBERG, E., LEHMAN, W., MURRAY, M. J. AND ROZENGURT, E. (1990). Early signals in the mitogenic response of Swiss 3T3 cells: a comparative study of purified PDGF homodimers. *Growth Factors* **3**, 83–95.

MENDOZA, S. A., SCHNEIDER, J. A., LOPEZ-RIVAS, A., SINNETT-SMITH, J. W. AND ROZENGURT, E. (1986). Early events elicited by bombesin and structurally related peptides in quiescent Swiss 3T3 cells. II. Changes in Na^+ and Ca^+ fluxes, Na^+/K^+ pump activity, and intracellular pH. *J. cell Biol.* **102**, 2223–2233.

MILLAR, J. B. A. AND ROZENGURT, E. (1988). Bombesin enhancement of cAMP accumulation in Swiss 3T3 cells: evidence of a dual mechanism of action. *J. cell. Physiol.* **137**, 214–222.

MORRIS, C. AND ROZENGURT, E. (1988). Purification of a phosphoprotein from rat brain closely related to the 80 kDa substrate of protein kinase C identified in Swiss 3T3 fibroblasts. *FEBS lett.* **231**, 311–316.

MUIR, J. G. AND MURRAY, A. W. (1987). Bombesin and phorbol ester stimulate phosphatidylcholine hydrolysis by phospholipase C: evidence for a role of protein kinase C. *J. cell. Physiol.* **130**, 382–391.

NANBERG, E. AND ROZENGURT, E. (1988). Temporal relationship between inositol polyphosphate formation and increases in cytosolic Ca^{2+} in quiescent 3T3 cells stimulated by platelet-derived growth factor, bombesin and vasopressin. *EMBO J.* **7**, 2741–2747.

POLAKIS, P. G., UHING, R. J. AND SNYDERMAN, R. (1988). The formylpeptide chemoattractant receptor copurifies with a GTP-binding protein containing a distinct 40 kDa pertussis toxin substrate. *J. biol. Chem.* **263**, 4969–4976.

RODRIGUEZ-PENA, A. AND ROZENGURT, E. (1985). Serum, like phorbol esters, rapidly activates protein kinase C in intact quiescent fibroblasts. *EMBO J.* **4**, 71–76.

RODRIGUEZ-PENA, A. AND ROZENGURT, E. (1986). Phosphorylation of an acidic mol. wt. 80 000 cellular protein in a cell-free system and intact Swiss 3T3 cells: a specific marker of protein kinase C activity. *EMBO J.* **5**, 77–83.

RODRIGUEZ-PENA, A., ZACHARY, I. AND ROZENGURT, E. (1986). Rapid dephosphorylation of a M_r 80 000 protein, a specific substrate of protein kinase C, upon removal of phorbol esters, bombesin and vasopressin. *Biochem. biophys. Res. Commun.* **140**, 379–385.

ROZENGURT, E. (1986). Early signals in the mitogenic response. *Science* **234**, 161–166.

ROZENGURT, E. (1988). Bombesin-induction of cell proliferation in 3T3 cells. Specific receptors and early signaling events. *Ann. N.Y. Acad. Sci.* **547**, 277–292.

ROZENGURT, E. (1989). Signal transduction pathways in mitogenesis. *Br. med. Bull.* **45**, 515–528.

ROZENGURT, E., BROWN, K. AND PETTICAN, P. (1981). Vasopressin inhibition of epidermal growth factor binding to cultured mouse cells. *J. biol. Chem.* **256**, 716–722.

ROZENGURT, E., ERUSALIMSKY, J., MEHMET, H., MORRIS, C., NANBERG, E. AND SINNETT-SMITH, J. (1988). Signal transduction in mitogenesis: further evidence for multiple pathways. *Cold Spring Harb. Symp. quant. Biol.* **53**, 945–954.

ROZENGURT, E., HIGGINS, T., CHANTER, N., LAX, A. J. AND STADDON, J. M. (1990). *Pasteurella multocida* toxin: a novel potent mitogen for cultured fibroblasts. *Proc. natn. Acad. Sci. U.S.A.* **87**, 123–127.

ROZENGURT, E., LEGG, A. AND PETTICAN, P. (1979). Vasopressin stimulation of 3T3 cell growth. *Proc. natn. Acad. Sci. U.S.A.* **76**, 1284–1287.

ROZENGURT, E., RODRIGUEZ-PENA, A., COOMBS, M. AND SINNETT-SMITH, J. (1984). Diacylglycerol stimulates DNA and cell division in mouse 3T3 cells: role of Ca^{2+}-sensitive, phospholipid-dependent protein kinase. *Proc. natn. Acad. Sci. U.S.A.* **81**, 5748–5752.

ROZENGURT, E., RODRIGUEZ-PENA, M. AND SMITH, K. A. (1983). Phorbol esters, phospholipase C, and growth factors rapidly stimulate the phosphorylation of a M_r 80 000 protein in intact quiescent 3T3 cells. *Proc. natn. Acad. Sci. U.S.A.* **80**, 7244–7248.

ROZENGURT, E. AND SINNETT-SMITH, J. (1983). Bombesin stimulation of DNA synthesis and cell division in cultures of Swiss 3T3 cells. *Proc. natn. Acad. Sci. U.S.A.* **80**, 2936–2940.

ROZENGURT, E. AND SINNETT-SMITH, J. W. (1987). Bombesin induction of c-*fos* and c-*myc* proto-oncogenes in Swiss 3T3 cells: significance for the mitogenic response. *J. cell. Physiol.* **131**, 218–225.

ROZENGURT, E. AND SINNETT-SMITH, J. (1988). Early signals underlying the induction of the c-*fos* and c-*myc* genes in quiescent fibroblasts: studies with bombesin and other growth factors. *Prog. nucl. acid Res. molec. Biol.* **35**, 261–295.

ROZENGURT, E. AND SINNETT-SMITH, J. (1990). Bombesin stimulation of fibroblast mitogenesis: specific receptors, signal transduction and early events. *Phil. Trans. R. Soc.* (in press).

SINNETT-SMITH, J., LEHMANN, W. AND ROZENGURT, E. (1990). Bombesin receptor in membranes from Swiss 3T3 cells. Binding characteristics, affinity labelling and modulation by guanine nucleotides. *Biochem. J.* **265**, 485–493.

SINNETT-SMITH, J., ZACHARY, I. AND ROZENGURT, E. (1988). Characterization of a bombesin receptor on Swiss mouse 3T3 cells by affinity cross-linking. *J. cell. Biochem.* **38**, 237–249.

STADDON, J. M., CHANTER, N., LAX, A. J., HIGGINS, T. E. AND ROZENGURT, E. (1990). *Pasteurella multocida* toxin, a potent mitogen, stimulates protein kinase C-dependent and -independent protein phosphorylation in Swiss 3T3 cells. *J. biol. Chem.* **265**, 11 841–11 848.

STUMPO, D. J., GRAFF, J. M., LABERT, K. A., GREENGARD, P. AND BLACKSHEAR, P. J. (1989). Molecular cloning, characterization, and expression of a cDNA encoding the '80- to 87-kDa' myristoylated alanine-rich C kinase substrate: a major cellular substrate for protein kinase C. *Proc. natn. Acad. Sci. U.S.A.* **86**, 4012–4016.

TAKUWA, N., TAKUWA, Y., BOLLAG, W. E. AND RASMUSSEN, H. (1987). The effects of bombesin on polyphosphoinositide and calcium metabolism in Swiss 3T3 cells. *J. biol. Chem.* **262**, 182–188.

TAKUWA, N., TAKUWA, Y., YANAGISAWA, M., YAMASHITA, K. AND MASAKI, T. (1989). A novel vasoactive peptide endothelin stimulates mitogenesis through inositol lipid turnover in Swiss 3T3 fibroblasts. *J. biol. Chem.* **264**, 7856–7861.

WILLIAMS, L. T. (1989). Signal transduction by the platelet-derived growth factor receptor. *Science* **243**, 1564–1570.

WOLL, P. J., COY, D. H. AND ROZENGURT, E. (1988). [Leu^{13}-ψ(CH_2NH)-Leu^{14}] bombesin is a specific bombesin receptor antagonist in Swiss 3T3 cells. *Biochem. biophys. Res. Commun.* **155**, 359–365.

WOLL, P. J. AND ROZENGURT, E. (1988). Two classes of antagonist interact with receptors for the mitogenic neuropeptides bombesin, bradykinin, and vasopressin. *Growth Factors* **1**, 75–83.

WOLL, P. J. AND ROZENGURT, E. (1989*a*). Neuropeptides as growth regulators. *Br. med. Bull.* **45**, 492–505.

WOLL, P. J. AND ROZENGURT, E. (1989*b*). Multiple neuropeptides mobilise calcium in small cell lung cancer: effects of vasopressin, bradykinin, cholecystokinin, galanin and neurotensin. *Biochem. biophys. Res. Commun.* **164**, 66–73.

YOKOTA, Y., SASAI, Y., TANAKA, K., FUJIWARA, T., TSUCHIDA, K., SHIGEMOTO, R., KAKIZUKA, A., OHKUBO, H. AND NAKANISHI, S. (1989). Molecular characterization of a functional cDNA for rat substance P receptor. *J. biol. Chem.* **264**, 17 649–17 652.

ZACHARY, I., MILLAR, J., NANBERG, E., HIGGINS, T. AND ROZENGURT, E. (1987*b*). Inhibition of bombesin-induced mitogenesis by pertussis toxin: dissociation from phospholipase C pathway. *Biochem. biophys. Res. Commun.* **146**, 456–463.

ZACHARY, I. AND ROZENGURT, E. (1985*a*). High-affinity receptors for peptides of the bombesin family in Swiss 3T3 cells. *Proc. natn. Acad. Sci. U.S.A.* **82**, 7616–7620.

ZACHARY, I. AND ROZENGURT, E. (1985*b*). Modulation of the epidermal growth factor receptor by mitogenic ligands: effects of bombesin and role of protein kinase C. *Cancer Surveys* **4**, 729–765.

ZACHARY, I. AND ROZENGURT, E. (1987*a*). Identification of a receptor for peptides of the bombesin family in Swiss 3T3 cells by affinity cross-linking. *J. biol. Chem.* **262**, 3947–3950.

ZACHARY, I. AND ROZENGURT, E. (1987*b*). Internalization and degradation of peptides of the bombesin family in Swiss 3T3 cells occurs without ligand-induced receptor down-regulation. *EMBO J.* **6**, 2233–2239.

ZACHARY, I., SINNETT-SMITH, J. W. AND ROZENGURT, E. (1986). Early events elicited by bombesin and structurally related peptides in quiescent Swiss 3T3 cells. I. Activation of protein kinase C and inhibition of epidermal growth factor binding. *J. cell Biol.* **102**, 2211–2222.

ZACHARY, I., WOLL, P. J. AND ROZENGURT, E. (1987*a*). A role for neuropeptides in the control of cell proliferation. *Devl Biol.* **124**, 295–308.

ZURIER, R. B., KOZMA, M., SINNETT-SMITH, J. AND ROZENGURT, E. (1988). Vasoactive intestinal peptide synergistically stimulates DNA synthesis in mouse 3T3 cells: role of cAMP, Ca^{2+}, and protein kinase C. *Expl cell Res.* **176**, 155–161.

J. Cell Sci. Suppl. 13, 57–74 (1990)
Printed in Great Britain © The Company of Biologists Limited 1990

The biochemistry and biology of the myeloid haemopoietic cell growth factors

C. M. HEYWORTH[1], S. J. VALLANCE[2], A. D. WHETTON[2] AND T. M. DEXTER[1]

[1]*Cancer Research Campaign Department of Experimental Haematology, Paterson Institute for Cancer Research, Christie Hospital and Holt Radium Institute, Manchester M20 9BX, UK*
[2]*Department of Biochemistry and Applied Molecular Biology, UMIST, Manchester M60 1QD, UK*

Summary

In the adult, blood cell production or haemopoiesis takes place mainly in the bone marrow. The blood cell types produced are a reflection of the needs of the organism at any moment, for example bacterial infection leads to a large increase in neutrophil production. The rate and scale of blood cell production *in vivo* are regulated, at least in part, by the synthesis and release of specific cytokines both within the bone marrow and also from other tissues. Here we detail the range of cytokines which act directly on haemopoietic stem cells and myeloid progenitor cells. Also cellular systems which will permit the elucidation of the specific interactions between these various cytokines which regulate stem cell self-renewal, differentiation and proliferation are described.

Introduction

The continuous production of mature blood cells is required for a wide variety of functions and is essential for survival. The physiology of haemopoiesis has been elucidated in part, and some of the complex regulatory networks which control blood cell development are now known. For example, the proliferation and development of multipotent and lineage-restricted myeloid progenitor cells are controlled by a series of haemopoietic growth factors which have recently been purified and molecularly cloned. The availability of sufficient quantities of these recombinant growth factors has enabled an elucidation of the specific target cells for these agents (Table 1). This in turn has encouraged studies aimed at identifying the intracellular mechanisms, stimulated by these growth factors, that facilitate the survival, proliferation and development of haemopoietic cells. Recent progress in these fields is considered here.

Haemopoiesis

All mature blood cells: erythrocytes, platelets, B lymphocytes, T lymphocytes, neutrophils, eosinophils, basophils and macrophages, are derived from pluripo-

Key words: self-renewal, differentiation, haemopoiesis, stem cells, haemopoietic growth factors.

Table 1. *Growth factors which stimulate myeloid cell development*

Growth factor	Responsive progenitor cell populations
Interleukin 3 (IL-3)	GM-CFC Eos-CFC Meg-CFC Bas-CFC BFU-E Multipotent Cells
Granulocyte Macrophage Colony Stimulating Factor (GM-CSF)	GM-CFC Meg-CFC Eos-CFC BFU-E
Macrophage Colony Stimulating Factor (M-CSF or CSF-1)	GM-CFC (Mainly macrophage development)
Granulocyte Colony Stimulating Factor (G-CSF)	GM-CFC (Mainly granulocyte development)
Erythropoietin (epo)	CFU-e
Interleukin 1 (IL-1)	Multipotent cells (see text)
Interleukin 6 (IL-6)	Multipotent cells (see text)
Interleukin 5 (IL-5)	Eos-CFC
Interleukin 4 (IL-4)	BFU-E Bas-CFC

tent stem cells which, in the adult, reside mainly in the bone marrow (Lajtha, 1982). The presence of stem cells is inferred by the ability of marrow cells to reconstitute lympho- and myelo-poiesis when transferred into irradiated hosts. However, several months are required before full and lasting reconstitution with donor cells can be unequivocally demonstrated, and for this reason, the shorter term quantitative assay of 'spleen colony formation', developed in the 1960's by Till and McCulloch, is consequently used as an assay for pluripotent haemopoietic cells (McCulloch and Till, 1964). The most primitive spleen colony forming cells (CFU-S) are multipotent, can produce mature cells representative of all the myeloid cell lineages and can undergo extensive self-renewal. Normally, only a minority of CFU-S are undergoing DNA synthesis, but this proportion increases dramatically during recovery from cytoreductive therapy or following bone marrow transplantation (Lajtha, 1982).

The recognition of CFU-S as a minimal-cycling population during steady-state haemopoiesis infers that the massive expansion in cell numbers between the CFU-S and the mature cells occurs *via* an intermediate population of proliferating cells. These intermediate progenitor cells have been characterised using *in vivo* assays such as for the erythropoietin responsive cell (ERC) (Bruce and McCulloch, 1964) and *in vitro* assays, which assess the ability of cells to give rise to colonies of mature haemopoietic cells when immobilised in soft gel media, such as agar or methylcellulose (Metcalf, 1984). The clonal nature of the colonies arising from the

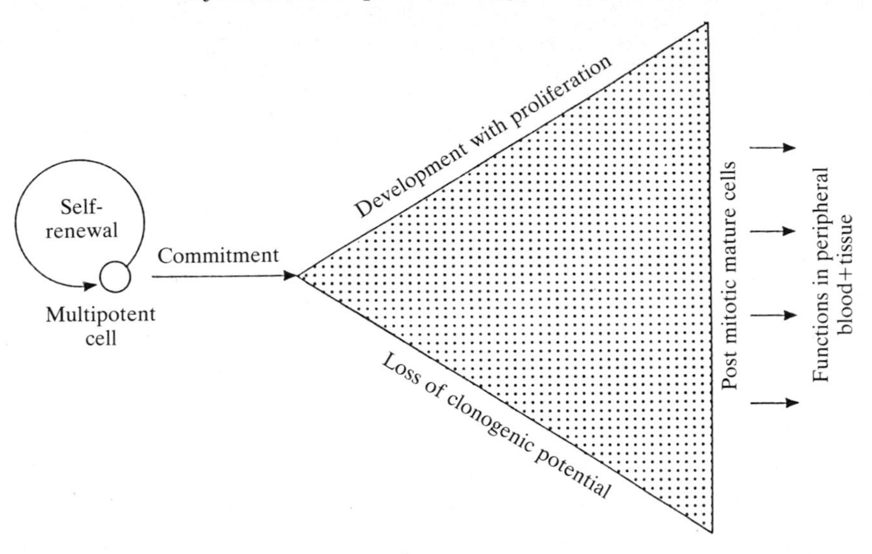

Fig. 1. Haemopoiesis. Cells involved in haemopoiesis can be divided into three main stages: the multipotent cells which have differing degrees of ability to reconstitute the bone marrow of irradiated mice, but nonetheless are multipotent and have the capacity to self-renew; progenitor cells, which are unipotent or bipotent cells committed to a programme of development to form the third stage, which consists of mature cells, displaying a range of grossly different morphologies and functional activities, yet all derived from a common pool of pluripotent stem cells. There are a number of assays which can be employed to identify committed progenitor cells and some of the multipotent cells found in the bone marrow. The CFU-S (Colony Forming Unit-Spleen) assay is an *in vivo* assay for multipotent, self-renewing stem cells, whilst the high proliferative potential colony forming cell (HPP-CFC Stanley *et al.* 1986) assay identifies a primitive, multipotential cell using soft agar colony formation as the parameter which is measured (see text). These cells can give rise to the colony forming cell-mix (CFC-Mix) which has a lower capacity for self-renewal but is still multipotent. The multipotent cell then differentiates to give rise to the lineage-restricted, committed progenitor cells. These include the Burst Forming Unit-Erythroid (BFU-E), Colony Forming Unit-Erythroid (CFU-E), Granulocyte Macrophage Colony Forming Cell (GM-CFC), Eosinophil Colony Forming Cell (Eos-CFC), Megakaryocyte-CFC (Meg-CFC) and Basophil-CFC (Bas-CFC).

progenitor cells has been demonstrated using karyotypic analysis, isoenzyme markers and a number of other techniques (Metcalf, 1984). From these assays it has become apparent that the provision of media containing the appropriate growth stimulator(s) is an essential element in the formation of such colonies and is necessary for the progenitor cells to survive, proliferate and develop. The absence of these 'colony stimulating factors' (CSFs) leads to the death of progenitor cells within a very short time scale (8–48 h) (Metcalf and Merchav, 1982).

Using these *in vitro* assays, a number of distinct, committed, progenitor cell populations have now been identified (see Fig. 1). Such cells are generally limited in their developmental potential to only one or two of the haemopoietic lineages,

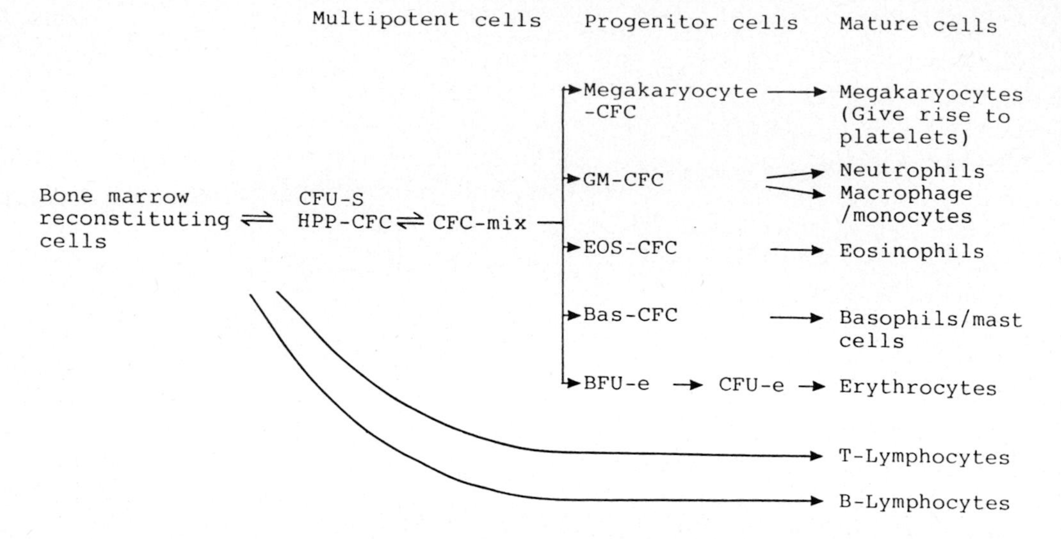

Fig. 2. The basic structure of haemopoiesis. Multipotent cells can either undergo self-renewal, or become committed to development. The development of haemopoietic progenitor cells is coupled to proliferation, such that one progenitor cell can give rise to thousands of mature cells. A corollary of this development is the loss of proliferative capacity until postmitotic but functionally active cells (with finite lifetimes) such as neutrophils, platelets, or erythrocytes are formed. The death of these cells requires that they are constantly replaced throughout the lifetime of the organism, thus haemopoiesis is an ongoing process in all healthy animals.

their proliferation is normally tightly coupled to development and the cells progressively display the antigenic, biochemical and morphological features characteristic of the mature cells of the appropriate lineage. Accompanying this process of development, there is a gradual loss of proliferative potential, resulting in the formation of the postmitotic mature cells (see Fig. 2): in the case of the erythrocyte and the platelet, no inheritable genetic material is retained by the mature cell, while in neutrophils the DNA is retained in a condensed form and the cells are unable to undergo replication. Thus, when cells from these three lineages leave the bone marrow, they possess little or no proliferative potential. Other haemopoietic cells, however, can migrate from the bone marrow and undergo proliferation and development at other specific sites in the body (e.g. pre T cells in the thymus) (Wood, 1982). Similarly, the mast cell and the monocyte can undergo further replication and development in the tissues (Stanley *et al.* 1983; Tsuji *et al.* 1990). Although the mechanism of acquisition of a postmitotic phenotype in myeloid cells is not yet understood, an insight into this process is clearly of value in understanding the cellular mechanisms underlying the formation of leukaemic cells from normal haemopoietic tissue. In this respect, identification of the CSFs and other growth factors, and the characterisation of their target cells and modes of action, has implications not only for development of normal haemopoietic cells, but for leukaemic cells also.

Growth factors which regulate myeloid cell production

Many growth factors responsible for initiating proliferation and development of haemopoietic progenitor cells *in vitro* have now been molecularly cloned and purified to homogeneity (see Clark and Kamen, 1987; Whetton and Dexter, 1989). The availability of large quantities of highly purified material has facilitated both *in vitro* and *in vivo* studies. Both animal and clinical trials have confirmed their physiological significance. The target cell populations which respond to each growth factor, and its effects *in vivo*, are summarised in the next section.

Interleukin 3 (IL-3)

Of the known haemopoietic growth factors acting on myeloid cells, IL-3 has activity on the greatest range of committed progenitor cells. It can maintain the viability of the CFU-S population in a highly enriched population of cells (precluding the possibility that IL-3 promotes the production of a secondary growth factor from other cell types in the bone marrow which then act on CFU-S) and also initiate DNA synthesis in these cells (Heyworth *et al.* 1988; Spirak *et al.* 1985; Schrader and Clark-Lewis, 1982). It can stimulate multipotent cells to develop into colonies containing mature cells from the macrophage, neutrophil, eosinophil, mast cell, megakaryocytic and erythroid lineages (see Whetton and Dexter, 1989). Similarly most of the committed progenitor cells from the myeloid cell lineages will proliferate and develop in response to IL-3. IL-3 can stimulate the development of mature cells from granulocyte macrophage colony forming cells (Cook *et al.* 1989), mast cell progenitors, eosinophil (Rothenberg *et al.* 1988), megakaryocyte (Bruno *et al.* 1988) and erythroid (Sonoda *et al.* 1988) progenitor cells. Although there is little evidence to suggest that cells committed to T and B lymphocyte development respond directly to IL-3, using haemopoietic reconstitution in irradiated mice as an assay, it has been demonstrated that cells which can give rise to T and B lymphocytes can respond to IL-3 (Clark-Lewis *et al.* 1985). In some instances there is a requirement for a second growth factor for the complete development of all the progenitor cells: late erythroid progenitor cells require the presence of erythropoietin to form mature erythroid cells (Iscove *et al.* 1974), and IL-5 is required for eosinophil maturation (Yamaguchi *et al.* 1988). However IL-3 is generally a powerful mitogen for the early progenitor cells in any given lineage: it is the more mature cells which are relatively less responsive. Thus IL-3 acts at several stages of development of haemopoietic cells, and on cells from several distinct lineages.

Furthermore, some mature cells are responsive to IL-3, although it is notable that erythrocytes and possibly human neutrophils no longer express cell surface receptors for this cytokine. Monocytes are still able to respond to IL-3: treatment with this growth factor not only stimulates survival and maintains the functional activity but also enhances the cytotoxic activity of these cells (Cannistra *et al.* 1988). This latter activity may be mediated by the potentiation of tumour necrosis factor expression (Cannistra *et al.* 1987). Similarly, mast cells exhibit an IL-3 potentiated histamine granule release, and connective tissue type mast cells are dependent on IL-3 for their survival and proliferation (Tsuji *et al.* 1990). Thus, like other myeloid growth factors (see below), IL-3 can potentiate mature cell function in addition to its role in the proliferation and development of haemopoietic progenitor cells.

Interleukin 1 (IL-1)

Some cytokines have no ability to stimulate directly the formation of colonies from normal bone marrow, but can markedly potentiate the proliferation and development of haemopoietic cells *in vitro* and *in vivo* in response to growth factors. For example, Bradley, Stanley and co-workers identified preparations of conditioned media from leukaemic cells which, when added to bone marrow cells individually, had no ability to stimulate colony formation; when added together large colonies of mature haemopoietic cells were formed (Stanley *et al.* 1986). Based on these observations, a biological activity named haemopoietin 1 was purified and characterised from the conditioned medium of leukaemic 5637 cells, on the basis of its ability to stimulate proliferation and differentiation of multipotent cells when combined with macrophage colony stimulating factor (M-CSF) (see later). These two agents synergistically promote the development of large colonies containing mature macrophages, but when added alone neither agent stimulates colony formation. Mochizinki *et al.* (1987) later demonstrated that haemopoietin 1 was identical to Interleukin 1 (IL-1). However, the experiments performed did not supply absolute proof of a direct synergy between these two growth factors acting on a single (multipotent) target cell population. Since Interleukin 1 is known to stimulate the production of a range of cytokines and haemopoietic growth factors from cells found in the bone marrow (Lee *et al.* 1987), it was possible that IL-1 was acting in a paracrine fashion to stimulate the release of growth factors from such cells which then act in concert with M-CSF to stimulate colony formation (see Bagby, 1989, for a review). Indeed, it has been proposed that such a model is responsible for the actions of IL-1 on multipotent haemopoietic cells (see Ikebuchi *et al.* 1988*a*). In more recent work, however, it has been shown that IL-1 can synergise with M-CSF to stimulate the formation of colonies from a cell population highly enriched for CFU-S (Heyworth *et al.* 1988). This indicates that, at least in some circumstances, IL-1 can directly influence the growth and development of multipotential cells. It is of course possible that IL-1 can act on multipotent cells to stimulate the production of other growth factors, which then act in an *intrinsic* fashion. With *in vivo* experiments the situation is, of course, even more complex. But in terms of the effects on haemopoietic cells perhaps the most significant discovery is that IL-1 can decrease the time taken to recover from leucopenia following treatment with cytotoxic drugs such as 5-fluorouracil (Moore *et al.* 1990).

Interleukin 4 (IL-4)

Originally described as a B cell growth factor (Widmer and Grabstein, 1987), IL-4 has recently been shown to act on a number of other cell types, including myeloid progenitor cells. Early studies suggested that IL-4 could synergise with IL-3 (to stimulate mast cell, but suppress neutrophilic cell, development), erythropoietin (erythroid cell development), G-CSF and M-CSF (neutrophil and macrophage development respectively) (Rennick *et al.* 1987; Peschel *et al.* 1987).

Further studies using bone marrow from 5-fluorouracil treated mice (which has few committed progenitor cells, but still contains multipotent, primitive, haemopoietic cells) suggest that IL-4 can promote the differentiation and development of multipotent cells (Kishi *et al.* 1989). The addition of IL-4 to soft agar cultures of such bone marrow preparations gives a few granulocyte macrophage and blast cell colonies. Further replating of such blast cell colonies resulted in the development of colonies containing cells from several distinct lineages, indicating that IL-4 had supported the proliferation of multipotent cells in the original blast cell colonies.

Interleukin 6 (IL-6)

Interleukin 6 is a pleiotropic cytokine that has the ability to modulate acute phase protein levels, stimulate the production of neutrophils and to stimulate B cell proliferation. Ogawa and co-workers have presented evidence that IL-6 can act on multipotential cells to reduce the G_0 period of the cell cycle, and that when the multipotent cells leave this phase of the cycle they then become responsive to IL-3 (Ikebuchi *et al.* 1988*a*). *In vivo* experiments also provided evidence that IL-6 can act on the primitive multipotent haemopoietic cells as well as megakaryocyte precursors; infusion of this growth factor into mice leads to a large increase in the number of CFU-S present in the spleen (Suzuki *et al.* 1989), and also to an increase in the platelet count. Although it has been suggested that the effects of IL-1 on multipotent cells can be attributed to its ability to stimulate the production of IL-6 from bone marrow stromal cells (Ikebuchi *et al.* 1988*a*), our investigations using highly enriched multipotent cells (where no helper cell population is present) suggest that both IL-6 and IL-1 (see above) can act directly on these cells (Heyworth *et al.* 1988).

The model presented by Sachs and co-workers (1990) may explain this enigma. They suggest that normal myeloid progenitor cells can be stimulated by growth factors such as IL-3, GM-CSF, M-CSF or G-CSF to produce IL-6. This autocrine production of IL-6 then acts to make the progenitor cells differentiate, while the growth factors promote proliferation; in this way proliferation and differentiation are coupled. Of course distinct growth factors may stimulate different levels of production of IL-6, thus the balance between proliferation and development would vary depending on the growth factor employed. Clearly the appropriate techniques are now available to test the validity of this hypothesis using highly enriched (growth factor responsive) progenitor cell populations coupled to an analysis of IL-6 gene expression. However, it should be noted that data do not support the concept of IL-6 acting solely as a differentiation inducer. Using enriched progenitor cells it has been demonstrated that IL-6 can act as a proliferative stimulus. Furthermore there is no evidence that IL-6 induces cellular maturation without accompanying proliferation of such cells (Ikebuchi *et al.* 1987; Hoang *et al.* 1988; Suda *et al.* 1988; Carcacciolo *et al.* 1989).

Transforming Growth Factor-1 (TGF-β)

TGF-β has a wide variety of effects including stimulation of osteoclast and Schwann cell proliferation, and the inhibition of proliferation of epithelial cells, fibroblasts and endothelial cells (Axelrad, 1990). Recently it has been demonstrated that TGF-β is produced by cells present in the bone marrow (Eaves *et al.* 1988) and the haemopoietic areas of foetal liver, and evidence from experiments *in vitro* (using highly enriched marrow progenitor cells) suggest that TGF-β can directly influence the growth of haemopoietic cells. TGF-β does not stimulate colony formation from progenitor cells, but can inhibit proliferation and development of fluorescence-activated, cell sorted (FACS) purified CFU-S stimulated with IL-3 (Hampson *et al.* 1989) and also inhibit the proliferation of granulocyte macrophage colony forming cell (GM-CFC) following stimulation with IL-3 and M-CSF. Similarly the addition of TGF-β to colony forming assays for BFU-E or CFU-E leads to inhibited growth (Del Rizzo *et al.* 1990; Hino *et al.* 1988; Axelrad *et al.* 1987). GM-CSF stimulated colony formation is, in some cases, inhibited (Sing *et al.* 1988) and in others, activated, by TGF-β.

These data firmly indicate a physiological role for TGF-β in the control of haemopoiesis.

Stem Cell Inhibitor (Macrophage Inflammatory Protein 1)

Crude extracts from several tissues, and also from the conditioned medium of cells such as macrophages, have been known for some time to contain an activity which has the ability to inhibit the progression of CFU-S into the cell cycle from G_0. This activity is present in haemopoietic tissues where CFU-S can be shown to be quiescent, but is much reduced in regenerating bone marrow, where CFU-S are actively cycling (Graham *et al.* 1990). Such data suggest that this 'inhibitor' may play a role in regulating stem cell proliferation.

Recent attempts to isolate, purify and obtain a partial sequence for this stem cell inhibitor by Pragnell and co-workers have met with success, and it has now been shown that the CFU-S inhibitor is probably the same molecular entity as Macrophage Inflammatory Protein 1α (MIP-1α) (Graham *et al.* 1990). At present the range of target cells on which this cytokine acts have not been established, although Broxmeyer and co-workers have shown a significant enhancement of granulocyte macrophage colony formation in the presence of MIP-1α, using highly enriched GM-CFC (Broxmeyer *et al.* 1989). It will now be of some interest to determine the *in vivo* effects of this novel cytokine, particularly on stem cell proliferation.

The negative regulator of CFU-S entry into DNA synthesis, characterised and purified by Guigon, Frindel and co-workers, has already been shown to be effective *in vivo* as a proliferation inhibitor (Lenfant *et al.* 1989). Surprisingly, this molecule is a simple, acetylated tetrapeptide, although this may be a fragment of a larger molecule found *in vivo* (Axelrad, 1990; Lenfant *et al.* 1989). Undoubtedly, though, this molecule, MIP-1α, TGF-β and other inhibitors of stem cell proliferation such as inhibin (Axelrad, 1990), will be the focus of a great deal of attention because of their possible use in the treatment of patients who may risk bone marrow failure as a result of chemotherapy, and also because the failure of stem cells to respond to such inhibitors may be one of the underlying causes of malignant disease (see Dexter and White, 1990).

Granulocyte Macrophage Colony Stimulating Factor

In *in vitro* assays, murine GM-CSF can stimulate the development of colonies containing predominantly neutrophils and/or macrophages, and also eosinophils. Higher concentrations of the growth factor can also stimulate the development of megakaryocytic cells, and stimulate the proliferation and development of early erythroid progenitor cells (BFU-E) (see Whetton and Dexter, 1989; Metcalf, 1984). There is also evidence that some multipotent cells can respond to GM-CSF. This ability to act on the immature myeloid progenitor cell does not, however, define the full range of biological activity of this growth factor.

The mature myeloid cells also exhibit a profound response to GM-CSF: peripheral blood monocytes and tissue-based macrophages proliferate in response to GM-CSF and also show an enhanced cytotoxic activity (Grabstein *et al.* 1986; Reed *et al.* 1987). Monocytes can also be stimulated by GM-CSF to release prostaglandin E, arachidonic acid, IL-1, tumour necrosis factor and other cytokines such as M-CSF (Di Persio *et al.* 1989; Arnaout *et al.* 1986; Cannistra *et al.* 1987). In addition to these effects, GM-CSF can prime the functional activity of mature, circulating neutrophils; both the phagocytic and cytotoxic activity of these cells is markedly enhanced by preincubation with this growth factor (Golde *et al.* 1990; Fleischmann *et al.* 1986). Primed neutrophils are more effective in the phagocytosis and killing of yeast, bacteria and opsonised tumour cells (Fleischmann, 1986; Villallta and Kierszenbaum, 1986). These effects are in part achieved by enhancing the superoxide anion production of the neutrophils, as well as enhancing the rate of phagocytosis.

Another facet of the biological activity of GM-CSF, which is perhaps just as physiologically relevant, is its effect on neutrophil locomotion and adhesion. GM-CSF is a chemotactic factor for phagocytic cells, and its production can be stimulated in macrophages (e.g. in response to lipopolysaccharide or interferon-γ) (Hamilton and Adams, 1987; Thorens *et al.* 1987; Piacibells *et al.* 1985; Brussollino *et al.* 1989) and endothelial cells (in response to agents such as lipopolysaccharide and tumour necrosis factor). This may be of significance in the movement of neutrophils to areas of infection. Recent reports also suggest that the adhesion of neutrophils to endothelium is enhanced following treatment of the cells with GM-CSF. These data indicate a major role for GM-CSF in response to infectious agents (Gamble *et al.* 1989).

Haemopoietic cells are not the only cell types which respond to GM-CSF. It can also stimulate the proliferation and chemotaxis of endothelial cells. Thus at sites of injury, GM-CSF can potentially attract both phagocytic cells and endothelial cells. This suggests that GM-CSF is a pleiotropic cytokine which can not only stimulate the development of myeloid progenitor cells, but also mediate the inflammatory response associated with infection and the process of wound healing.

Granulocyte Colony Stimulating Factor (G-CSF)

Originally, Granulocyte Colony Stimulating Factor was purified on the basis of its ability to induce the differentiation of the WEHI-3B myelomonocytic leukaemia cell line, and to stimulate the development of mature neutrophils from normal bone marrow progenitor cells in colony forming assays (Burgess and Metcalf, 1980; Platzer *et al.* 1983). Recent data has extended the range of activity of this growth factor to a number of other cells, including endothelial cells, mature, postmitotic neutrophils (see below) and multipotent haemopoietic stem cells. Like IL-6, the serum levels of this protein rise markedly in mice after the injection of endotoxin, suggesting a role for these agents in the response to infections.

There is evidence that, like IL-6, G-CSF can shorten the G_0 period of the dormant multipotent cell in the bone marrow (Ikebuchi *et al.* 1988*b*). In the presence of G-CSF, some progenitor cells present in bone marrow from 5-fluorouracil treated mice form blast cell colonies, but when IL-3 is also added there is a synergistic interaction between these factors, leading to the development of colonies containing mature cells from several lineages. These synergistic effects are also seen using highly enriched populations of multipotent cells, inferring that the effects are not mediated through the paracrine production of secondary growth factors (see above) (Heyworth *et al.* 1988).

In addition to the effects of G-CSF on multipotent cells, and as a stimulus for the development of neutrophilic cells, there are a number of other biological effects of this growth factor. In many respects, these are similar to the effects seen with GM-CSF. For example, G-CSF can increase the lifetime of circulating neutrophils, and enhance their antibody-dependent cellular cytotoxicity (Begley *et al.* 1986; Vadas *et al.* 1983; Tsuchiya *et al.* 1986), although it is of interest to note that this effect is additive with that of GM-CSF, suggesting that they do not share a common priming mechanism (Vadas *et al.* 1983, 1985). However, both GM-CSF and G-CSF enhance the production of reactive oxygen intermediates by neutrophils in response to the chemotactic peptide formyl-methionyl-leucyl-phenylalanine, and can also stimulate endothelial cell proliferation and chemotaxis. The dramatic elevation in GM-CSF and G-CSF levels in bacterially-infected mice (Burgess and Nicola, 1983), and the ability of G-CSF treatment to reduce the incidence of infection in patients (Bronchud and Dexter, 1989), suggest that G-CSF has a role in host defence mechanisms *via* its effects on myeloid cell production and functional activity.

Interleukin 5

Human Interleukin 5 (IL-5) can promote the development of eosinophilic progenitor cells, whilst its murine counterpart (which is 70 % homologous) can also promote the proliferation of B cell precursors. Evidence from murine model systems suggests that IL-5 has a role *in vivo*, in the systemic response to parasitic infections (Strath and Sanderson, 1986). Serum levels of this cytokine are markedly elevated in mice infected with parasites, and this is followed by an increase in eosinophil production from the bone marrow, and eosinophil numbers in the peripheral blood (Strath and Sanderson, 1986). Infusion of exogenous IL-5 has a similar effect (Sanderson *et al.* 1986).

Furthermore, the survival of eosinophils, and their cytotoxic activity, is markedly potentiated by IL-5, which is also a chemotactic factor for these cells (Yamaguchi *et al.* 1988*b*). Thus, like G-CSF and GM-CSF, IL-5 not only facilitates the proliferation and development of eosinophils but also the capacity of the mature cells to perform their specific functions. However, unlike the effects of GM-CSF or G-CSF on phagocytic cells, IL-5 does not *prime* eosinophils to display enhanced levels of superoxide production (see above) from eosinophils, it *directly* activates superoxide production by eosinophils (Yamaguchi *et al.* 1988*a*).

Macrophage Colony Stimulating Factor (M-CSF)

In soft agar cultures of normal murine bone marrow, M-CSF (or Colony-Stimulating Factor 1) stimulates the formation of predominantly macrophage colonies, derived from granulocyte macrophage colony forming cells. The response of multipotent cells to M-CSF is limited unless a second cytokine such as IL-1 is present. Interestingly, human M-CSF, although a potent growth factor for murine GM-CFC, is relatively poor in its ability to promote the development of colonies from human bone marrow; the reason for this is unknown.

M-CSF does, however, support the survival and proliferation of tissue based macrophages and monocytes. M-CSF can also affect the functional activity of mature macrophages. For example, the production of prostaglandins, plasminogen activator, interferons, tumor necrosis factor, and GM-CSF have all been shown to increase in M-CSF treated macrophages. Similarly M-CSF promotes the cytotoxic and phagocytic capacities of these cells, as well as priming them to produce greater quantities of reactive oxygen intermediates (see above).

Model systems to study haemopoietic stem and progenitor cell development

It has been suggested that the effects of the different growth factors on haemopoietic stem cells resemble those of the so-called competence and progression factors in fibroblastic cell proliferation (Rozengurt, 1986). Agents such as IL-6, G-CSF and IL-1 can stimulate the transition of the stem cells from a G_0 cell cycle state into G_1, and as such can be thought of as 'competence factors'. These cells are then able to respond to a second set of factors (such as IL-3 and also

GM-CSF) which can stimulate the progression of these cells into S phase, and development into committed progenitor cells. While the growth factors mentioned above undoubtedly have the ability to influence the proliferation and development of multipotent haemopoietic cells, there is little information on the biochemical mechanisms mediated by these factors and leading to regulation of cell cycle progression, stem cell differentiation, lineage restriction and the development of the postmitotic phenotype.

The most appropriate population for such studies are stem cells or committed progenitor cells prepared from *in vivo* sources. However, there are difficulties in the preparation of such populations, in that the bone marrow contains a heterogenous population of which immature blood cells represent a minor proportion. The use of drugs such as cyclophosphamide or thiamphenicol can increase the proportion or numbers of the progenitor cells required. This, and recent advances in cell purification procedures (such as centrifugal elutriation, or cell surface antigen-based selection techniques) have produced relatively pure populations of stem and committed progenitor cells (Lord and Spooncer, 1986; Williams *et al.* 1987). Unfortunately, these procedures give a relatively low yield of the required cells, often insufficient for molecular, biological or biochemical analysis. However, given the very low number of cells required for colony forming assays, sufficient cells can usually be obtained to determine if there is a direct (as opposed to a paracrine) effect of mitogenic agents on enriched cells (see above). Enriched populations of CFU-S, GM-CFC, CFU-E or BFU-E have also been employed to investigate the types of cytokines that can act directly on immature haemopoietic cells (Heyworth *et al.* 1990; Williams *et al.* 1987; Cook *et al.* 1989; Mitler *et al.* 1989), and the possible biochemical events they elicit within the cell (Mitler *et al.* 1989; Cook *et al.* 1989; Imagawa *et al.* 1989). Although the results obtained give an insight into the nature of the target cell populations for the various cytokines, the limited numbers of cells that can be prepared precludes investigation of the molecular mechanisms regulating stem cell commitment, lineage restriction, and development to mature cells. For this reason, many of these events are presently being studied using a variety of cell lines.

Cell lines for the study of myeloid growth factor stimulated development

Perhaps the most frequently employed cell line for biochemical studies on the maturation of myeloid cells is the human promyelocytic leukaemic cell line, HL-60. In response to a wide variety of extracellular stimuli (such as retinoic acid phorbol esters and dimethylsulphoxide), HL-60 cells will develop into granulocyte-, macrophage- and also eosinophil-like cells, although differences do exist between these cells and normal granulocytes and macrophages (Collins, 1987). Although the developmental stimuli which have been used in the past to induce HL-60 cells have no physiological role in haemopoiesis *in vivo*, there is evidence that GM-CSF can also induce differentiation (Metcalf, 1983; Begley *et al.* 1987). Another important distinction between HL-60 cells and normal myeloid precursors is that HL-60 cells will grow in simple defined medium in the absence of any haemopoietic growth factor (Collins, 1987), whereas normal progenitor cells rapidly die in such conditions. In this respect, HL-60 cells can be distinguished from the majority of primary myeloid leukaemia cells, which also require growth factors for survival and proliferation.

These marked distinctions between normal progenitor, primary myeloid leukaemia and HL-60 cells, suggest that the latter may not be an appropriate model to study many aspects of myeloid cell development. Similar criticism applies to the human leukaemic cell line K562, which can develop along several distinct lineages in response to a variety of non-physiological stimuli (Ohlsson-Wilhelm *et al.* 1987; Leary *et al.* 1987; Nishimura, 1988).

There are other cell lines, however, which can differentiate to mature cells in response to specific haemopoietic growth factors. For example, the M1 and WEHI-3B and 32DC13 cell lines will undergo differentiation to mature cells in response to G-CSF (Valtieri *et al.* 1987). Also the IL-3 dependent, multipotential, LyD9 (Kinachi *et al.* 1989) and FDCP-Mix (Spooncer *et al.* 1986) cell lines can be stimulated by a number of distinct cytokines to develop into the appropriate mature cell type.

We have employed the FDCP-Mix cells to study the balance between self-renewal and differentiation in haemopoietic cells. These events are apparently regulated not only by the type of growth factors in which the cells are cultured, but also by the concentrations of these factors. When combined with a low concentration of IL-3, the cells respond to GM-CSF, M-CSF, G-CSF or erythropoietin by developing into mature postmitotic cells. The type(s) of mature cells produced are governed by which growth factor(s) are added (e.g. erythropoietin stimulates erythroid development). At a high concentration of IL-3, the cells proliferate and self-renew; they do not differentiate, irrespective of the presence of other growth factors. In the absence of IL-3 the cells show little or no response to other growth factors.

Initially we have investigated the mechanisms regulating the development of FDCP-Mix cells to neutrophils. In the presence of high concentrations of IL-3, FDCP-Mix (A4) cells undergo self-renewal, irrespective of the presence of other haemopoietic growth factors such as GM-CSF or G-CSF. At a low concentration of IL-3, in the presence of GM-CSF plus G-CSF, FDCP-Mix cells become committed to development and produce predominantly neutrophils. The concentration of IL-3 dictates whether the cells self-renew or differentiate, that is to say there is an antagonism between IL-3 and stimulators of neutrophilic development. IL-3 has an ability to promote self-renewal, but how does it antagonise the developmental response given to multipotent A4 cells by growth factors such as GM-CSF?

The FDCP-Mix cell lines offer the opportunity to examine the biochemical events elicited by haemopoietic growth factors leading to differentiation and development. Using these cells and the preparations of highly enriched, normal, bone marrow-derived progenitor cells such as GM-CFC (Cook *et al.* 1989) and CFU-S (Heyworth *et al.* 1988) it will now be possible to assess the molecular mechanisms which regulate the survival, proliferation, differentiation and development of apparently normal haemopoietic stem and progenitor cells. This combination of cellular biological and biochemical approaches will not only allow the mechanisms regulating normal haemopoietic cell development to be identified, but also reveal how these mechanisms are restricted during leukaemic transformation.

Work in the authors' laboratories is funded by the Cancer Research Campaign (T.M.D.) and the Leukaemia Research Fund (A.D.W.).

References

AXELRAD, A. A. (1990). Some hemopoietic negative regulators. *Expl Hematol.* **18**, 143–150.

AXELRAD, A. A., CROIZAT, H., DEL RIZZO, D., ESKINAZI, D., PEZZUTTI, G., STEWART, S. AND VAN DER GOAG, H. (1987). Properties of a protein NRP that negatively regulates DNA synthesis of the early erythropoietic progenitor cells, BFU-E. In *The inhibitors of haematopoiesis* (ed. A. Najman, M. Guigon, N.-C. Gorin and J.-Y. Mary), vol. **162**, p. 79. Paris: Cataloque INSERM/John Libbey Eurotext.

ARNAOUT, M. A., WANG, E. A., CLARK, S. C. AND SIEFF, C. A. (1986). Human recombinant GM-CSF increases cell–cell adhesion and surface expression of adhesion-promoting surface glycoproteins on mature granulocytes. *J. clin. Invest.* **78**, 597–601.

BAGBY, G. C. (1989). Interleukin-1 and Hematopoiesis. *Blood Reviews* **3**, 152–161.

BEGLEY, C. G., LOPEZ, A. F., NICOLA, N. A., WARREN, D. J., VADAS, M. A., SANDERSON, C. J. AND METCALF, D. (1986). Purified CSFs enhance survival of human neutrophils and eosinphils *in vitro*: a rapid and sensitive microassay for CSFs. *Blood* **68**, 162–166.

BEGLEY, C. G., METCALF, D. AND NICOLA, N. A. (1987). Purified colony stimulating factors (G-CSF and GM-CSF) induce differentiation in human HL60 leukaemic cells with suppression of clonogenicity. *Int. J. Cancer.* **39**, 99–105.

BRONCHUD, M. H. AND DEXTER, T. M. (1989). Clinical use of growth factors. *Br. med. Bull.* **45** (2) 590–599.

BROXMEYER, H. E., SHERRY, B., LU, L., COOPER, S., CAROW, C., WOLPE, S. D. AND CERAMI, A. (1989). Myelopoietic enhancing effects of murine macrophage inflammatory proteins 1 and 2 on colony formation *in vitro* by murine and human bone marrow granulocyte/macrophage progenitor cells. *J. exp. Med.* **170** (5), 1583–94.

BRUCE, W. R. AND MCCULLOCH, E. A. (1964). The effect of erythropoietic stimulation on the haemopoietic colony-forming cells of mice. *Blood* **23**, 216–221.

BRUNO, E., BRIDDELL, R. AND HOFFMAN, R. (1988). Effect of recombinant and purified haemopoietic growth factors on human megakaryocyte colony formation. *Expl Haematol.* **16**, 371–377.

BURGESS, A. AND NICOLA, N. (1983). *Growth Factors and Stem Cells* , pp. 93–124. Academic Press: New York.

BURGESS, A. W. AND METCALF, D. (1980). Characterisation of a serum factor stimulating the differentiation of myelomonocytic leukaemic cells. *Int. J. Cancer.* **39**, 647–654.

BUSSOLINO, F., WANG, J. H., DEFILIPPI, P., TURRINI, F., SANAVIO, F., EGDELL, C-J. S., AGLIETTA, M., ARESE, P. AND MANTOVANI, A. (1989). Granulocyte- and granulocyte-macrophage colony stimulating factors induce human endothelial cells to migrate and proliferate. *Nature* **337**, 471–473.

CANNISTRA, S. A., RAMBALDI, A., SPRIGGS, D. R., HERRMANN, F., KUFE, D. AND GRIFFIN, J. D. (1987). Human GM-CSF induce expression of the TNF gene by the V937 cell line. *J. clin. Invest.* **79**, 1720–1728.

CANNISTRA, S. A., VELLENGA, E., GROSHEK, P., RAMBALDI, A. AND GRIFFIN, J. D. (1988). Human GM-CSF and IL-3 stimulate monocyte cytotoxicity through a tumour necrosis factor-dependent mechanism. *Blood* **71**, 672–676.

CARCACCIOLO, D., CLARK, S. C. AND RONERA, G. (1989). Human IL-6 supports granulocytic differentiation of hematopoietic progenitor cells and acts synergistically with GM-CSF. *Blood* **73** (3), 666–70.

CLARK, S. C. AND KAMEN, R. (1987). The human haemopopietic colony stimulating factors. *Science* **236**, 1229–1237.

CLARK-LEWIS, I., CRAPPER, R. M., LEDIE, K., SCHRADER, S. AND SCHRADER, J. W. (1985). In *Cellular and Molecular Biology of Lymphokines* (ed. C. Song and A. Schimpl) pp. 455–459. Academic Press: Orlando, Florida.

COLLINS, S. J. (1987). The HL60 promyelocytic leukaemia cell line: proliferation, differentiation and oncogene expression. *Blood* **70**, 1233–1244.

COOK, N., DEXTER, T. M., LORD, B. I., CRAGOE, E. J. AND WHETTON, A. D. (1989). Identification of a common signal associated with cellular proliferation stimulated by 4 haemopoietic growth factors in a highly enriched population of granulocyte/macrophage colony-forming cells. *EMBO J.* **8**, 2967–2974.

DEL RIZZO, D. F., ESKINAZI, D. AND AXELRAD, A. A. (1990). IL-3 opposes the action of negative regulatory protein (NRP) and of transforminggrowth factor-beta (TGF-β) in their inhibition of DNA synthesis of the erythroid stem cell BFU-E. *Expl Hematol.* **18**, 138–142.

DEXTER, T. M. AND WHITE, H. (1990). Growth factors: growth without inflation. *Nature* **344**, 380–381.

DI PERSIO, J. F., HEDNAT, C. AND GASSON, J. C. (1989). GM-CSF indirectly down-regulates high-affinity LTB_4 receptor expression by directly stimulating neutrophil LTB_4 synthesis. *J. cell. Biochem.* (Suppl. 13c) 9.

EAVES, C. J., CASHMAN, J. D., KAY, K. H., DOUGHERTY, G. J., GABOURY, L. A., EAVES, A. C. AND HUMPHRIES, R. K. (1988). Evidence that human marrow stromal cells produce TGF-β and thereby arrest the cycling of primitive populations of normal hemaopopietic progenitors. *Blood* (Suppl. 1) **72**, 84a.

FLEISHMANN, J., GOLDE, D. W., WEIBART, R. G. AND GASSON, J. C. (1986). GM-CSF enhances phagocytosis of bacteria by human neutrophils. *Blood* **68**, 708–711.

GAMBLE, J. R., ELLIOTT, M. J., JAIPARGAS, E., LOPEZ, A. F. AND VADAS, M. A. (1989). Regulation of human monocyte adherence by GM-CSF. *Proc. Natn. Acad. Sci. U.S.A.* **86**, 7022–6.

GOLDE, D. W., BALDWIN, C. G. AND WEISBART, R. H. (1990). Molecular control of haemopoesis. *CIBA Foundation Symposium 148*, pp. 62–75. Wiley: Chichester.

GRABSTEIN, K. H., URDAL, D. L., TUSHINSKI, R. J., MOCHIZUKI, D. Y., PRICE, V. L., CANTRELL, M. A., GILLIE, S. AND CONLAN, P. J. (1986). Induction of macrophage tumoricidal activity by GM-CSF. *Science* **232**, 506–508.

GRAHAM, G. J., WRIGHT, E. G., HEWICK, R., WOLPE, S. D., WILKIE, N. M., DONALDSON, D., LORIMORE, S. AND PRAGNELL, I. B. (1990). Identification and characterisation of an inhibitor of haemopoietic stem cell proliferation. *Nature* **344** (6265), 442–4.

HAMILTON, J. A. AND ADAMS, D. O. (1987). Molecular mechanism of signal transduction in macrophages. *Immunol. Today* **8**, 151–158.

HAMPSON, J., PONTING, I. L. O., COOK, N., VODINELICH, L., REDMOND, S., ROBERTS, A. B. AND DEXTER, T. M. (1989). The effects of TGF-β on haemopoietic cells. *Growth Factor* **1**, 193–202.

HEYWORTH, C. M., DEXTER, T. M., KAN, O. AND WHETTON, A. D. (1990). The role of haemopoietic growth factors in self-renewal and differentiation of IL-3-dependent multipotential stem cells. *Growth Factors* **22**, 197–211.

HEYWORTH, C. M., PONTING, I. L. O. AND DEXTER, T. M. (1988). The response of haemopoietic cells to growth factors: developmental implications of synergistic interactions. *J. Cell. Sci.* **91**, 239–247.

HINO, M., TOJO, A., MIYAZONA, K., URABI, A. AND TAKAKU, F. (1988). Effects of type beta transforming growth, factors on haemopoietic progenitor cells. *Br. J. Haemat.* **70**, 143.

HOANG, T., HAMAN, A., GONCALVES, O., WONG, G. G. AND CLARK, S. C. (1988). Interleukin-6 enhances growth factor dependent proliferation of the blast cells of acute myeloblastic leukaemia. *Blood* **72**, 823–826.

IKEBUCHI, K., IHLE, J. N., HIRAI, Y., WONG, G. G., CLARK, S. C. AND OGAWA, M. (1988*a*). Syngestic factors for stem cell proliferation. Further studies of the target stem cells and the mechanism of stimulation by Interleukin-1, Interleukin-6 and Granulocyte-colony-stimulating factor. *Blood* **72**, 2007–2014.

IKEBUCHI, K., CLARK, S. C., IHLE, J. N., SOUZA, L. M. AND OGAWA, M. (1988*b*), GM-CSF enhances

IL-3-dependent proliferation of multipotential haemopoietic progenitors. *Proc. natn. Acad. Sci. U.S.A.* **85**, 3445–3449.

IKEBUCHI, K., WONG, G. G., CLARK, S. C., IHLE, J. N., HIRAI, Y. AND OGAWA, M. (1987). IL-6 enhancement of IL-3 dependent proliferation of multipotential haemopoietic progenitors. *Proc. natn. Acad. Sci. U.S.A.* **84**, 9035–9.

IMAGAWA, S., SMITH, B. R., PALMER-CROCKER, R. AND BUNN, H. F. (1989). The effect of erythropoietin on intracellular free calcium in erythropoietin-responsive cells. *Blood* **73**, 1452–1457.

ISCOVE, N. N., SIEBER, F. AND WINTERHALTER, K. H. (1974). Erythroid colony formation in cultures of mouse and human bone marrow: Analysis of the requirement for erythropoietin by gel filtration and affinity chromatography on agarose-concavalin A. *J. cell. Physiol.* **83**, 309–316.

ISHIBASHI, T., KIMURA, H., SHIKANA, Y., UCHIDA, T., KARIYONE, S., TURAIO, T., KISHIMOTO, T., TATATSUKI, F. AND AKIYANA, Y. (1989). Interleukin-6 is a potent thrombopoietic factor *in vivo* in mice. *Blood* **74**, 1241–1244.

KINACHI, T., TOUHIRO, K., INABA, K., TAKEDA, T., PALACUIS, R. AND HANJO, T. (1989). An IL-4-dependent precursor clone differentiates into myeloid cells as well as B-lymphocytes. In *Lymphokine Receptor Interactions* (ed. D. Fradeliji and J. Bertoglio), vol. **179**, pp. 119–126. Collaque INSERM/John libbey Eurotext Ltd.

KISHI, K., IHLE, J. N., URDAL, D. L. AND OGAWA, M. (1989). Murine B-cell stimulatory factor-1 (BSF-I)/Interleukin-4 (IL-4) is a multilineage colony-stimulating factor that acts directly on primitive haemopoietic progenitors. *J. cell Physiol.* **139** (3), 463–8.

LAJTHA, L. G. (1982). In *Blood and its Disorders* (ed. R. M. Hardisty and D. J. Weatherall), pp. 57–75. Blackwell: London.

LEARY, J. F., FARLEY, B. A., GUILANO, R., KOSCIOLEK, B. A., LABELLA, S. AND ROWLEY, P. T. (1987). Induction of megakaryocytic characteristics in human leukaemic cell line K562: polyploidy inducers and secretion of mitogenic activity. *J. Biol. Regul. Hameost. Agents* **1**, 73–80.

LEE, M., SEGAL, G. M. AND BAGBY, G. C. (1987). Interleukin-1 induces human bone marrow derived fibroblasts to produce multilineage haemopoietic growth factors. *Expl. Haematol.* **15**, 983–988.

LENFANT, M., WDZIECZAK-BAKALA, J., GUITTET, E., PROME, J. C., SOTTY, D. AND FRINDEL, E. (1989). Inhibitor of haemopoietic pluripotent stem cell proliferation: purification and determination of its structure. *Proc. natn. Acad. Sci. U.S.A.* **86**, 779–785.

LORD, B. I. AND SPOONCER, E. (1986). Isolation of haemopoietic spleen colony forming cells. *Lymphokine Res.* **5**, 97–104.

MCCULLOCH, E. A. AND TILL, J. E. (1964). Proliferation of haemopoietic colony-forming cells transplanted into irradiated mice. *Radiation Res.* **22**, 383–389.

METCALF, D. (1983). Clonal analysis of the response of HL60 human myeloid leukaemia cells to biological regulators. *Leuk. Res.* **7**, 117–132.

METCALF, D. (1984). *Haemopoietic colony-stimulating factors.* Elsevier: Amsterdam.

METCALF, D. AND MERCHOV, S. (1982). Effects of GM-CSF deprivation on precursors of granulocytes and macrophages. *J. cell. Physiol.* **112**, 411–418.

MITLER, B. A., CHEUNG, S. Y., TILLOTSON, D. L., HOPE, S. M. AND SCADUTO, R. C. (1989). Erythropoietin stimulates a rise in intracellular free calcium concentration in single BFU-E derived erythroblasts at specific stages of differentiation. *Blood* **73**, 1188–1194.

MOCHIZINKI, D. Y., EISENMAN, J. K., CONLON, P. J., LARSEN, A. D. AND TUCHINSKI, R. J. (1987). IL-1 regulates haemopoietic activity, a role previously ascribed to haemopoietin 1. *Proc. natn. Acad. Sci. U.S.A.* **84**, 5267–5271.

MOORE, M. A. S., MUENCH, M. O., WARREN, D. J. AND LOOVER, J. (1990). Molecular control of haemopoiesis. *CIBA Foundation Symposium* 148. pp. 43–61. Wiley: Chichester.

NISHIMURA, J., TAKAHIRA, H., SHIBATA, K., MUTA, K., YAMAMOTO, M., IDEGUCHI, H., UMEMURA, T. AND NAWATA, H. (1988). Regulation of biosynthesis and phosphorylation of p210 bcr/abl protein during differentiation induction of K562 cells. *Leukaemic Research* **12**, 875–885.

OHLSSON-WILHELM, B. M., FARLEY, B. A. AND ROWLEY, P. J. (1987). Erythroid differentiation of K562 cells: mixed colonies as an index of delayed expression of commitment. *Expl Hematol.* **15**, 817–821.

PESCHEL, C., PAUL, W. E., O'HARA, J. AND GREEN, I. (1987). Effects of B-cell stimulatory factor 1/IL-4 on haemopoietic progenitor cells. *Blood* **70**, 254–263.

PIACIBELLS, W., LU, L., WACHTER, M., RUBIN, B. AND BROXMEYER, H. E. (1985). Release of GM-CSF from major histocompatability complex class II antigen-positive monocytes is enhanced by human gamma-interferon. *Blood* **66**, 1343–1351.

PLATZER, E., WELTE, K., GABRILORE, J., LU, L., HARRIS, P., MERTILSMANN, K. AND MOORE, M. A. S. (1983). Biological activities of a human pluripotent haemopoietic colony stimulating factor on normal and leukaemic cells. *J. Exp. Med.* **162**, 1788–1801.

REED, S. G., NATHAN, C. F., PIHL, O. L., RODRICKS, P., SHANEBECK, K., CONLON, P. J. AND GRABSTEIN, K. H. (1987). Recombinant GM-CSF activates macrophages to inhibit *Typanosoma cruzi* and release hydrogen peroxide. *J. Expl Med.* **166**, 1734–1746.

RENNICK, D., YANG, G., MULLER-SIEBEURG, C., SMITH, C., ARAI, N., TAKABE, Y. AND GEMMELL, L. (1987). IL-4 (B-cell stimulatory factor 1) can enhance or antagonise the factor-dependent growth of haemopoietic progenitor cells. *Proc. natn. Acad. Sci. U.S.A.* **84** 6889–6893.

ROTHENBERG, M. E., OWEN, W. F., SILBERSTEIN, D. S., WOODS, J., SILVERMAN, R. J. AUSTEN, K. F. AND STEVENS, R.L. (1988). Human eosinophils have prolonged survival, enhanced functional properties and become hypodense when exposed to human Interleukin-3. *J. clin. Invest.* **81**, 1986–1992.

ROZENGURT, E. (1986). Early signals in the mitogenic response. *Science* **234**, 161–166.

SACHS, L. (1990). Molecular control of haemopoiesis. CIBA Foundation Symposium *148*, 5–24. Wiley: Chichester.

SANDERSON, C. J., O'GARRA, A., WARREN, D. S. AND KLAUS, G. G. (1986). Eosinophil differentiation factor also has B-cell growth factor activity. Proposed name Interleukin-4. *Proc. natn. Acad. Sci. U.S.A.* **83**, 437–440.

SCHRADER, J. W. AND CLARK-LEWIS, I. (1982). A T-cell-derived factor, stimulating multipotential haemopoietic stem cells – molecular weight and distribution from T cell growth factor and T-cell-derived GM-CSF. *J. Immunol.* **129**, 30–35.

SING, G. K., KELLER, J. R., ELLINGSWORTH, J. R. AND RUSCETTI, F. W. (1988). TFG selectivity inhibits normal and leukaemic human bone marrow cell growth *in vitro*. *Blood* **72**, 1504–1510.

SONODA, Y., YANG, Y. G., WONG, G. G., CLARK, S. C. AND OGAWA, M. (1988). Erythroid burst-promoting activity of purified recombinant human GM-CSF and interleukin-3 sera and studies in serum-free culture. *Blood* **72**, 1381–1387.

SPIRAK, J. L., SMITH, R. R. AND IHLE, J. N. (1985). IL-3 promotes the *in vitro* proliferation of murine, pluripotent hematopoietic stem cells. *J. clin. Invest.* **76**, 1613–1621.

SPOONCER, E., HEYWORTH, C. M., DUNN, A. AND DEXTER, T. M. (1986). Self-renewal and differentiation of IL-3 dependent multipotential stem cells are modulated by stromal cells and serum factors. *Differentiation* **31**, 111–118.

STANLEY, E. R., BARTOLIN, A., PATINKIN, D., ROSENDAAL, M. AND BRADLEY, T. R. (1986). Regulation of very primitive multipotential haemopoietic cells by haemopoietin 1. *Cell* **45**, 667–674.

STANLEY, E. R., GUILBERT, L. J., TUSHINSKI, R. J. AND BARTELMEZ, S. H. (1983). CSF-1 – a mononuclear phagocyte lineage-specific haemopoietic growth factor. *J. cell. Biochem.* **21**, 151–159.

STRATH, M. AND SANDERSON, C. J. (1986). Detection of eosinophil differentiation factor and its relationship to eosinophils in mesocestoides corti-infected mice. *Exptl Hematol.* **14**, 16–20.

SUDA, T., YAMAGUCHI, Y., SUDA, J., MIURA, Y., OKANE, A. AND AKIYAMA, Y. (1988). Effect of IL-6 on the differentiation and proliferation of murine and human haemopoietic progenitors. *Exptl Hematol.* **16**, 891–5.

SUZUKI, C., OKANO, A., TAKATSUKI, F., MIYOSAKA, Y., HIRANO, T., KISHIMOTO, T., EJIMA, D. AND AKIYAMA, Y. (1989). Continuous perfusion with IL-6 enhances production of hemotopoietic stem cells (CFU-s). *Biochem. biophys. Res. Commun.* **159**, 933–938.

THORENS, B., MERMOD, J. J. AND VASSALLI, P. (1987). Phagocytosis and inflammatory stimuli induce GM-CSF mRNA in macrophages through post-transcriptional regulation. *Cell* **48**, 671–679.

TSUCHIYA, M., ASANO, S., KAZZIRO, Y. AND NAGATA, S. (1986). Isolation and characterisation of the cDNA for murine granulocyte colony-stimulating factor. *Proc. natn. Acad. Sci. U.S.A.* **83**, 7633–7637.

TSUJI, K., NAKAHATA, T., TAKAGI, M., KOBAYASHI, T., ISHIGURO, A., KIKUCHI, T., NAGANUMA, K., KOIKE, X., MIYAJIMA, A., ARAI, K-I AND AKABANE, T. (1990). Effects of IL-3 and IL-4 on the development of 'connective tissue-type' mast cells: IL-3 supports their survival and IL-4 triggers and supports their proliferation synergistically with IL-3. *Blood* **75**, 421–7.

VADAS, M. A. AND LOPEZ, A. F. (1985). Regulation of granulocyte function by colony stimulating factors and monoclonal antibodies. *Lymphokines* **12**, 179–200.

VADAS, M. A., NICOLA, N. A. AND METCALF, D. (1983). Activation of antibody dependent cell-mediated cytoxicity of human neutrophils and eosinophils by separate colony stimulating factors. *J. Immunol.* **130**, 795–799.

VALTIERI, M., TWEARDY, D. J., CARACCIOLO, D., JOHNSON, K., MAVILIO, F., ALTMANN, S., SANTOLI, D. AND RONERA, G. (1987). Cytokine-dependent granulocytic differentiation. *J. Immunol.* **138**, 3829–3835.

VILLALLTA, F. AND KIERSZENBAUM, F. (1986). Effects of human colony-stimulating factor on the uptake and destruction of a pathogenic parasite (*Trypanosoma cruzi*) by human neutrophils. *J. Immunol.* **137**, 1703–1709.

WHETTON, A. D. AND DEXTER, T. M. (1989). Myeloid haemopoietic growth factors. *Biochim. Biophys. Acta* **989**, 111–132.

WIDMER, M. B. AND GRABSTEIN, K. H. (1987). Regulation of cytolytic T-lymphocyte generation by B-cell stimulatory factor. *Nature* **326**, 795–798.

WILLIAMS, D. E., STRANEVA, J. E., COOPER, S., SHADDUCK, R. K., WAHEAD, A., GILLIS, S., URDAL, D. AND BROXMEYER, H. E. (1987). Interactions between purified murine colony stimulating factors (natural CSF-1, recombinant GM-CSF and recombinant IL-3) on the *in vitro* proliferation of purified murine granulocyte-macrophage progenitor cells. *Expl Haematol.* **15**, 1007–1012.

WOOD, D. G. (1982). In *Blood and its Disorders* (ed. R. M. Hardisty and D. J. Weatherall), pp. 55–74. Blackwell: London.

YAMAGUCHI, Y., HAYASHI, Y., SUGAMA, Y., MIURA, Y., KASAHGRA, T., KITAMURA, S., TORISHU, M., MITA, S., TOMINAGA, A., TAKATSU, K. AND SUDA, T. (1988*a*). Highly purified murine Interleukin-5 (IL-5) stimulates purified murine Interleukin-5 (IL-5 stimulates eosinophil function and prolongs *in vitro* survival: IL-5 as an eosinophil chemotatic factor. *J. Expl Med.* **167**, 1737–1742.

YAMAGUCHI, Y., SUDA, T., SUDA, J., EGUCHI, M., MIURA, Y., HARADA, N., TOMINGA, A. AND TAKATGSKU, K. (1988). Purified interleukin 5 supports the terminal differentiation and proliferation of murine eosinophilic precursors. *J. Expl Med.* **167**, 43–56.

J. Cell Sci. Suppl. 13, 75–85 (1990)

Printed in Great Britain © The Company of Biologists Limited 1990

Growth and differentiation factors of pluripotential stem cells

JOHN K. HEATH, AUSTIN G. SMITH, LI-WEI HSU
AND PETER D. RATHJEN

CRC Growth Factor Group, The Department of Biochemistry, University of Oxford, South Parks Rd., Oxford OX1 3QU, UK

Summary

The mammalian embryo develops as a quasi-stem cell system whose differentiation and pluripotentiality *in vitro* is controlled by a single regulatory factor, Differentiation Inhibiting Activity/Leukemia Inhibitory Factor (DIA/LIF). DIA/LIF is expressed in two distinct functional forms, derived from the use of alternate transcriptional start sites, one of which is freely diffusible and the other tightly associated with the extracellular matrix. The dissemination of the DIA/LIF signal is therefore under specific molecular control. The expression of DIA/LIF *in vitro* is both developmentally programmed and controlled by the action of other growth factors, the most notable of which are members of the fibroblast growth factor family expressed by the stem cells themselves. This indicates that differentiation and proliferation in early development of the mouse are controlled, at least in part, by an interactive network of specific growth and differentiation regulatory factors.

Introduction

The development of most organisms is characterised by two features: the progressive elaboration of different cell types from common precursors and the generation of tissue and body architecture in a defined pattern. These processes in many situations are thought to depend upon specific cellular responses to environmental cues, relating either to relative position within the developing organism or interactions with other cells or groups of cells. Two key issues in modern embryology are accordingly the molecular identity of such environmental signals and the mechanisms by which such signals are interpreted into changes in gene expression and cell behaviour.

There is now very good evidence that intercellular signalling in many developmental systems is mediated by specific polypeptide growth and differentiation factors, often initially identified by their biological activity in adult or pathological situations. This evidence comes from three main sources. Firstly, analysis of expression *in vivo* by *in situ* hybridisation and immunohistochemistry techniques has revealed that many growth factors are principally expressed in the embryonic stages of development. Where detailed information is available, e.g. for TGFβ (Lehnert and Ackhurst, 1988; Heine *et al.* 1987), TGFβ and BMP2A (Lyons *et al.* 1989), IGF2 (Beck *et al.* 1987) and Int-2 (Wilkinson *et al.* 1988), expression is often found in multiple sites and at different stages of development.

Key words: embryonic stem cells, growth factors, leukemia inhibitory factor, fibroblast growth factor, differentiation.

This evidence suggests that growth factors have some kind of activity in embryological development, and the often complex patterns of expression observed suggest that either individual growth factors perform some specific function in a diversity of situations or that they have diverse functions which depend upon the cellular context in which they act. The second type of evidence comes from experimental situations in which it has been possible to demonstrate that exogenously applied growth factors can mimic the effects of normal intercellular interactions in controlling developmental decisions. Examples would include the effect of members of the FGF gene family in inducing ventral-type mesoderm differentiation in the animal pole cells of early *Xenopus* embryos (Slack *et al.* 1987), or the action of TGFβ and the TGFβ-like XTC mesoderm inducing factor in the induction of dorsal mesoderm differentiation in the same system (Smith, 1987; Rosa *et al.* 1988). The third line of evidence comes from the characterisation of developmental mutants (particularly in *Drosophila*) where genes which can be shown, by genetic methods, to control specific pathways of differentiation and tissue pattern are found to encode molecules which are very similar to growth factor receptors (eg. *sevenless* and *torpedo*, reviewed by Hafen and Basler, this volume) or growth factors themselves (e.g. *wingless*, Rijsewick *et al.* 1987).

In this paper we shall be concerned with the identity and function of growth factors controlling developmental decisions and gene expression in the early development of the mouse. The mouse offers certain advantages for this problem which principally arise from the strategy of early mouse development. Abundant experimental evidence (reviewed by Gardner and Beddington, 1988) has shown that early mouse development occurs as a quasi-stem cell system whereby the major differentiated cell types formed in the early mouse embryo are withdrawn by differentiation from a pool of cells, leaving a residual 'stem cell' population which retains multipotential properties and from whom subsequent differentiated cell types are formed. This multipotent stem cell population only exists transiently during the course of normal development, although it has been hypothesised that the residue of this multipotent stem cell pool goes on to form the germ cell lineage (Heath, 1978). Each successive differentiation event in this situation occurs as a consequence of cellular responses to environmental cues such as relative location or proximity to other cell types. A historical difficulty with the direct molecular analysis of these signals has been the small size and inaccessibility of the embryo as development unfolds. However, a major experimental advantage is the availability of cultured cell lines derived either by tumourgenic conversion of the pluripotent quasi-stem cell population present in the normal embryo, embryonal carcinoma (EC) cells, or directly from it, embryonic stem (ES) cells, as reviewed by Robertson (1989). The key feature of ES cell lines, in particular, is that upon reintroduction into a host preimplantation embryo they will participate in normal development and give rise to progeny in all the tissues of the adult mouse, including the germ cell lineage (Bradley *et al.* 1984).

This phenomenon has three significant implications. Firstly, it permits the use of ES cells as a cellular vector for genetic manipulation of the mouse genome, providing the means to generate defined gain of function or loss of function mutations for experimental analysis of gene function (reviewed by Frohman and Martin, 1989). Secondly, ES cells provide an experimentally accessible resource for the molecular characterisation of growth factor expression and action: the demonstrated pluripotentiality of ES cells argues that ES cells (and their differentiated progeny) are capable of both generating and responding correctly to all the environmental signals that occur in the course of normal development. Thirdly, the extent of tissue colonisation occurring upon reintroduction of ES cells into the embryo provides a rigorous biological assay for the action of regulatory agents on ES cell function.

Differentiation Inhibitory Activity/Leukaemia Inhibitory Factor (DIA/LIF)

1. *Control of stem cell function*

An empirical finding of the first workers to isolate ES cells, Martin (1981) and Evans and Kaufman (1981), was, that like many EC cell lines (Martin and Evans, 1975), effective propagation and maintenance of ES cell pluripotentiality required that the cells be grown in association with a heterologous feeder cell layer (usually irradiated fibroblasts). Whilst this was technically inconvenient it strongly implied that maintenance of ES cell pluripotentiality and inhibition of their differentiation depended upon some interaction with the feeder cell layer; in other words some type of feeder-derived signal was continuously required to actively maintain ES cell function *in vitro*. A number of workers (Smith and Hooper, 1983; Koopman and Cotton, 1984; Smith and Hooper, 1987) observed that the requirement for a functional feeder cell layer to suppress initially EC, and subsequently ES, cell differentiation could be replaced by culture media conditioned by a variety of cell types such as STO fibroblasts (Smith and Hooper, 1983) and the BRL rat liver cell line (Smith and Hooper, 1987). The differentiation of ES cells *in vitro* can therefore be suppressed by soluble factors secreted by heterologous cell types.

The principal agent responsible for these effects of feeder-conditioned media on ES cell function is a single polypeptide regulatory factor DIA/LIF. This was conclusively demonstrated by the finding that recombinant DIA/LIF from either bacterial or mammalian sources can completely substitute for heterologous feeder cells or feeder-conditioned medium, both for maintenance of ES cell pluripotentiality in continuous culture (Smith *et al.* 1988; Williams *et al.* 1988) and for isolation of pluripotent ES cell lines *de novo* from the normal mouse embryo (L. Williams and S. Pease, personal communication; A. G. Smith and J. Nichols, unpublished observations). These observations suggest that DIA/LIF is a major determinant of pluripotent stem cell function in the normal embryo, and the pluripotent stem cell state is maintained as long as DIA/LIF is present. It follows that an understanding of the intercellular mechanisms of stem cell differentiation require analysis of DIA/LIF action at the molecular level and that the transient existence of the stem cell state in the course of normal development (Gardner and Beddington, 1988) is likely to be controlled by the exact timing and form of the delivery of the DIA/LIF signal to the cells *in vivo*.

2. *Molecular and biological characteristics*

DIA/LIF is a molecule of about 43×10^3 M_r derived by extensive glycosylation of a core 19×10^3 M_r polypeptide chain (Smith *et al.* 1988). Although the extent of glycosylation can depend upon the identity of the cellular source, it does not appear that the carbohydrate moiety is essential for biological function since activity is manifest in non-glycosylated material derived from expression in bacteria (Gearing *et al.* 1989). It is present in 'natural' sources, biologically active at sub-nanomolar concentrations and exerts its biological activity through association with specific, high affinity cell surface receptors expressed at between 300–1500 sites per cell (Smith *et al.* 1988; Hilton *et al.* 1988). The structural characterisation of DIA/LIF receptor(s) will be an important step towards defining the intercellular mechanism controlling ES cell differentiation and function. However, the molecular nature of the DIA/LIF receptor(s) is at present unclear, although indirect arguments (see below) suggest that they may be of the PIN-type form typical of certain lymphokines such as Il-6 (reviewed by Guy *et al.* 1990) rather than the tyrosine kinase or G-Protein linked serpentine receptors associated with many growth factors and mitogens.

DIA/LIF exemplifies the multiple biological functions ascribed to polypeptide regulatory factors described above. It was initially observed as an activity which induced macrophage differentiation in the M1 mouse leukaemia cell line (Tomida *et al.* 1984) and this property enabled isolation of mouse and human cDNA clones encoding DIA/LIF (Gearing *et al.* 1987; Gough *et al.* 1988). It was independently observed that the activity present in BRL-conditioned media which acted to suppress differentiation of ES cells was active as a maintenance factor for the DA1a murine leukaemia cell line. Exploitation of this latter activity led to the isolation of human cDNA clones encoding DIA/LIF (Smith *et al.* 1988; Moreau *et al.* 1988). These findings created a link between stem cells of the early mouse embryo and cell differentiation and proliferation in the haemopoetic system. Following these initial discoveries, DIA/LIF has emerged as a molecule with distinct bioregulatory actions in a diversity of cellular systems including bone, liver and the nervous system (Table 1). This strongly indicates that, specifically in the case of DIA/LIF, the exact biological effects of a polypeptide regulatory agent *in vivo* depend on the precise circumstances in which the signal is delivered to the target cell and the cellular context in which the signal is interpreted. A second feature of the range of DIA/LIF biological activity is an intriguing congruity between the actions of DIA/LIF and Interleukin 6 in systems such as MI, DA1a cells and hepatocytes (although Il-6 does not affect ES cell differentiation *in vitro*, A. G. Smith, unpublished results). This may be coincidental, but could indicate

Table 1. *Bioregulatory actions of DIA/LIF in various cellular systems*

Name	Cellular target	Activity	Reference
Differentiation Inhibitory Activity (DIA)	ES cells	Differentiation inhibition	a
Differentiation Retarding Factor (DRF)	PSA4 EC cells	Differentiation inhibition	b
Human Interleukin for DA cells (HILDA)	DA1a myeloid leukemia cells	Factor dependence	c
Leukemia Inhibitory Factor (LIF)	M1 monocytic leukemia cells	Differentiation induction	d
D factor	M1 monocytic leukemia cells	Differentiation induction	e
Osteoclast Activating Factor (OAF)	Calvarial osteoclasts	Activation	f
Hepatocyte stimulatory factor III (HSFIII)	R53 hepatoma/hepatocytes	Induction of APR proteins	g
Melanoma-derived LPLipase inhibitor	Adipocytes	Suppression of LPLipase	h
Cholinergic Neural Differentiation Factor	Neurons	Transmitter selection	i

References: (a) Smith and Hooper, 1987; (b) Koopman and Cotton, 1984; (c) Moreau *et al.* 1988; (d) Gearing *et al.* 1987; (e) Tomida *et al.* 1984; (f) Abe *et al.* 1986; (g) Baumann and Wong, 1989; (h) Mori *et al.* 1989; (i) Yamamori *et al.* 1989.

that both agents operate through analogous signal transduction pathways in which ligand specificity is determined by expression of appropriate receptors.

Multiple functional forms

It has been argued above that a key issue in understanding the role of DIA/LIF action *in vivo* is the way in which the signal is delivered. It is of considerable significance therefore that recent findings indicate that DIA/LIF exists in two functional forms which are distinguished by their range of action *in vivo* (Rathjen *et al.* 1990). Detailed examination of DIA/LIF gene transcription in cultured cells has shown that two classes of DIA/LIF mRNAs exist which diverge in sequence at their 5′ ends. Molecular cloning of cDNAs encoding these two forms of DIA/LIF mRNA shows that they encode two distinct forms of DIA/LIF protein which differ in sequence at the amino terminus (Fig. 1). The 5′ sequence divergence is generated by initiation of transcription at two distinct promoters, thereby creating two distinct first exons which are subsequently spliced onto common second and third exons to generate the two distinct mRNA species. The functional significance of this arrangement is that one form of the DIA/LIF protein (containing the amino-terminal sequence MKVLA) is expressed in a form which is freely diffusible and the alternative form (containing the amino-terminal sequence MRCR) is expressed in a form which is specifically associated with the extracellular matrix of the expressing cells.

Although it is not clear how the amino-terminal sequence determines the localisation of the parent protein, the use of alternative transcripts dictates whether a DIA/LIF signal can act over a distance or whether its sphere of action is confined to cells in intimate physical contact. In essence, the way a DIA/LIF signal is delivered is of paramount importance in understanding its function *in*

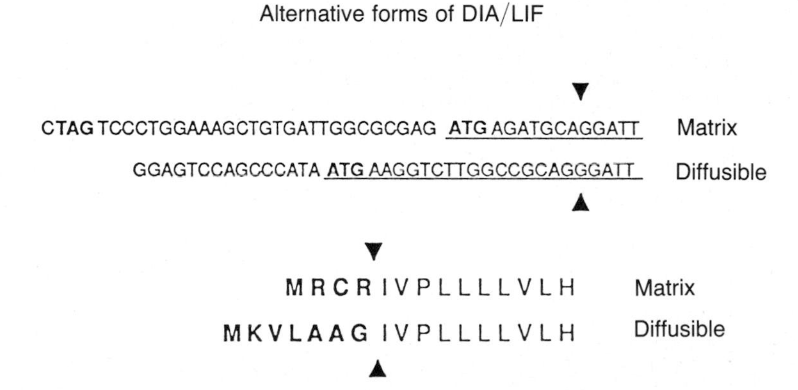

Fig. 1. Nucleotide sequence (upper) and predicted amino acid sequence (lower) of 5′ ends of alternative DIA/LIF cDNA clones. Coding regions are underlined, the position of exon1/exon2 boundaries are indicated by arrows and amino-terminal differences in bold type.

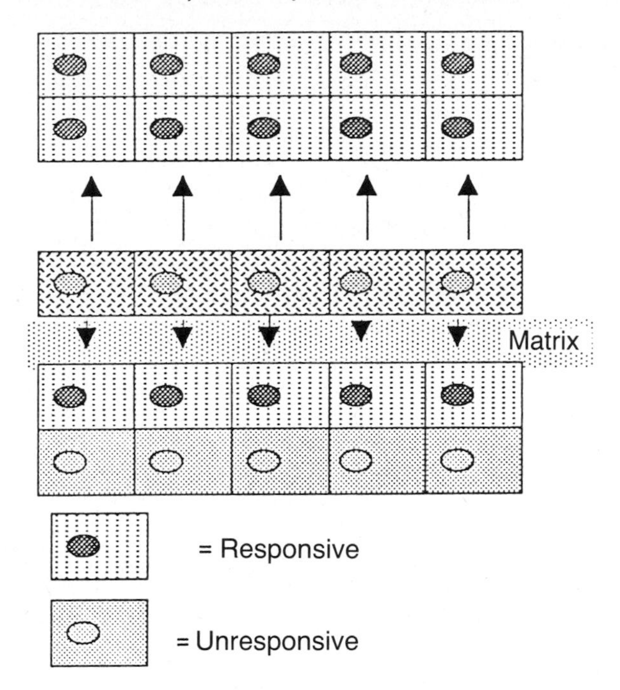

Fig. 2. Different forms of DIA/LIF have different range of actions. MKVLA-DIA/LIF (upward facing arrows) is freely diffusible and can affect the behaviour of cells remote from the site of expression. MRCR-DIA/LIF (downward facing arrows) is deposited in the extracellular matrix of expressing cells, and only those cells in direct physical contact are capable of responding.

vivo and a molecular mechanism exists to control tightly the delivery of the signal by exploitation of the extracellular matrix to limit dissemination.

The biological importance of this distinction can be envisaged by consideration of a hypothetical multilayer tissue (Fig. 2). Expression of the ECM-associated form of DIA/LIF would confine its action to cells in the adjacent layer, whereas expression of the diffusible form would permit control of cell behaviour in cell layers remote from the signal source. This behaviour represents a simple mechanism for creating pattern in a multilayered tissue, and is intriguingly reminiscent of the concept of stem cell 'niche' (reviewed by Lajtha, 1979) which proposes that stem cells are inhibited from differentiation into more mature cell types by sequestration in a specific physical compartment. According to this view, differentiation induction occurs as a consequence of cell movement out of the 'niche' (or away from the source of signal). It should be emphasised that the use of multiple protein forms to control the dissemination of a signal is not confined to DIA/LIF. Recent studies of int-2 expression, for example, have revealed the existence of multiple amino termini whose identity directs localisation of the int-2 molecule to the nucleus, endoplasmic reticulum or extracellular matrix (Acland *et al.* 1990; Dickson *et al.* this volume).

Stem cell derived signals

Embryonic stem cells not only modify their behaviour in response to environmental cues such as DIA/LIF, but are themselves a source of potent regulatory factors. It has been known for some time that EC cells and ES cells express multiple growth factor species with important functional consequences. The major growth factor species expressed by ES and EC cells have been characterised by protein purification and recombinant DNA techniques (Heath *et al.* 1989). These prove to be secreted (and soluble) members of the FGF family of growth factors: K-FGF, also called *Ks*, (Delli-bovi *et al.* 1987) or *hst-1* (Taira et. al, 1987) and a second FGF-like growth factor, closely related to K-FGF in biochemical properties and biological function, whose primary sequence is at present undefined (Heath *et al.* 1989). It has been established (mainly from analysis of EC cell differentiation systems) that these stem cell-derived growth factors are capable of controlling the multiplication of their differentiated progenitors (Heath and Rees, 1985) suggesting the existence of a form of 'feedforward control' whereby stem cells control the behaviour of the differentiated progeny *via* the action of these secreted factors (reviewed by Heath and Smith, 1988).

The importance of stem cell-derived growth factors goes beyond their mitogenic action on differentiated progeny. The expression of DIA/LIF (both soluble and matrix associated forms) in a variety of cell types has been found to depend upon the presence of exogenous growth factors, including members of the FGF family (Rathjen *et al.* 1990). Furthermore, expression in many cases is enhanced in the presence of modulators of the TGFβ family which are also known to be expressed (albeit in the latent form) by EC cells and their differentiated derivatives (Mummery *et al.* 1990). DIA/LIF expression (soluble and matrix associated forms) is also found to increase greatly as ES cells undergo differentiation *in vitro* in the absence of exogenous DIA/LIF (Rathjen *et al.* unpublished data).

These findings have two implications. Firstly, they suggest that the natural source of DIA/LIF made available to stem cells *in vivo* is their immediate differentiated progeny; the stem cell 'pool' described above would in this view be maintained as a consequence of initial differentiation events. This provides further support for a 'feedback' relationship between differentiated progeny and stem cell parents (Heath and Smith, 1988). Secondly, the inducibility of DIA/LIF expression by stem cell-derived factors in certain cell types provides a potential patterning mechanism, since stem cell differentiation (and manifestations of DIA/LIF action in other sites) would be controlled by the inducibility of other cell types in the immediate vicinity. The specific case of the feeder effect in ES cells may well be due to this phenomenon where ES cells locally induce responsive heterologous feeder cells to express DIA/LIF.

Conclusion

This review has focused on the role of differentiation regulatory factors in early mouse development. The exploitation of cultured stem cell lines has led to the identification of both a stem cell regulatory factor (DIA/LIF) and stem cell-derived growth factors such as K-FGF. A closer examination of the action of these agents has revealed two specific issues which may have broader application outside the early phases of mammalian development. Firstly, the way a regulatory signal is delivered to a responding cell and how it is subsequently interpreted are key issues in fully understanding the role of bioregulatory factors in controlling patterning and the generation of form. In the case of DIA/LIF (and other embryonic growth and differentiation factors) mechanisms exist which control the dissemination of the regulatory signal. Secondly, individual regulatory factors participate in a network of interactions between parental cells and their differentiated progeny. The specific design of such a regulatory network will have great influence on the form and pattern of the differentiation events that ensue.

The authors' research is supported by the Cancer Research Campaign. L–W H is supported by a Taiwanese Ministry of Education postgraduate scholarship.

References

Abe, E., Tanaka, H., Ishimi, Y., Miyaura, C., Hayashi, T., Nagasawa, H., Tomida, M., Yamaguchi, Y., Hozumi, M. and Suda, T. (1986). Differentiation-inducing factor purified from conditioned medium of mitogen-treated spleen cell cultures stimulates bone resorption. *Proc. natn. Acad. Sci. U.S.A.* **83**, 5958–5962.

Acland, P., Dixon, M., Peters, G. and Dickson, C. (1990). Subcellular fate of the Int-2 oncoprotein is determined by the choice of initiation codon. *Nature* **343**, 662–665.

Baumann, H. and Wong, G. G. (1989). Hepatocyte-stimulating factor III shares structural and functional identity with leukemia-inhibitory factor. *J. Immunol.* **143**, 1163–7.

Beck, F., Samani, N. J., Penschow, J., Thorley, B., Tregear, G. and Coghlan, J. (1987). Histochemical localisation of IGF-I and IGF-II mRNA in the developing rat embryo. *Development.* **101**, 175–184.

Bradley, A., Evans, M., Kaufman, M. H. and Robertson, E. (1984). Formation of germ-line chimaeras from embryo-derived teratocarcinoma cell lines. *Nature* **309**, 255–256.

Delli Bovi, P., Curatola, A., Kern, F., Greco, A., Ittman, M. and Basilico, C. (1987). An oncogene isolated by transfection of Kaposis's sarcoma DNA encodes a growth factor that is a member of the FGF family. *Cell* **50**, 729–737.

Dickson, C., Acland, P., Smith, R., Dixon, M., Deed, R., MacAllan, D., Walther, W., Fuller-Pace, F., Kiefer, P. and Peters, G. (1990). Characterization of int-2: a member of the fibroblast growth factors. *J. Cell Sci. (Suppl.)* **13**, 87–96.

Evans, M. J. and Kaufman, M. (1981). Establishment in culture of pluripotential cells from mouse embryos. *Nature* **292**, 154–156.

Frohman, M. A. and Martin, G. R. (1989). Cut paste and save: new approaches to altering specific genes in mice. *Cell* **56**, 145–148.

Gardner, R. L. and Beddington, R. S. P. (1988). Multilineage 'stem cell' in the mammalian embryo. *J. Cell Sci. Supp.* **10**, 11–28.

Gearing, D. P., Gough, N. M., King, J. A., Hilton, D. J., Nicola, N. A., Simpson, R. J., Nice, E. C., Kelso, A. and Metcalf, D. (1987). Molecular cloning and expression of cDNA encoding a murine myeloid leukaemia inhibitory factor (LIF). *EMBO J.* **6**, 3995–4002.

Gearing, D. P., Nicola, N. A., Metcalf, D., Foote, S., Willson, T. A., Gough, N. M. and Williams, R. L. (1989). Production of leukemia inhibitory factor in *Escherichia coli* by a novel procedure and its use in maintaining embryonic stem cells in culture. *Bio-Technology* **7**, 1157–1161.

GOUGH, N. M., GEARING, D. P., KING, J. A., WILLSON, T. A., HILTON, D. J., NICOLA, N. A. AND METCALF, D. (1988). Molecular cloning and expression of the human homologue of the murine gene encoding myeloid leukemia-inhibitory factor. *Proc. natn. Acad. Sci. U.S.A.* **85**, 2623–7.

GUY, G. R., BEE, N. S. AND PENG, C. S. (1990). Lymphokine signal transduction. *Prog. Growth factor Res.* **2**, in press.

HAFEN, E. AND BASLER, K. (1990). Mechanisms of positional signalling in the developing eye of *Drosophila* studied by ectopic expression of *sevenless* and *rough*. *J. Cell Sci. (Suppl.)* **13**, 157–168.

HEATH, J. K. (1978). Mammalian Primordial Germ Cells. *Development in Mammals* **3**, 267.

HEATH, J. K., PATERNO, G. D., LINDON, A. C. AND EDWARDS, D. R. (1989). Expression of multiple heparin binding growth factor species by murine embryonal carcinoma and embryonic stem cells. *Development* **107**, 113–122.

HEATH, J. K. AND REES, A. R. (1985). Growth factors in mammalian embryogenesis. In *Growth Factors in Biology and Medicine* (ed. D. Evered and M. Stoker). *Ciba Symp.* **116**, 21–29.

HEATH, J. K. AND SMITH, A. G. (1988). Regulatory factors of embryonic stem cells. *J. Cell Sci. Supp.* **10**, 257–266.

HEINE, U., MUNOZ, E., FLANDERS, K., ELLINGSWORTH, L., LAM, H-Y., THOMPSON, L., ROBERTS, A. AND SPORN, M. (1987). The role of TGF-β in the devlopment of mouse embryo. *J. Cell Biol.* **105**, 2861–2876.

HILTON, D. J., NICOLA, N. A. AND METCALF, D. (1988). Specific binding of murine leukemia inhibitory factor to normal and leukemic monocytic cells. *Proc. natn. Acad. Sci. U.S.A.* **85**, 5971–5.

KOOPMAN, P. AND COTTON, R. (1984). A factor produced by feeder cells which inhibits embryonal carcinoma differentiation. *Expl Cell Res.* **154**, 233–242.

LAJTHA, L. G. (1979). Stem cell concepts. *Differentiation* **14**, 23–34.

LEHNERT, S. AND ACKHURST, R. (1988). Embryonic pattern of TGF beta type-1 RNA suggests both paracrine and autocrine mechanisms of action. *Development* **104**, 263–273.

LYONS, K. M., PELTON, R. W. AND HOGAN, B. L. M. (1989). The patterns of expression of murine vgr-1 and BMP-2a RNA suggest that transforming growth factor-beta-like genes coordinately regulate aspects of embryonic development. *Genes and Development* **3**, 1657–1668.

MARTIN, G. R. (1981). Isolation of a pluripotent cell line from early mouse embryos cultured in medium conditioned by teratocarcinoma stem cells. *Proc. natn. Acad. Sci. U.S.A.* **78**, 7634–7638.

MARTIN, G. R. AND EVANS, M. (1975). Differentiation of clonal teratocarcinoma cells: formation of embryoid bodies *in vitro*. *Proc. natn. Acad. Sci. U.S.A.* **78**, 7634–7638.

MOREAU, J. F., DONALDSON, D. D., BENNETT, F., WITEK-GIANNOTTI, J., CLARK, S. C. AND WONG, G. G. (1988). Leukaemia inhibitory factor is identical to the myeloid growth factor human interleukin for DA cells. *Nature* **336**, 690–2.

MORI, M., YAMAGUCHI, K. AND ABE, K. (1989). Purification of a lipoprotein lipase-inhibiting protein produced by a melanoma cell line associated with cancer cachexia. *Biochem. biophys. Res. Commun.* **160**, 1085–92.

MUMMERY, C. L., SLAGER, H., KRUIJER, W., FEIJEN, A., FREUND, E., KOORNEFF, I. AND VAN-DER-EIJDEN-VAN-RAAIJ, A. J. M. (1990). Expression of transforming growth factor beta-2 during the differentiation of murine embryonal carcinoma and embryonic stem cells. *Devl. Biol.* **137**, 161–170.

RATHJEN, P. D., TOTH, S., EDWARDS, D. R., HEATH, J. K. AND SMITH, A. G. (1990). Expression of the developmental regulatory factor Differentiation Inhibiting Activity/Leukaemia Inhibitory Factor (DIA/LIF) and control of embryo stem cell populations. *Cell* **62**, 1105–1114.

RIJSEWICK, F., SCHUERMANN, M., WAGENAAR, E., PARREN, P., WEIGEL, D. AND NUSSE, R. (1987). The *Drosophila* homologue of the mouse mammary oncogene int-1 is identical to the segment polarity gene *wingless*. *Cell* **50**, 649–658.

ROBERTSON, E. J. (1989). Developmental potential of embryonic stem cells. *Molecular Genetics of Early* Drosophila *and Mouse Development* (ed. M. R. Capecchi), pp. 39–44. New York: Cold Spring Harbour Laboratory Press.

ROSA, F., ROBERTS, A., DANIELPOUR, D., DART, L., SPORN, M. AND DAWID, I. (1988). Mesoderm induction in amphibians: the role of TGF-β2-like factors. *Science* **239**, 783–785.

SLACK, J., DARLINGTON, B., HEATH, J. AND GODSAVE, S. (1987). Mesoderm induction in early *Xenopus* embryos by heparin-binding growth factors. *Nature* **326**, 197–200.

SMITH, A. G., HEATH, J. K., DONALDSON, D. D., WONG, G. G., MOREAU, J., STAHL, M. AND ROGERS, D. (1988). Inhibition of pluripotential embryonic stem cell differentiation by purified polypeptides. *Nature* **336**, 688–90.

SMITH, A. G. AND HOOPER, M. L. (1987). Buffalo rat liver cells produce a diffusible activity which inhibits the differentiation of murine embryonal carcinoma and embryonic stem cells. *Devl Biol.* **121**, 1–9.

SMITH, J. (1987). A mesoderm-inducing factor is produced by a *Xenopus* cell line. *Development* **99**, 3–14.

SMITH, T. A. AND HOOPER, M. L. (1983). Medium conditioned by feeder cells inhibts the differentiation of embryonal carcinoma cell cultures. *Expl cell Res.* **145**, 458–462.

TAIRA, M., YOSHIDA, T., MIYAGAWA, K., SAKAMOTO, H., TERADA, M. AND SIGMURA, T. (1987). cDNA sequence of human transforming gene *hst* and identification of the coding sequence required for transforming activity. *Proc. natn. Acad. Sci. U.S.A.* **84**, 2980–2984.

TOMIDA, M., YAMAMOTO-YAMAGUCHI, Y. AND HOZUMI, M. (1984). Characterization of a factor inducing differentiation of mouse myeloid leukaemic cells purified from conditioned medium of mouse Ehrlich ascites tumor cells. *FEBS. Lett.* **178**, 291–296.

WILKINSON, D., PETERS, G., DICKSON, C. AND MACMAHON, A. (1988). Expression of of the FGF-related proto-oncogene *int-2* during gastrulation and neurulation in the mouse. *EMBO. J.* **7**, 691–695.

WILLIAMS, R. L., HILTON, D. J., PEASE, S., WILLSON, T. A., STEWART, C. L., GEARING, D. P., WAGNER, E. F., METCALF, D., NICOLA, N. A. AND GOUGH, N. M. (1988). Myeloid leukaemia inhibitory factor maintains the developmental potential of embryonic stem cells. *Nature* **336**, 684–7.

YAMAMORI, T., FUKADA, K., AEBERSOLD, R., KORSCHING, S., FANN, M. J. AND PATTERSON, P. H. (1989). The cholinergic neuronal differentiation factor from heart cells is identical to leukemia inhibitory factor. *Science* **246**, 1412–6.

J. Cell Sci. Suppl. 13, 87–96 (1990)
Printed in Great Britain © The Company of Biologists Limited 1990

Characterization of *int*-2: a member of the fibroblast growth factor family

CLIVE DICKSON[1], PIERS ACLAND[1], ROSALIND SMITH[1], MARK DIXON[1], RICHARD DEED[2], DAVID MacALLAN[2], WOLFGANG WALTHER[2], FRANCES FULLER-PACE[1], PAUL KIEFER[1] AND GORDON PETERS[2]

Laboratories of Viral Carcinogenesis[1] and Molecular Oncology[2], Imperial Cancer Research Fund, Lincoln's Inn Fields, London WC2A 3PX, UK

Summary

int-2 was discovered as a proto-oncogene transcriptionally activated by MMTV proviral insertion during mammary tumorigenesis in the mouse. Sequence analysis showed *int*-2 to be a member of the fibroblast growth factor family of genes. In normal breast and most other adult mouse tissues, *int*-2 expression was not detected except for low levels in brain and testis. However, using *in situ* hybridization, expression was found at a number of sites during embryonic development, from day 7 until birth. An analysis of the *int*-2 transcripts found in embryonal carcinoma cells revealed six major classes of RNA initiating at three promoters and terminating at either of two polyadenylation sites. Despite the transcriptional complexities, all size classes of RNA encompass the same open reading frame. Using an SV40 early promoter to drive transcription of an *int*-2 cDNA in COS-1 cells, several proteins were observed. These were shown to be generated by initiation from either of two codons: One, a CUG, leads to a product which localizes extensively to the cell nucleus and partially to the secretory pathway. In contrast, initiation at a downstream AUG codon results in quantitative translocation across the endoplasmic reticulum and the accumulation of products ranging in size from $27.5\times10^3 M_r$ to $31.5\times10^3 M_r$ in organelles of the secretory pathway. These proteins represented glycosylated and non-glycosylated forms of the same primary product with or without the signal peptide removed. These findings suggest the potential for a dual role of *int*-2; an autocrine function acting at the cell nucleus, and a possible paracrine action through a secreted product.

Introduction

The *int*-2 gene encodes a member of the fibroblast growth factor (FGF) family of which the archetype is basic FGF (bFGF) (Dickson and Peters, 1987). To date, there are seven known members of this family which over a central core region share approximately 20% amino acid identity (for review see Burgess and Maciag, 1989; Dickson *et al.* 1989). At the gene level, the FGF family also share a common architecture, with three coding exons and a highly conserved middle exon of 104 nucleotides (Fig. 1). Substantial sequence differences reside at the amino and carboxyl terminal regions which are often extended compared with the prototypic factors, bFGF and the related acidic FGF (aFGF). Perhaps the most biologically significant difference is that all members of the family, apart from aFGF and bFGF, contain a signal peptide at their amino terminus, which directs passage through the secretory pathway. In contrast, the mechanisms by which

Key words: *int*-2, FGF-family, growth factor, MMTV, embryogenesis.

```
bFGF                                                                        MAAGSITTLPALPEDGGSG
aFGF                                                                          MAEGEITTFTALTEKF
INT-2                                                        MGLIWLLLLSLLEPGWPAAGPGARLRRDAGGRGGV
HST/kFGF     MSGPGTAAVALLPAVLLALLAPWAGRGGAAAPTAPNGTLEAELERRWESLVALSLARLPVAAQPKEAAVQSGA
FGF-5     MSLSFLLLLFFSHLILSAWAHGEKRLAPKGQPGPAATDRNPRGSSSRQSSSSAMSSSSASSSPAASLGSQGSGLEQS
FGF6
KGF                                MHKWILTWILPTLLYRSCFHIICLVGTISLACNDMTPEQMATNVNCSSPERHTRSY

bFGF      AFPPGHFKDPKRLYCKNG-GFFLRIHPDGRVDGVREKSDPHIKLQLQAEERGVVSIKGVCANRYLAMKEDGRLLAS
aFGF      NLPPGNYKKPKLLYCSNG-GHFLRILPDGTVDGTRDRSDQHIQLQLSAESVGEVYIKSTETGQYLAMDTDGLLYGS
INT-2     YEHLGGAPRRRKLYCATK--YHLQLHPSGRVNGSLENSAYSI-LEITAVEVGIVAIRGLFSGRYLAMNKRGRLYAS
HST/kFGF  GDYLLGIKRLRRLYCNVGIGFHLQALPDGRIGGAHADTRDSL-LELSPVERGVVSIFGVASRFFVAMSSKGKLYGS
FGF-5     SFQWSLGARTGSLYCRVGIGFHLQIYPDGKVNGSHEANMLSV-LEIFAVSQGIVGIRGVFSNKFLAMSKKGKLHAS
FGF-6     .....GIKRQRRLYCNVGIGFHLQGLPDGRISGTHEENPYSL-LEISTVERGVVSLFGVRSALFVAMNSKGRLYAT
KGF       DYMEGGDIRVRRLFCRTQW-Y-LRIDKRGKVKGTQEMKNNYNIMEIRTVAVGIVAIKGVESEFYLAMNKEGKLYAK

bFGF      KCVTDECFFFERLGSNNYNTYRSRKYTS----------------WYVALKRTGQYKLG--SKTGPGQKAILFLPMS
aFGF      QTPNEECLFLERLEENHYNTYISKKHAEKN--------------WFVGLKKNGSCKRG--PRTHYGQKAILFLPLP
INT-2     EHYSAECEFVERIHELGYNTYASRLYRTVSSTPGARRQPSAERLWYVSVNGKGRPRRGFKTRRT--QKSSLFLPRV
HST/kFGF  PFFTDECTFKEILLPNNYNAYESYKYPG----------------MFIALSKNGKTKKG--NRVSPTMKVTHFLPRL
FGF-5     AKFTDDCKFRERFQENSYNTYASAIHRTEKTGRE----------WYVALNKRGKAKRGCSPRVKPQHISTHFLPRF
FGF-6     PSFQEECKFRETLLPNNYNAYESDLYQG----------------TYIALSKYGRVKRG--SKVSPIMTVTHFLPRI
KGF       KECNEDCNFKELILENHYNTYASAKWTHNGGE------------MFVALNQKGIPVRGKKTKK---QKTAHFLPMA

bFGF      AKS                                                      155
aFGF      VSSD                                                     155
INT-2     LDHRDHEMVRQLQSGKPRPPGKGVQPRRRRQKQSPDNLEPSHVQASRLGSQLEASAH  239
HST/kFGF                                                           206
FGF-5     KQSEQPELSFTVTVPEKKNPPSPIKSKIPLSAPRKNTNSVKYRLKFRFG        267
FGF-6                                                              ?
KGF       IT                                                       194
```

Fig. 1. The human homologues of the FGF family. The amino acid sequences are taken from the literature and aligned to maximize the homologies between family members (Abraham *et al.* 1986; Jaye *et al.* 1986; Yoshida *et al.* 1987; Brookes *et al.* 1989; Marics *et al.* 1989; Finch *et al.* 1989; Zhan *et al.* 1988). The residues which are identical in all seven proteins are highlighted with black shading. The position of the conserved 104 nucleotide exon is indicated by the bar above and below the sequences. The boxed amino acids show the position of the N-linked glycosylation motifs.

aFGF and bFGF are released from cells remain enigmatic. However, it is clear that many cells possess high affinity cell surface receptors which interact with both aFGF and bFGF. It is assumed that these receptors mediate the various activities ascribed to aFGF and bFGF, including the ability to induce mitogenesis, angiogenesis, chemotaxis, and differentiative responses (reviewed in Burgess and Maciag, 1989; Gospodarowicz, 1985; Rifkin and Moscatelli, 1989).

Members of the FGF family were discovered by a number of different approaches. For example, aFGF and bFGF, and the recently described keratinocyte growth factor (KGF) were identified as mitogens in extracts from bovine brain, pituitary and human embryonic lung fibroblasts respectively (Gospodarowicz *et al.* 1974; Finch *et al.* 1989). Two other members, kFGF (also known as hst, hstf1, or KS3) and FGF-5 were detected as dominantly acting transforming genes following transfection of human tumour DNA into NIH3T3 cells (Sakamoto *et al.* 1986; Yoshida *et al.* 1987; Delli Bovi *et al.* 1987*a*,*b*; Zhan *et al.* 1988). The FGF-6 gene was isolated by low stringency hybridization with a kFGF probe and has similar *in vitro* transforming properties to kFGF and FGF-5 (Marics *et al.* 1989). It is probably no coincidence that the three FGF genes that transform cells encode actively secreted products. Indeed, bFGF can be converted into an effective transforming gene if a signal peptide sequence is appended to the amino terminus (Rogelj *et al.* 1988; Blam *et al.* 1988). The most likely explanation would be that, at least in cell culture conditions, the synthesis and secretion of these factors can lead to morphological transformation *via* an autocrine loop.

Although these secreted FGFs are potent transforming agents, and their cognate genes are therefore classed as oncogenes, there is little evidence implicating them in the genesis of a naturally occurring tumour. In contrast, the *int*-2 gene was originally discovered through its transcriptional activation in spontaneous mammary tumours in mice infected by mouse mammary tumour virus (Peters *et al.* 1983; Dickson *et al.* 1984). The remainder of this report will summarize the known properties of the *int*-2 gene and its products in relation to other members of the FGF family.

Int-2 as a proto-oncogene in mammary cancer

Int-2 is one of a group of genes found to lie at common integration sites for the mouse mammary tumour virus (MMTV) in murine breast tumours (reviewed in Nusse, 1988; Peters and Dickson, 1987). Insertion of viral DNA within several kilobases either side of the *int*-2 gene results in its transcriptional activation by a mechanism that is thought to involve the regulatory elements of the viral promoter acting in *cis*. As integration of viral DNA appears to occur at random, it is only rarely in infected cells that an appropriate insertion occurs next to *int*-2. Presumably the transcriptionally active *int*-2 gene, which is apparently silent in the normal mammary gland, endows the afflicted cell with a proliferative advantage. The result is an initially clonal hyperplasia that, through further mutations, will eventually develop into frank neoplasia. Despite the random nature of the causal integration event, up to 70 % of the mammary tumours that develop in certain inbred strains of mice show proviral insertion and transcriptional activation of *int*-2.

Embryonic expression of *int*-2

As determined by Northern blotting and RNAase protection analysis, adult mouse tissues do not express detectable levels of *int*-2 RNA, with the exception of very small amounts in brain and testes. However, the same techniques readily

Table 1. *Timing of* int-2 *transcription during mouse development*

Tissue/Cell type	Days of development
Parietal endoderm	7.5–9
Migrating mesoderm	8–9.5
Hindbrain/neuroepithelium	8.5–9.5
Pharyngeal pouches	9.5
Inner ear	10.5–17.5
Cerebellum/Purkinje cells	14.5–21
Retina	14.5–21
Tooth bud/mesenchyme	17.5–18.5

Taken from: Wilkinson *et al.* 1988, 1989.

detect *int*-2 transcripts in mid-gestation mouse embryos (Jakobovits *et al.* 1986; Wilkinson *et al.* 1988). Further detailed studies, using *in situ* RNA hybridization, have confirmed that embryonic expression of *int*-2 is quite extensive, from around day 7 until parturition (see Table 1), but is restricted to a few specific sites in a distinct temporal sequence (Wilkinson *et al.* 1988, 1989). The number and diversity of these sites suggest multiple roles, consistent with functions such as mitogenesis, chemotaxis or induction of differentiation.

Structure of the *int*-2 gene

One of the early sites of *int*-2 expression is the parietal endoderm, which has an *in vitro* counterpart in the form of F9 or PCC4 embryonal carcinoma cells (reviewed in Hogan *et al.* 1983). Following exposure to retinoic acid and dibutyryl cAMP, these cultures differentiate to a parietal endoderm-like cell, and concordantly start to synthesize *int*-2 RNA (Jakobovits *et al.* 1986; Smith *et al.* 1988). These cell lines have been particularly useful in characterizing the normal transcripts from the *int*-2 gene (Mansour and Martin, 1988; Smith *et al.* 1988). From a combination of approaches, including sequencing of *int*-2 cDNAs and RNAase protection assays, six distinct classes of RNA have been resolved (Fig. 2). These transcripts derive from three different promoter regions and terminate at either of two polyadenylation sites. While different embryonal carcinoma cells express different subsets of these RNAs, mammary tumours, in which *int*-2 has been activated by the insertion of proviral DNA, generally express all six transcripts (Dickson *et al.* 1990). Since the longest open reading frame appears to be the same in all these RNAs, it seems reasonable to speculate that the multiple promoters are required for regulating the expression of *int*-2 in different tissue types and at precise times during embryogenesis.

Synthesis and processing of the *int*-2 protein

The sequence of the mouse *int*-2 gene predicts a primary translation product of $27\times10^3 M_r$, which is distinctly basic and hydrophilic, apart from a short hydrophobic region at the amino terminus (Moore *et al.* 1986; Mansour and

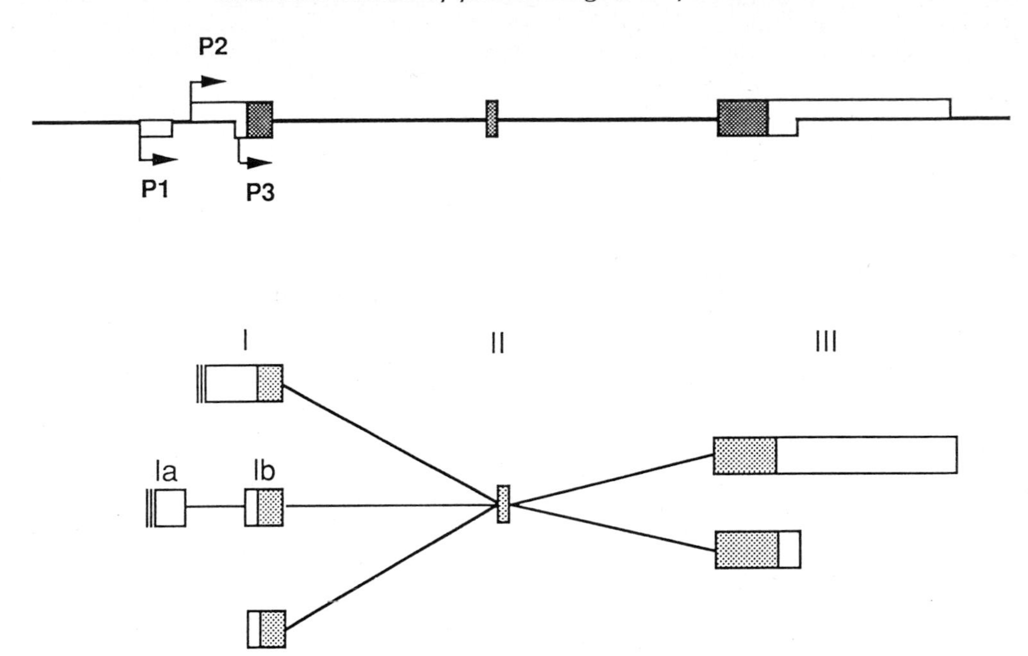

Fig. 2. Structure of the mouse *int*-2 gene and RNA transcripts. The upper part of the figure schematically shows the structure of the *int*-2 gene with the exons marked as boxes. The alternative use of 5′ exons and 3′ polyadenylation sites is indicated by the depth of the boxes. The shaded areas highlight the extent of the open reading frame. The position of the cap sites for the three promoters (P1, P2 and P3) are marked by arrows. The lower part of the illustration shows the possible combinations of 5′ and 3′ exons with the invariant middle exon which give rise to six RNA classes.

Martin, 1988). This hydrophobic region has hallmarks of a signal peptide, and there is a single consensus site for N-linked glycosylation, suggesting that the protein may belong to the secreted class of FGFs. Antisera raised against synthetic peptides, based on the predicted amino acid sequence, have been used in a variety of immunological procedures to search for *int*-2 products. However, attempts to detect *int*-2 protein in mammary tumours or embryonal carcinoma cells have met with limited success, and these naturally occurring systems have not been amenable to further biochemical analyses. To examine the properties of *int*-2 in mammalian cells, we therefore chose to introduce the cloned cDNA into cultured cells in a plasmid vector capable of high levels of expression.

The vector chosen was based on the SV40 early promoter, since it undergoes amplification following transient transfection into COS-1 cells. An immunoblot analysis of total cell extract prepared 60 h after transfection shows multiple *int*-2-related proteins in the molecular weight range of around 30×10^3 M_r, as shown in Fig. 3A (Dixon *et al.* 1989). Analogous products can also be detected by translating synthetic *int*-2 RNA in a cell free system (Fig. 3B). The latter experiments provide an explanation for the multiplicity of proteins in terms of carbohydrate addition and signal peptide cleavage. Thus, translation of *int*-2 in a

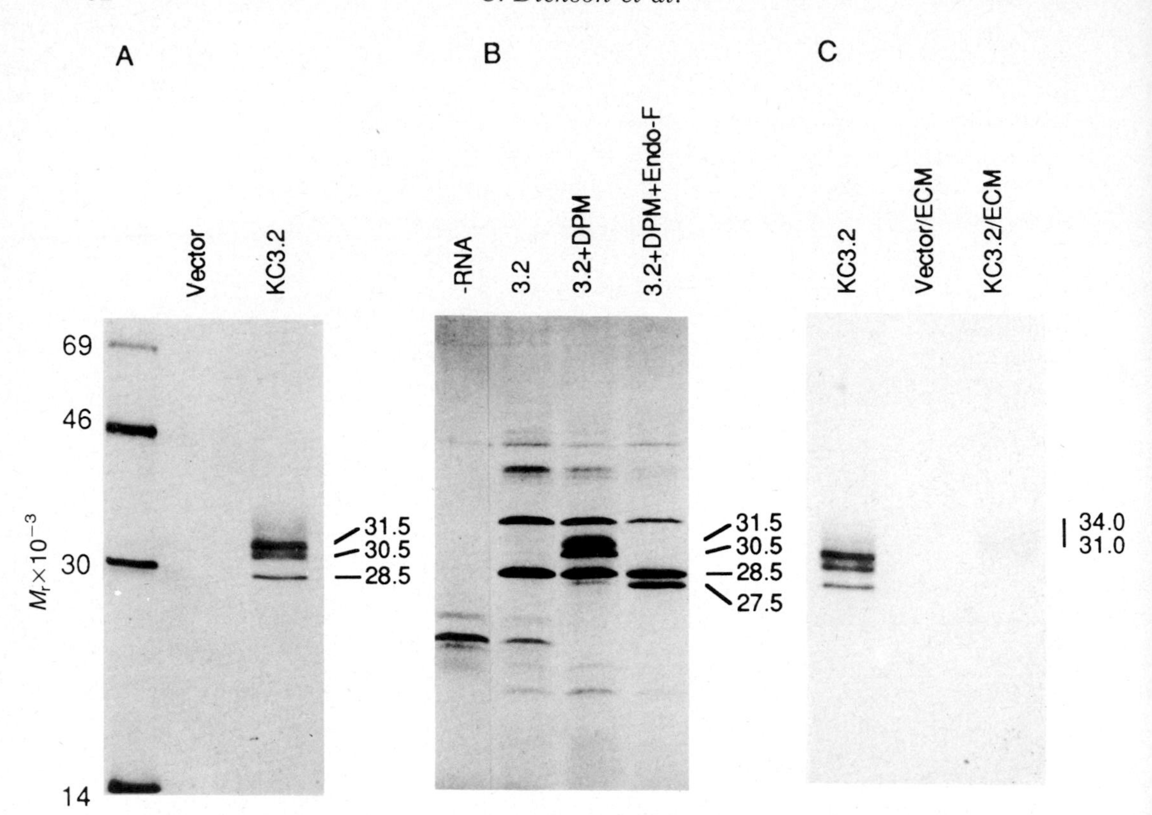

Fig. 3. Synthesis and processing of the major *int*-2 proteins in COS-1 cells, and a cell free system. (A) An immunoblot showing multiple forms of the *int*-2 proteins synthesized in COS-1 cells. Cells were transfected with vector DNA or a plasmid DNA containing the *int*-2 open reading frame starting at the first in-frame AUG codon. Four days after transfection, a total cell extract was prepared from the cultures, separated on a polyacrylamide gel and transferred to nitrocellulose. *Int*-2 proteins were detected using anti-peptide serum as described in Dixon *et al.* 1989. (B) *In vitro* translation in a rabbit reticulocyte lysate containing ^{35}S-methionine and programmed with complimentary RNA. The panel shows an autoradiograph of a polyacrylamide gel of the proteins synthesized in the absence of added RNA (−RNA) or following addition of *int*-2 cRNA alone (3.2), or 3.2 in the presence of dog pancreas microsomes to facilitate N-linked glycosylation (3.2+DPM). A further sample of 3.2+DPM was treated with endoglycosidase-F which removes added sugars as an additional control (3.2+DPM+endo-F). The polypeptides with molecular weights higher than the marked *int*-2 proteins are apparently artifactual as they disappear upon treatment with ribonuclease. (C) An immunoblot showing the *int*-2 proteins synthesized in COS-1 cells transfected with KC3.2 as in A, and the *int*-2 related protein found associated with the extracellular matrix in control vector and KC3.2 transfected cells (vector/ECM and KC3.2/ECM respectively).

rabbit reticulocyte lysate yields a single polypeptide of 28.5×10^3 M_r, representing the primary translation product. If canine pancreatic microsomes are included in the system, three other species are generated, one of lower molecular weight that corresponds to the non-glycosylated product after signal peptide cleavage, and two higher molecular weight forms that are glycosylated, as indicated by their sensitivity to endoglycosidase F (Fig. 3B). These two glycosylated species are consistent with carbohydrate addition at the single consensus site, with and without signal peptide cleavage.

The processing and secretion of *int*-2 protein in transfected COS-1 cells appears to be relatively inefficient, as judged by the proportion of unprocessed forms observed in cell lysates and the difficulty in detecting *int*-2 proteins in the medium. Nevertheless, small quantities of slightly higher molecular weight forms of *int*-2 do appear to be associated with the extracellular matrix. This was demonstrated by removing the cells from the monolayer with non-ionic detergent or EDTA, and then dissolving the matrix material remaining on the dish with a buffer containing SDS. As shown in Fig. 3C, the amount of matrix-bound material detected by immunoblotting is small compared with the intracellular pool. It is not clear at present whether the inefficiency of processing in COS-1 cells is a reflection of the system, or an intrinsic feature of the *int*-2 product.

Alternative initiation codons for *int*-2

We considered whether our inability to detect *int*-2 protein in embryonal carcinoma cells or mammary tumours was in part a result of low translational efficiency. This suspicion was based on the presence of an AUG codon in the +1 reading frame immediately preceding the predicted AUG start site (Dixon *et al.* 1989). In an experiment where this upstream AUG codon was mutated, there was indeed a small but significant increase in the level of translation, both in COS-1 cells and by cell free synthesis. To further explore the possible effects of 5′ leader sequences on the efficiency of translation, various forms of *int*-2 RNA were synthesized and tested in an *in vitro* system. Surprisingly, it was found that inclusion of the presumed leader sequences resulted in a qualitative rather than quantitative effect, producing a larger translation product (Acland *et al.* 1990). From the DNA sequence, it was apparent that the open reading frame extended 5′ of the initiator methionine codon, but that there were no alternative AUG codons at which translation might begin. The most likely candidate was therefore an in-frame CUG codon, an idea that was tested using site-directed mutagenesis. In this study, point mutations were placed in each of the proposed initiation codons and the mutants subsequently tested in the COS-1 cell expression system. To simplify the analysis, the glycosylation site was also mutated (Asn to Gln) to prevent the formation of multiple glycosylated forms of the *int*-2 proteins. The results of such an analysis, depicted in Fig. 4, confirmed that both CUG and AUG codons can be used to initiate *int*-2 protein synthesis (Acland *et al.* 1990).

The use of alternative initiation codons raised the question of what effect the 29 amino acid N-terminal extension might have on the function of the signal peptide, encoded immediately downstream of the AUG codon. The answer was most clearly shown by immunofluorescence of COS-1 cells transfected with the various mutated froms of *int*-2 cDNA (Acland *et al.* 1990). As expected, the AUG-initiated products were located in the endoplasmic reticulum and Golgi, characteristic of a

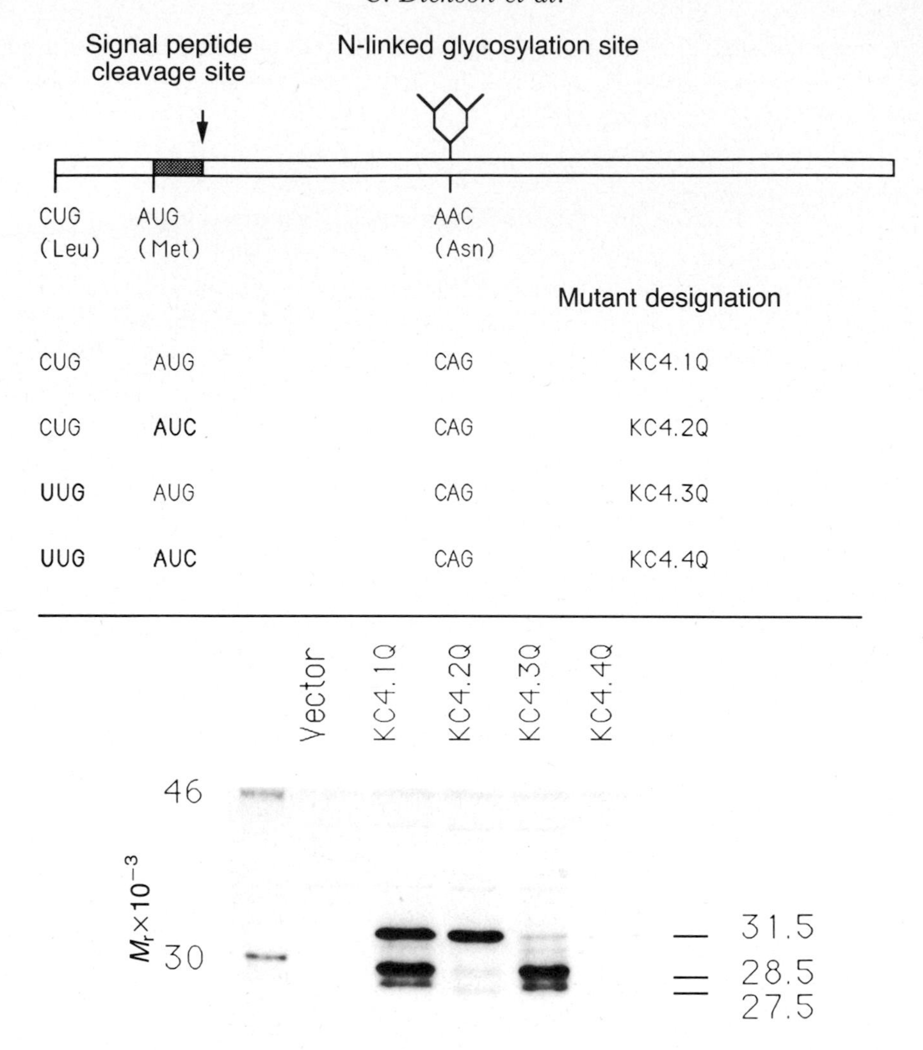

Fig. 4. Protein initiation and processing motifs used for *int*-2 synthesis. The top part of the figure schematically shows the two protein synthesis initiation codons, the position of the signal peptide and the site of N-linked carbohydrate addition. Below the diagram is a table showing the position of site specific mutations used to identify these motifs. The lower third of the diagram shows an immunoblot illustrating the results of a COS-1 transfection experiment using the mutations described above (reproduced from Acland *et al.* 1990).

secreted protein, but a significant proportion of the CUG-initiated protein was found in the nucleus. Thus, it would seem that the subcellular fate of the *int*-2 protein is dependent on the choice of initiation codon. This observation raises the possibility that *int*-2 may have dual functions, operating *via* alternative intracellular pathways: one as a secreted product interacting with high affinity cell surface receptors, the other acting directly in the nucleus. To determine whether such a hypothesis is viable, it will be necessary to show that a nuclear form of the protein has measurable biological effect and is not simply a consequence of a highly basic sequence forming a fortuitous nuclear localization signal. In addition, it will be essential to ascribe an independent activity to the secreted form of the protein.

Functions of the *int*-2 protein

In general, it has been difficult to associate measurable biological functions with the *int*-2 protein, at least in *in vitro* assays, such as cell transformation and mitogenicity. This is in sharp contrast to several other members of the FGF family, such as FGF-5 and kFGF, where we can readily demonstrate mitogenic and transforming activities in parallel experiments. One of the reasons may be the choice of indicator cells and the possibilty that FGFs interact with a corresponding family of specific receptors. It may be necessary to identify specialized cell types that respond to *int*-2. Nevertheless, the development of mice carrying *int*-2 as a transgene has clearly shown that *int*-2 can induce extensive hyperplasia of the mammary epithelium, a property consistent with its known participation in mammary tumorigenesis (Muller *et al.* 1990). Future studies must be aimed at understanding the biological properties of the multiple forms of *int*-2, and their potential role in animal development.

References

ABRAHAM, J. A., WHANG, J. L., TUMULO, A., MERGIA, A., FRIEDMAN, J., GOSPODAROWICZ, D. AND FIDDES, J. C. (1986). Human basic fibroblast growth factor: nucleotide sequence and genomic organization. *EMBO J.* **5**, 2523–2528.

ACLAND, P., DIXON, M., PETERS, G. AND DICKSON, C. (1990). Subcellular fate of the int-2 oncoprotein is determined by choice of initiation codon *Nature* **343**, 662–665.

BLAM, S. B., MITCHELL, R., TISCHER, E., RUBIN, J. S., SILVA, M., SILVER, S., FIDDES, J. C., ABRAHAM, J. A. AND AARONSON, S. A. (1988). Addition of growth hormone secretion signal to basic fibroblast growth factor results in cell transformation and secretion of aberrant forms of the protein. *Oncogene* **3**, 129–136.

BROOKES, S., SMITH, R., CASEY, G., DICKSON, C. AND PETERS, G. (1989). Sequence organization of the human *int*-2 gene and its expression in teratocarcinoma cells. *Oncogene* **4**, 429–436.

BURGESS, W. H. AND MACIAG, T. (1989). The heparin binding (fibroblast) growth factor family proteins. *A. Rev. Biochem.* **58**, 575–606.

DELLI-BOVI, P. AND BASILICO, C. (1987*a*). Isolation of a rearranged human transforming gene following transfection of Kaposi sarcoma DNA. *Proc. natn. Acad. Sci. U.S.A.* **84**, 5660–5664.

DELLI-BOVI, P., CURATOLA, A. M., KERN, F. G., GRECO, A., ITTMANN, M. AND BASILICO, C. (1987*b*). An oncogene isolated by transfection of Kaposi's sarcoma DNA encodes a growth factor that is a member of the FGF family. *Cell* **50**, 729–737.

DICKSON, C. AND PETERS, G. (1987). Potential oncogene product related to growth factors. *Nature* **326**, 833.

DICKSON, C., SMITH, R., BROOKES, S. AND PETERS, G. (1984). Tumorigenesis by mouse mammary tumor virus: proviral activation of a cellular gene in the common integration region *int-2*. *Cell* **37**, 529–536.

DICKSON, C., SMITH, R., BROOKES, S. AND PETERS, G. (1990). Proviral insertions within the *int-2* gene can generate multiple anomalous transcripts but leave the protein coding domain intact. *J. Virol.* **64**, 784–793.

DIXON, M., DEED, R., ACLAND, P., MOORE, R., WHYTE, A., PETERS, G. AND DICKSON, C. (1989).

Detection and characterization of the fibroblast growth factor-related oncoprotein INT-2. *Molec. cell Biol.* **9**, 4896–4902.

FINCH, P. W., RUBIN, J. S., MIKI, T., RON, D. AND AARONSON, S. A. (1989). Human KGF is FGF-related with properties of a paracrine effector of epithelial cell growth. *Science* **245**, 752–755.

GOSPODAROWICZ, D. (1985). Biological activity *in vivo* and *in vitro* of pituitary and brain fibroblast growth factor. In *Mediators in Cell Growth and Differentiation* (ed. R. J. Ford and A. L. Maizel), pp. 109–134. Raven Press, New York.

GOSPODAROWICZ, D., JONES, K. L. AND SATO, G. (1974). *Proc. natn. Acad. Sci. U.S.A.* **71**, 2295–2299.

HOGAN, B. L. M., BARLOW, D. P. AND TILLY, R. (1983). F9 teratocarcinoma cells as a model for the differentiation of parietal endoderm in the mouse embryo. *Cancer Surveys* **2**, 115–140.

JAKOBOVITS, A., SHACKLEFORD, G. M., VARMUS, H. E. AND MARTIN, G. R. (1986). Two proto-oncogenes implicated in mammary carcinogenesis, *int-1* and *int-2*, are independently regulated during mouse development. *Proc. natn. Acad. Sci. U.S.A.* **83**, 7806–7810.

JAYE, M., HOWK, R., BURGESS, W., RICCA, G. A., CHIU, I., RAVERA, M. W., O'BRIEN, S. J., MODI, W. S., MACIAG, T. AND DROHAN, W. N. (1986). Human endothelial cell growth factor: cloning, nucleotide sequence, and chromosome localization. *Science* **233**, 541–545.

MANSOUR, S. L. AND MARTIN, G. R. (1988). Four classes of mRNA are expressed from the mouse *int-2* gene, a member of the FGF gene family. *EMBO J.* **7**, 2035–2041.

MARICS, I., ADELAIDE, J., RAYBAUD, F., MATTEI, M-G., COULIER, C., PLANCHE, J., DE LAPEYRIERE, O. AND BIRNBAUM, D. (1989). Characterization of the *HST*-related *FGF*. 6 gene, a new member of the fibroblast growth factor gene family. *Oncogene* **4**, 335–340.

MOORE, R., CASEY, G., BROOKES, S., DIXON, M., PETERS, G. AND DICKSON, C. (1986). Sequence, topography and protein coding potential of mouse *int-2*: a putative oncogene activated by mouse mammary tumour virus. *EMBO J.* **5**, 919–924.

MULLER, W. J., LEE, F. S., DICKSON, C., PETERS, G., PATTENGALE, P. AND LEDER, P. (1990). The *int*-2 gene product acts as an epithelial growth factor in transgenic mice. *EMBO J.* **9**, 907–913.

NUSSE, R. (1988). The activation of cellular oncogenes by proviral insertion in murine mammary cancer. In *Breast Cancer: Cellular and Molecular Biology* (ed. M. E. Lippman and R. Dickson), pp. 283–306. Martinus Nijhoff, Boston.

PETERS, G., BROOKES, S., SMITH, R. AND DICKSON, C. (1983). Tumorigenesis by mouse mammary tumor virus: evidence for a common region for provirus integration in mammary tumors. *Cell* **33**, 369–377.

PETERS, G. AND DICKSON, C. (1987). On the mechanism of carcinogenesis by mouse mammary tumor virus. In *Cellular and Molecular Biology of Mammary Cancer* (ed. D. Medina, W. Kidwell, G. Heppner and E. Anderson), pp. 307–319. Plenum, New York.

RIFKIN, D. B. AND MOSCATELLI, D. (1989). Recent developments in the cell biology of basic fibroblast growth factor. *J. Cell Biol.* **109**, 1–6.

ROGELJ, S., WEINBERG, R. A., FANNING, P. AND KLAGSBRUN, M. (1988). Basic fibroblast growth factor fused to a signal peptide transforms cells. *Nature* **331**, 173–175.

SAKAMOTO, H., MORI, M., TAIRA, M., YOSHIDA, T., MATSUKAWA, S., SHIMIZU, K., SEKIGUCHI, M., TERADA, M. AND SUGIMURA, T. (1986). Transforming gene from human stomach cancers and a noncancerous portion of stomach mucosa. *Proc. natn. Acad. Sci. U.S.A.* **83**, 3997–4001.

SMITH, R., PETERS, G. AND DICKSON, C. (1988). Multiple RNAs expressed from the *int-2* gene in mouse embryonal carcinoma cell lines encode a protein with homology to fibroblast growth factors. *EMBO J.* **7**, 1013–1022.

WILKINSON, D. G., BHATT, S. AND MCMAHON, A. P. (1989). Expression pattern of the FGF-related proto-oncogene *int*-2 suggests multiple roles in fetal development. *Development* **105**, 131–136.

WILKINSON, D. G., PETERS, G., DICKSON, C. AND MCMAHON, A. P. (1988). Expression of the FGF-related proto-oncogene *int-2* during gastrulation and neurulation in the mouse. *EMBO J.* **7**, 691–695.

YOSHIDA, T., MIYAGAWA, K., ODAGIRI, H., SAKAMOTO, H., LITTLE, P. F. R., TERADA, M. AND SUGIMURA, T. (1987). Genomic sequence of *hst*, a transforming gene encoding a protein homologous to fibroblast growth factors and the *int-2*-encoded protein. *Proc. natn. Acad. Sci. U.S.A.* **84**, 7305–7309.

ZHAN, X., BATES, B., HU, X. AND GOLDFARB, M. (1988). The human FGF-5 oncogene encodes a novel protein related to fibroblast growth factors. *Molec. cell. Biol.* **8**, 3487–3495.

J. Cell Sci. Suppl. 13, 97–117 (1990)
Printed in Great Britain *© The Company of Biologists Limited 1990*

Basic fibroblast growth factor (bFGF), a multifunctional growth factor for neuroectodermal cells

REINER WESTERMANN*, CLAUDIA GROTHE AND KLAUS UNSICKER

Department of Anatomy and Cell Biology, Philipps University, Marburg, Fed. Rep. Germany

Summary

Basic fibroblast growth factor (bFGF), a heparin-binding mitogen for mesoderm-derived cells, also acts as a mitogen, differentiation inducing and maintenance factor for many neuroectodermal cells including glial cells, neurons, paraneurons, and their tumor counterparts. The molecule is expressed in several types of neuroectodermal cells *in vitro* and *in vivo*. Furthermore, bFGF occurs in many neuronal target tissues, and can prevent ontogenetic as well as lesion-induced neuron death. Thus, in terms of its wide range of functions, bFGF is apparently more than a 'classical' neurotrophic factor. Some of its essential features, such as regulation of expression, local availability and transport in the nervous system remain to be studied.

Neurotrophic factors

During a distinct phase of nervous system development, 20–70 % (depending on the location) of the neurons originally established are eliminated. This naturally occurring neuron death is thought to be regulated by target-derived proteins called neurotrophic factors (NTFs, recently reviewed by Davies, 1988; Barde, 1989; Hefti *et al.* 1989, Oppenheim, 1989). The biological features of NTFs have been deduced from those that can be attributed to nerve growth factor (NGF, for recent review see Levi-Montalcini, 1987), which, for a long time, was the one and only NTF. Several definitions for NTFs have been proposed, but we shall focus on a very stringent one. Barde (1988) defined an NTF as a protein that is (1) able to prevent the ontogenetic neuron death (to be shown by *in vivo* administration of the NTF and inhibiting antibodies), (2) synthesized in the respective neuronal target tissue, (3) present in limiting amounts and (4) taken up by receptor-mediated retrograde axonal transport. So far, only NGF completely fulfills these critera (Barde, 1988, 1989). However, since the neuronal specificity of NGF is restricted to sympathetic and neural crest-derived sensory neurons in the peripheral nervous system (PNS) and magnocellular cholinergic neurons of the forebrain in the central nervous system (CNS) (Barde, 1989; Hefti *et al.* 1989), it is safe to postulate the existence of additional NTFs.

Recently, two further proteins (brain-derived neurotrophic factor, BDNF, and neurotrophin-3, NT-3) with 50–60 % sequence homology to NGF have been

* Author for correspondence.

Key words: FGF, NGF, growth factor, neurotrophic factor, nervous system.

described (Barde *et al.* 1982; Leibrock *et al.* 1989; Hohn *et al.* 1990; Maisonpierre *et al.* 1990).

Several other proteins (recently reviewed by Barde, 1989; Hefti *et al.* 1989; Walicke, 1989) can also promote survival, induce fiber outgrowth and influence transmitter metabolism of various neurons. Whether these proteins match NGF in terms of the criteria defining an NTF remains to be investigated. Even for BDNF it still has to be shown that (1) anti-BDNF antibodies are able to enhance the amount of dying neurons during ontogenetic cell death, (2) BDNF is present in the target organs of all neuron populations being affected and (3) BDNF is retrogradely transported.

Recently, bFGF has emerged as a protein that is synthesized in the nervous system and exerts a variety of effects on neural cells *in vivo* and *in vitro*. The aim of this article is to summarize our present knowledge on neural functions of bFGF, compare them with those of NGF and discuss the pros and cons of bFGF as a neurotrophic factor.

Basic fibroblast growth factor

The molecular properties, tissue distribution and functions of bFGF outside the nervous system have been extensively studied and reviewed (*e.g.* by Gospodarowicz *et al.* 1986*a*; Thomas, 1987; Lobb, 1988; Boehlen, 1989; Rifkin and Moscatelli, 1989; Baird and Boehlen, 1990).

The FGF-family

The FGF-family at present consists of seven members, which share approximately 40–60 % sequence homology. Best characterized are acidic FGF (aFGF) and basic FGF (bFGF), which will be discussed in more detail. The five other members are the oncogene products kFGF, also designated hst and ksFGF, (Taira *et al.* 1987; Delli-Bovi *et al.* 1987; Paterno *et al.* 1989), int-2 (Dickson and Peters, 1987; Paterno *et al.* 1989), FGF-5 (Zhan *et al.* 1988), FGF-6 (Marics *et al.* 1989) and KGF (Rubin *et al.* 1989), which were identified as members of the FGF family by their sequence. They also share some functions (mitogenic activity, mesoderm induction, see below) with the well-characterized aFGF and bFGF.

Molecular properties

Basic FGF, like aFGF, varies in size ($15.6–17.8 \times 10^3 M_r$) due to N-terminal truncation. This, possibly tissue-specific, heterogeneity does not seem to affect its functions (see reviews). In its full form, bFGF is a peptide composed of 154 amino acid residues (Ueno *et al.* 1986).

Sequence analysis has revealed a 5′-extended open reading frame (Abraham *et al.* 1986*a*,*b*), suggesting the existence of precursor molecules. Indeed, several bFGF forms of higher molecular mass have been isolated or tentatively identified using immunological techniques (see reviews and below).

Levels of bFGF mRNA are very low in most cell types. This may be a consequence of mRNA instability (Jaye *et al.* 1986; Abraham *et al.* 1986*a*). Protein levels, however, are relatively high in several cell types, especially neuroectodermal tissues such as brain, pituitary and adrenal gland (Gospodarowicz, 1975; Gospodarowicz *et al.* 1984, 1986*b*).

Sequence data for the 18×10^3 M_r bFGF have revealed the absence of a typical signal sequence (Abraham *et al.* 1986*a*), and, accordingly, the issue as to whether or not bFGF is released from cells is controversial (see reviews and below).

The three-dimensional structure of bFGF has not yet been established, but it seems that disulfide bonds may not be essential for activity (Seno *et al.* 1988). From results obtained with synthetic peptides it has been concluded that the N-terminal as well as C-terminal domains of bFGF are necessary for binding to heparin and to the bFGF receptor (Baird *et al.* 1988).

Receptors for bFGF have been described, and signal transduction events are very similar to those known for several other growth factors, e.g. oncogene induction, protein kinase C activation, Ca^{2+} influx, ribosomal protein S6 phosphorylation (see reviews).

Cells and tissues expressing bFGF

Many types of cells and tissues contain heparin-binding mitogenic factors, and in most cases, these factors have been identified as FGFs (see reviews). It was therefore concluded that bFGF is present in most, if not all tissues, and that bFGF can be expressed by most cell types (for a list of tissues and cells containing bFGF, see Lobb, 1988).

The expression of bFGF seems to be regulated differentially depending on cell type and developmental age (Munaim *et al.* 1988; Grothe and Unsicker, 1990; Grothe *et al.* 1990*b*).

Functions outside the nervous system

Basic FGF has been shown to exert a variety of effects on many mesenchymal and neuroectodermal cells, the most prominent being an induction or increase in proliferation. Other effects include survival, morphological changes, anchorage-independent growth, differentiation and delayed senescence (see reviews). These effects may be responsible for the putative participation of bFGF in many essential processes such as mesoderm-induction (Slack *et al.* 1987; Knoechel and Tiedemann, 1989), angiogenesis and neovascularization, tissue differentiation and regulation of tissue-specific functions, wound healing and other regenerative processes. In addition, bFGF may also affect tumor-growth and vascularization as well as diseases associated with abnormal cell proliferation and differentiation (see reviews). The precise roles of bFGF in these events are only poorly understood, but its essential participation is clear and possible clinical applications are beginning to be explored (for reviews see Lobb, 1988; Westphal and Herrmann, 1989; Hefti *et al.* 1989).

Molecular properties of neuroectodermal bFGF

Basic FGF has been purified from bovine pituitary (Gospodarowicz, 1975), brain (bovine, rat, guinea pig, human) (Gospodarowicz *et al.* 1984; Boehlen *et al.* 1985; Moscatelli *et al.* 1987; Presta *et al.* 1988*b*), as well as eye and retina (Baird *et al.* 1985; Courty *et al.* 1987). Sequence data are available, and demonstrate that these organs contain the $18\times10^3 M_r$ bFGF and N-terminal truncated forms (Boehlen *et al.* 1985; Esch *et al.* 1985; Abraham *et al.* 1986*a*, Gimenez-Gallego *et al.* 1986; Gospodarowicz *et al.* 1986*b*, Klagsbrun *et al.* 1987; Ho *et al.* 1988) similar to those from other tissues. In addition, bFGFs of higher molecular mass have been identified. Biologically active bFGF-like peptides of $21–25\times10^3 M_r$ have been isolated from rat and guinea pig brain, respectively (Presta *et al.* 1988*b*, Moscatelli *et al.* 1987). Such bFGF forms may have been translated by unusual initiation codons (i.e. CUG instead of AUG) (Florkiewicz and Sommer, 1989; Prats *et al.* 1989). Anti-bFGF antibodies recognize additional bFGF-like peptides of up to $55\times10^3 M_r$ in neuroectoderm-derived tissues (Presta *et al.* 1988*a*, Grothe *et al.* 1990*a*, Westermann *et al.* 1990). Whether these immunoreactive proteins are bFGF-forms or immunologically crossreactive proteins different from bFGF has not yet been resolved. An interesting feature, however, is that as well as the $18\times10^3 M_r$ bFGF, these larger proteins are expressed in a species and tissue specific pattern. In the rat, Presta *et al.* (1988*a*) found that in different regions of the brain a $29\times10^3 M_r$ peptide was more abundant than the $18\times10^3 M_r$ bFGF, while in the pituitary approximately equal amounts of 18, 27 and $29\times10^3 M_r$ proteins were detected. In contrast, bovine pituitary contained 18, 24, 29–33 and $46\times10^3 M_r$ immunoreactive proteins (Grothe *et al.* 1990*a*). For the adrenal gland, different patterns were found not only in different species (bovine, pig and rat), but also in whole glands as compared to medullary chromaffin cells (Westermann *et al.* 1990), which are of neuroectodermal origin (Unsicker *et al.* 1989).

Another problem still unresolved is the release of bFGF. While the missing signal peptide (Abraham *et al.* 1986*a*) argues against a release, there is sufficient evidence to indicate that bFGF may be released from intact cells: (1) bFGF is released by cultured cells (Sato *et al.* 1989; van Zoelen *et al.* 1989; Werner *et al.* 1989; Maier *et al.* 1990), (2) anti-bFGF antibodies inhibit autocrine growth of endothelial cells (Schweigerer *et al.* 1987*b*), (3) bFGF is present in secretory organelles of adrenal chromaffin cells (Westermann *et al.* 1990), (4) many effects of bFGF seem to be mediated by specific, membrane-bound surface receptors (see reviews), (5) bFGF is located in the extracellular matrix *in vitro* and *in vivo* (Baird and Ling, 1987; Jeanny *et al.* 1987; Vlodavski *et al.* 1987; Moscatelli, 1988; Bashkin *et al.* 1989; DiMario *et al.* 1989; Ingber and Folkmann, 1989), and, (6) bFGF is found in body fluids (Chodak *et al.* 1988; Smith *et al.* 1989). Thus, bFGF resembles interleukins and some other proteins that can be released from cells

CARL A. RUDISILL LIBRARY
LENOIR-RHYNE COLLEGE

despite the lack of a signal peptide (Muesch *et al.* 1990). Basic FGF may therefore be secreted by a mechanism other than the regulated pathway (Muesch *et al.* 1990), or *via* a so far unknown precursor containing a signal peptide. In this context, it is of interest that transfection with a bFGF-signal peptide construct results in the transformation of these cells (Thomas, 1988; Rogelj *et al.* 1988, 1989).

By its molecular and biological properties, bFGF from the nervous system appears to be identical with bFGF from non-neuronal sources.

Neurons and glial cells

Expression in neurons and/or glial cells and their tumor counterparts

Only few and controversial data are available on bFGF expression by neurons and glial cells *in vitro*. In primary cultures of brain cells and peripheral ganglia, Pettmann *et al.* (1987) and Janet *et al.* (1988) found bFGF immunoreactivity exclusively in neurons, but never in glial cells. Ferrara *et al.* (1988) reported that cultured astrocytes from adult bovine corpus callosum contain bFGF mRNA and protein. Gonzales *et al.* (1989) showed that *in vitro* only astrocytes express bFGF mRNA, while *in situ* neurons of several brain regions contain the mRNA. A neuronal rather than glial location of bFGF *in situ* was also shown by other authors (Pettmann *et al.* 1986; Janet *et al.* 1987; Grothe *et al.* unpublished). Nonetheless, the satellite glial cells of rat dorsal root ganglia also contain anti-bFGF immunoreactivity *in situ* (C. Grothe, unpublished data).

In the adult rat brain, bFGF-like immunoreactivity is restricted to neurons of discrete loci, *e.g.* the hippocampus (Gonzales *et al.* 1989), parietal cortex (P. Walicke, personal communication) and several, but not all, brain stem nuclei (Grothe *et al.* 1990*b*). Furthermore, developmental alterations to this pattern have been observed (Grothe *et al.* 1990*b*).

In the retina, bFGF is expressed in photoreceptor cells (Noji *et al.* 1990). From its specific, possibly light-dependent location in/on rod outer segments, an essential role in phototransduction has been postulated (Mascarelli *et al.* 1989).

Several types of neuroectodermal tumor cells can express bFGF, e.g. neuroblastoma cells (Huang *et al.* 1987; Heymann *et al.* 1988), glioma cell lines (Rogister *et al.* 1988; Sato *et al.* 1989, Okumura *et al.* 1989; Westermann and Unsicker, 1990) and diverse glial tumors (Paulus *et al.* 1990) as well as pheochromocytoma cells (see below).

Differences between these tumors have been found with regard to regulation and release of bFGF. For example, bFGF has been reported to be released from human astrocytoma cells (Sato *et al.* 1989), but not from rat glioma cells (Okumura *et al.* 1989; Westermann and Unsicker, 1990). Regulation of its expression in C6 glioma cells is controversial. Okumura *et al.* (1989) have shown that increasing cell density results in an increase of intracellular bFGF protein. In contrast, we have found that bFGF protein decreases with increasing cell

CARL A. RUDISILL LIBRARY
LENOIR-RHYNE COLLEGE

density and, furthermore, an external stimulus is essential for initiating bFGF expression (Westermann and Unsicker, 1990). The latter results were also obtained with astrocytoma cells (Sato *et al.* 1989). Furthermore, FGF-receptor expression also seems to be regulated by cell density (Veomett *et al.* 1989).

Thus the question as to whether neurons or glial cells, or both, physiologically express bFGF, remains to be answered. Further analyses in this regard should consider that (1) immunological localization of bFGF protein in a cell may reflect uptake from an exogenous source rather than its expression by this cell (*in situ* hybridizations required), (2) in different regions of the nervous system, bFGF may be expressed by different types of cells and (3) bFGF expression may qualitatively and quantitatively vary during development or different physiological states.

The above data indicate that bFGF expression *in vitro* is modulated not only by a variety of physiological but also by experimental conditions. Nonetheless, it seems that all neuroectoderm-derived cells have the capacity to express bFGF under distinct circumstances.

Effects on neurons in vitro

The most pronounced effects of bFGF on neurons are the promotion of *in vitro* survival and neurite outgrowth. Basic FGF maintains neurons of the cerebral cortex (Morrison *et al.* 1986; Walicke, 1988), hippocampus (Walicke *et al.* 1986; Mattson *et al.* 1989), thalamus (Walicke, 1988), striatum (Walicke, 1988), septum (Walicke, 1988; Grothe *et al.* 1989), mesencephalon (Ferrari *et al.* 1989) and spinal cord (Unsicker *et al.* 1987) in culture. The only peripheral nervous system neurons supported by bFGF are those of the chick ciliary ganglion (Unsicker *et al.* 1987; Giulian *et al.* 1988).

In explant cultures of rat retinae, bFGF enhanced survival and induced fiber outgrowth of ganglion cells (Baehr *et al.* 1989; Thanos and van Boxberg, 1989). Highly purified retinal ganglion cells of the chick embryo however did not survive in the presence of bFGF (Lehwalder *et al.* 1989).

Like NGF, bFGF does not indiscriminately address all neuron populations. Neurons which are obviously not targets for bFGF are located in the subiculum (Walicke *et al.* 1988), rat nodose and superior cervical ganglia as well as chick sympathetic and dorsal root ganglia (Unsicker *et al.* 1987).

Besides the survival and neurite outgrowth activities, other effects on neurons have been observed. Basic FGF is a mitogen for neuroblasts (Gensburger *et al.* 1987) and neuroblastoma cells (Luedecke and Unsicker 1990*a*,*b*). In ciliary ganglion neurons (Unsicker *et al.* 1987; Vaca *et al.* 1989), septal neurons (Grothe *et al.* 1989) and spinal cord neurons (McManaman *et al.* 1989), bFGF increases choline-acetyltransferase (ChAT) activity.

Interestingly, bFGF is able to overcome the toxic activities of glutamate on *in vitro* survival and neurite outgrowth of cultured hippocampal neurons (Mattson *et al.* 1989).

Thus, *in vitro*, bFGF has a variety of effects on neurons, comparable to those of the neurotrophic factor NGF.

Effects on glial cells in vitro

In vitro studies have revealed that bFGF has a variety of effects on glial cells. Several groups have reported on the mitogenic activity of bFGF on astrocytes (Eccleston *et al.* 1985; Pettmann *et al.* 1985, 1987; Sensenbrenner *et al.* 1985; Perraud *et al.* 1987, 1988*a*; Delaunoy *et al.* 1988; Kniss and Burry, 1988; Loret *et al.* 1989), oligodendrocytes (Eccleston and Silberberg, 1985; Eccleston *et al.* 1985; Perraud *et al.* 1987, 1988*a*; Delaunoy *et al.* 1988; Besnard *et al.* 1989) and glioblasts (Delaunoy *et al.* 1988). Yong *et al.* (1988*a*,*b*) found that bFGF is a mitogen for fetal, but not adult, human astrocytes, and not for oligodendrocytes and Schwann cells.

Another prominent effect of bFGF on astrocytes is to cause a morphological change. Astrocytes grown under different culture conditions mostly appear as flat epithelial cells. In the presence of bFGF the cell bodies become smaller and rounded, and extend several long processes. Simultaneously, a reorganization of intermediate filaments (Weibel *et al.* 1985) and an increased synthesis of alpha- and beta-tubulins can be observed (Weibel *et al.* 1987). The intensity of these morphological changes varies between astrocytes from different brain regions (Perraud *et al.* 1990).

Other effects include quantitative as well as qualitative changes in the synthesis of many different proteins, an increase in S-100 peptide, glutamine synthetase, free ribosomes, Na^+ and K^+ uptake, S6 protein kinase activity, and plasminogen activator release (Pettmann *et al.* 1985; Weibel *et al.* 1985; Latzkovits *et al.* 1988; Loret *et al.* 1988; Rogister *et al.* 1988; Gavaret *et al.* 1989). Numbers of Orthogonal array particles (OAPs, possibly a structural equivalent of K^+-channels) decrease (Wolburg *et al.* 1986). Furthermore, bFGF acts as a chemoattractant for astroglial cells (Senior *et al.* 1986). Most of these events can be correlated with the maturation of astrocytes.

Thus, bFGF appears to be a mitogen and maturation factor for astrocytes. Similar, but not identical, effects have been reported for acidic FGF, cyclic AMP and other growth factors (Weibel *et al.* 1987; Perraud *et al.* 1988*a*,*b*).

Oligodendrocytes cultured in the presence of bFGF also change their morphology (mostly showing outgrowths of two, long processes) and have increased carboanhydrase activity (Delaunoy *et al.* 1988), decreased myelin basic protein and inhibition of the enzyme CNP (Besnard *et al.* 1989). Thus bFGF apparently inhibits the maturation of oligodendrocytes.

Glioblasts cultured in the presence of bFGF proliferate, but fail to express differentiation markers for astro- or oligodendrocytes (Perraud *et al.* 1988*a*).

Comparable effects have been reported for glial tumor cells (Rogister *et al.* 1988; Westphal *et al.* 1988, Okumura *et al.* 1989; Westermann and Unsicker, 1990).

As is known for bFGF activities on cells of mesenchymal origin (for review see

Lobb, 1988), heparin and other glucosaminoglycans are potent enhancers of the bFGF effects on glial cells (Perraud *et al.* 1988*a*,*b*).

In summary, bFGF exerts various effects on glial cells, mainly affecting proliferation and differentiation. These effects appear to be different for glioblasts (proliferation, inhibition of differentiation), astrocytes (proliferation and maturation) and oligodendrocytes (possibly proliferation, inhibition of maturation) and further depend upon the site of origin of the cells.

With regard to glial cells, bFGF therefore differs from NGF, which does not influence glial cell morphology and physiology.

To our knowledge, there is only one documentation of an *in vivo* effect of bFGF on glial cells; this is an increase in number of reactive astrocytes around an axotomy wound, possibly due to the mitogenic activity of bFGF (Barotte *et al.* 1989).

Chromaffin cells

Chromaffin cells are modified sympathetic neurons which are scattered in peripheral ganglia and particularly prominent in the adrenal medulla (Unsicker *et al.* 1989).

Expression

Basic FGF is present in adrenal chromaffin cells (Grothe and Unsicker, 1989; Westermann *et al.* 1990). Immunohistochemistry using rat chromaffin cells has revealed that only a subpopulation, the noradrenergic, and not the adrenergic, cells contain bFGF (Grothe and Unsicker, 1990). Since cultured bovine chromaffin cells also contain anti-bFGF immunoreactive proteins (Blottner *et al.* 1989) it may be inferred that bFGF can be synthezised by chromaffin cells. Basic FGF immunoreactivity is subcellularly localized in the secretory granules of chromaffin cells (Westermann *et al.* 1990), and may therefore be released together with catecholamines, chromogranins and the other soluble molecules of the vesicle content.

Effects

Basic FGF induces proliferation, neurite outgrowth and NGF-dependence of adrenal chromaffin precursor cells *in vitro* (Stemple *et al.* 1988). A tumor counterpart of chromaffin cells, the PC12 rat pheochromocytoma cells, respond to bFGF in a manner similar to NGF. Basic FGF induces neurite outgrowth on PC12 cells (Togari *et al.* 1985; Wagner and D'Amore, 1986; Neufeld *et al.* 1987; Rydel and Greene, 1987; Schubert *et al.* 1987; Sigmund *et al.* 1990). The bFGF effect is different from that of NGF: the neurite network is less dense, protein kinase C inhibition enhances fiber outgrowth and bFGF-induced neurites disappear after about 6 days in culture (Togari *et al.* 1985; Rydel and Greene, 1987; Sigmund *et al.* 1990). The effects of NGF and bFGF on protein phosphorylation are identical, including reduced phosphorylation of the NGF-sensitive protein Nsp100, and enhanced phosphorylation of a microtubule-associated protein MAP 1.2, tyrosine hydroxylase and a nonhistone protein SMP (Togari *et al.* 1985; Rydel and Greene, 1987). The expression of the NGF-inducible protein NILE, Thy-1, acetylcholine-esterase (Rydel and Greene, 1987) and NGF-receptors (Doherty *et al.* 1988), as well as c-fos and SCG10, a growth cone membrane protein (Sigmund *et al.* 1990), are also affected. Differences in the actions of bFGF and NGF include (1) additive effects on the induction of ornithine decarboxylase (Togari *et al.* 1985), (2) an increase in the release of the β-amyloid precursor proteins NGF (2-fold) and FGF (7-fold) Schubert *et al.* 1989), (3) inability of bFGF to induce neurite outgrowth and expression of the protease transin in the NGF-responsive PC12/RG-5 subclone (Machida *et al.* 1989) and (4) involvement of protein kinase C in bFGF-dependent, but not NGF-dependent, c-fos induction (Sigmund *et al.* 1990). As known for other bFGF-activities, heparin and other glucosaminoglycans are able to enhance the bFGF effects on PC12 cells (Wagner and D'Amore, 1986; Damon *et al.* 1988) but probably in a way different from their modulation of the effects of NGF (Neufeld *et al.* 1987).

Basic FGF also affects the interaction between PC12 cells. In the presence of bFGF and heparin or chondroitinsulfate as culture matrices, PC12 cells form ring-like aggregates (Schubert *et al.* 1987).

In conclusion, these data demonstrate that bFGF and NGF, acting by at least partially different mechanisms, have similar, but not identical effects on chromaffin and pheochromocytoma cells.

Actions of bFGF on neuroectodermal cells *in vivo*

Presence of bFGF in neuronal target organs

An important feature of a neurotrophic factor is its expression by the neuronal target tissue. Since bFGF has been shown to be present in almost any tissue and organ (for a review, see Lobb, 1988), this criterion seems to be fulfilled. The relatively large amounts of bFGF present in most tissues, seem to conflict with the NTF-definition of Barde (1988). On the other hand, bFGF in these tissues may have additional functions requiring higher concentrations, and, as discussed by Oppenheim (1989), not 'limiting amounts' of factor, but 'limited access' (number of axonal branches and synapses) of factor, may control neuron death.

Basic FGF receptors in the nervous system

A further criterion for NTFs is their uptake by specific receptors and a retrograde axonal transport. So far, no retrograde transport of bFGF by neurons has been reported. In the case of embryonic chick ciliary and superior cervical ganglia, bFGF is apparently not retrogradely transported (Hendry and Belford, 1990).

Basic FGF receptors in the nervous system are well established. Courty *et al.* (1988) have described one type of bFGF receptor in bovine brain; others (Imamura *et al.* 1988; Ledoux *et al.* 1989; Mereau *et al.* 1989) have found two receptor types, differing in molecular mass and affinity for bFGF, in rat, guinea pig and bovine brain. In the chick brain, bFGF receptors are highly expressed during a long developmental period, as compared to other organs (Olwin and Hauschka, 1990).

Up to now, bFGF receptor mapping studies in the nervous system are missing, but from the effects of bFGF on neuroectodermal cells (see above) one may conclude that most of these cells express bFGF receptors.

Ontogenetic neuron death

The most important feature of NTFs is their ability to prevent the ontogenetic neuron death.

Dreyer *et al.* (1989) and Hendry *et al.* (1990) systemically applied bFGF and aFGF, respectively, to chick embryos and analyzed the number of ciliary ganglion neurons (which are known to survive, *in vitro*, in the presence of bFGF) (Unsicker *et al.* 1987; Giulian *et al.* 1988). They reported that neuron death, which occurs between embryonic day 8 (100% neurons) and 14 (56% neurons), is almost completely prevented by bFGF (Dreyer *et al.* 1989; Hendry *et al.* 1990).

Studies in other systems will be necessary to further substantiate this *in vivo* neuronal survival-promoting effect of bFGF; in particular, data on the *in vivo* effects of blocking anti-bFGF antibodies on nervous system development are required (for aFGF, see Hendry *et al.* 1990).

Central nervous system lesions

Well established lesions in the CNS are the fimbria-fornix and the optic nerve transection.

Basic FGF has been shown to maintain embryonic septal neurons *in vitro* (Grothe *et al.* 1989). Moreover, the protein is present in the hippocampus, an important target tissue of medial septal neurons (Pettmann *et al.* 1986). In adult rats, unilateral fimbria-fornix transection results in the loss of approximately 87% of the neurons, as visualised by Nissl staining (Otto *et al.* 1989) or about 60% of the cholinergic neurons in the medial septum (Anderson *et al.* 1988), and approximately 50% of the cholinergic neurons in the diagonal band of Broca on the ipsilateral side (Anderson *et al.* 1988). Application of bFGF in gelfoam or by intraventricular infusion significantly reduces this axotomy-induced neuron death. In the medial septum, approximately 20% of the total and 60% of the cholinergic neurons that otherwise would have died are rescued by bFGF. In the diagonal band of Broca about 80% of the cholinergic neurons survive in the presence of bFGF (Anderson *et al.* 1988). Comparable effects have been obtained with NGF (Otto *et al.* 1989). Further consequences of bFGF treatment are a reduction of the lesion-induced decrease in ChAT activity in the hippocampus, and, around the wound, an increase in the number of reactive astrocytes (Barotte *et al.* 1989).

In retinal explant cultures, neurite induction and survival of retinal ganglion cells can be induced by bFGF (Baehr *et al.* 1989; Thanos and van Boxberg, 1989). The presence of bFGF in the tectum, the target for the optic nerve fibers, however, has not been documented. Following transection of the rat optic nerve, bFGF partially prevents the death of retinal ganglion neurons (approximately 25 % of the otherwise dying neurons) (Sievers *et al.* 1987).

Peripheral nervous system lesions

In the PNS, sciatic nerve transection is a widely used lesion model. While bFGF is present in the skin of embryonic and adult rats (Gonzales *et al.* 1990; C. Grothe, unpublished data), a target tissue for the sensory nerve fibers, it has no effect on the *in vitro* survival of dorsal root ganglion neurons of chick embryos (Unsicker *et al.* 1987; see, however, Watters and Hendry, 1987; Hendry *et al.* 1990). Even so, local application of bFGF (as well as NGF) at the stump of a transected sciatic nerve rescues approx. 70–80 % (100 % in the case of NGF) of the dorsal root ganglion neurons (Otto *et al.* 1987). Basic FGF not only maintains these neurons, but also results in regeneration of the nerve fibers over a distance of up to 15 mm (Aebischer *et al.* 1989).

Neuron death induced by target ablation can also be prevented by bFGF. Blottner *et al.* (1989) induced the death of approximately 25 % of neurons in the intermediolateral column of the spinal cord by destruction of the adrenal medulla, which is known to contain bFGF (Grothe and Unsicker, 1989; Westermann *et al.* 1990). Substitution with bFGF results in the survival of approximately 85 % of the otherwise dying neurons (NGF 0 %) (Blottner *et al.* 1989). Splanchnic nerve transection abolishes these effects of bFGF (Blottner and Unsicker, 1990), suggesting that a retrograde axonal transport of bFGF is necessary for its function. Further evidence for the NTF-like function of bFGF in this system is the developmental correlation between the time of the first appearance of FGF-like neurotrophic activity and immunoreactivity and the onset of the functional innervation of the rat adrenal medulla (Blottner and Unsicker, 1989; Grothe and Unsicker, 1990).

For all the above lesion systems, it remains to be shown whether bFGF acts directly on the rescued neurons, or indirectly by inducing glial or other cells to synthesize and secrete a trophic factor.

Thus, bFGF resembles NGF in that it is able to prevent the physiological as well as lesion-induced neuron death of distinct neuronal populations, and to accelerate the regeneration of nerve connections *in vivo*.

Possible clinical applications of bFGF

Growth factors in general are of considerable interest for their clinical potential (for recent reviews, see Goustin *et al.* 1986; Deuel, 1987; Foster, 1988; Lobb, 1988; Westphal and Herrmann, 1989). Basic FGF, with its broad spectrum of functions, is a prominent candidate in this regard. Outside the nervous system, bFGF has been postulated to be associated with several pathological processes of enhanced or reduced cell growth (e.g. tumors, range of vascularization). Moreover, prominent roles in soft and hard tissue repair as well as immunomodulation have been observed. In addition, bFGF may be of diagnostic interest, because increased levels in different body fluids have been found in patients with tumors and retinopathies (discussed by Lobb, 1988).

By analogy, one may therefore speculate that bFGF may also be involved in neurological disorders.

Neural trauma

To date, therapy of neuronal trauma is mainly restricted to surgery, and especially in the CNS, only poor or no reconstitution is possible. The results obtained in central and peripheral lesion studies (*e.g.* axonal regeneration, increase of reactive astrocytes, prevention of neuron death, see above) however, provide evidence that bFGF may be used as a pharmacological tool to initiate or enhance regeneration processes after neuronal trauma.

Neuronal degeneration

Degenerative diseases of the CNS, such as Morbus Alzheimer, M. Parkinson and others, are characterized by neuron losses and connections in which bFGF may have an indirect role.

It has been shown that glioma and pheochromocytoma cells, as a consequence of bFGF-treatment, drastically enhance the synthesis and release of the β-amyloid precursor protein (Schubert *et al.* 1989; Quon *et al.* 1990) which is also enhanced and extracellularly deposited in Alzheimer's disease (Anderton *et al.* 1988; Glenner, 1988).

In Parkinson's disease, grafting of chromaffin tissue can result in at least partial reconstitution of the normal physiological status (see *e.g.* Olson, 1988), which may be due to liberation of growth factors rather than dopamine from the grafted cells. Consistent with the presence and possible release of bFGF by chromaffin cells is the observation that in an animal model of Parkinson's disease, bFGF is capable of partially reversing the degeneration, inducing nerve fiber growth and restoring dopamine and tyrosine hydroxylase levels (Otto and Unsicker, 1990).

Neuronal tumors

Screening of neuroectodermal tumors has shown that most of them not only contain, but also respond to, bFGF (Westphal *et al.* 1988, 1989; Paulus *et al.* 1990). Accordingly, possible participation of bFGF in initiation as well as growth and invasion of brain tumors may be conceived (for the transforming potential of FGF, see e.g. Thomas, 1988). Because of its angiogenic effects, bFGF may also be involved in the strong vascularization of brain tumors (Brem, 1976). Furthermore, increased expression of bFGF by brain tumors may result in increased levels of bFGF in the cerebrospinal fluid (CSF). In fact, growth factor activites resembling bFGF have been detected in the CSF of brain tumor patients but not in normal CSF (Lopez-Pousa *et al.* 1981; Brem *et al.* 1983).

Conclusion

Facts outlined above provide an indication of the putative functions bFGF may have in the nervous system. It may also have become clear, however, that we are far away from fully understanding the physiological role of bFGF in the nervous system.

Basic FGF shares a variety of functions with the fully established NTF, NGF. According to the NTF-definition of Barde (1988) two points remain to be clarified: (1) *in vivo* effects of inhibiting antibodies and (2) retrograde axonal transport. The issue of limiting amounts of factor available to nerve endings will be difficult to settle.

In addition to its NTF-like effects, bFGF, unlike NGF, also affects non-neuronal (glial, endothelial, etc.) cells of the nervous system and, furthermore, may be involved in growth and vascularization of brain tumors.

Thus, bFGF appears to be a physiological growth factor for neuroectodermal cells with a broader spectrum of functions than a classical NTF.

References

Abraham, J. A., Mergia, A., Whang, J. L., Tumolo, A., Friedman, J., Hjerrild, K. A., Gospodarowicz, D. and Fiddes, J. C. (1986*a*). Nucleotide sequence of a bovine clone encoding the angiogenic protein, basic fibroblast growth factor. *Science* **233**, 545–548.

Abraham, J. A., Whang, J. L., Tumolo, A., Mergia, A., Friedman, J., Gospodarowicz, D. and Fiddes, J. C. (1986*b*). Human basic fibroblast growth factor: nucleotide sequence and genomic organization. *EMBO J.* **5**, 2523–2528.

Aebischer, P., Salessiotis, A. N. and Winn, S. R. (1989). Basic fibroblast growth factor released from synthetic guidance channels facilitates peripheral nerve regeneration across long nerve gaps. *J. Neurosci. Res.* **23**, 282–289.

Anderson, K. J., Dam, D., Lee, A. and Cotman, C. W. (1988). Basic fibroblast growth factor prevents death of lesioned cholinergic neurons *in vivo*. *Nature* **332**, 360–361.

Anderton, B., Brian, J. P. and Power, D. (1988). The protein constituents of paired helical filaments and senile plaques in Alzheimer's disease. *ISI Atlas Sci. Biochem.* **1**, 81–87.

Baehr, M., Vanselow, J. and Thanos, S. (1989). Ability of adult rat ganglion cells to regrow axons *in vitro* can be influenced by fibroblast growth factor and gangliosides. *Neurosci. Lett.* **96**, 197–201.

Baird, A. and Boehlen, P. (1990). Fibroblast growth factors. *Handb. Expl Pharmacol.* 95/I, 369–418.

Baird, A., Esch, F., Gospodarowicz, D. and Guillemin, R. (1985). Retina- and eye-derived endothelial cell growth factors: partial molecular characterization and identity with acidic and basic fibroblast growth factors. *Biochemistry, Wash.* **24**, 7855–7860.

Baird, A. and Ling, N. (1987). Fibroblast growth factors are present in the extracellular matrix produced by endothelial cells *in vitro*: implications for a role of heparinase-like enzymes in the neovascular resp. *Biochem. biophys. Res. Commun.* **142**, 428–435.

Baird, A., Schubert, D., Ling, N. and Guillemin, R. (1988). Receptor and heparin-binding domains of basic fibroblast growth factor. *Proc. natn. Acad. Sci. U.S.A.* **85**, 2324–2328.

Barde, Y. A. (1988). What, if anything, is a neurotrophic factor? *Trends Neurosci.* **11**, 343–346.

Barde, Y. A. (1989). Trophic factors and neuronal survival. *Neuron* **2**, 1525–1534.

Barde, Y. A., Edgar, D. and Thoenen, H. (1982). Purification of a new neurotrophic factor from mammalian brain. *EMBO J.* **1**, 549–553.

Barotte, C., Eclancher, F., Ebel, A., Labourdette, G., Sensenbrenner, M. and Will, B. (1989). Effects of basic fibroblast growth factor (bFGF) on choline acetyltransferase activity and astroglial reaction in adult rats after partial fimbria transection. *Neurosci. Lett.* **101**, 197–202.

Bashkin, P., Doctrow, S., Klagsbrun, M., Svahn, C. M., Folkman, J. and Vlodavsky, I. (1989). Basic fibroblast growth factor binds to subendothelial extracellular matrix and is released by heparitinase and heparin-like molecules. *Biochemistry, Wash.* **28**, 1737–1743.

Besnard, F., Perraud, F., Sensenbrenner, M. and Labourdette, G. (1989). Effects of acidic and basic fibroblast growth factors on proliferation and maturation of cultured rat oligodendrocytes. *Int. J. Dev. Neurosci.* **7**, 401–409.

Blottner, D. and Unsicker, K. (1989). Spatial and temporal patterns of neurotrophic activities in rat adrenal medulla and cortex. *Dev. Brain Res.* **48**, 243–253.

Blottner, D. and Unsicker, K. (1990). Maintenance of intermediolateral spinal cord neurons by fibroblast growth factor administered to the medullectomized rat adrenal gland: dependence on intact adrenal innervation and cellular organization of implants. *Eur. J. Neurosci.* **2**, 378–382.

Blottner, D., Westermann, R., Grothe, C., Boehlen, P. and Unsicker, K. (1989). Basic fibroblast growth factor in the adrenal gland – possible trophic role for preganglionic neurons *in vivo*. *Eur. J. Neurosci.* **1**, 471–478.

Boehlen, P. (1989). Fibroblast growth factor. *Cytokines* **1**, 204–228

Boehlen, P., Esch, F., Baird, A., Jones, K. L. and Gospodarowicz, D. (1985). Human brain fibroblast growth factor: isolation and partial chemical characterization. *FEBS Lett.* **185**, 177–181.

Brem, S. (1976). The role of vascular proliferation in the growth of brain tumors. *Clin. Neurosurg.* **23**, 440–453.

Brem, S., Patz, J. and Tapper, D. (1983). Detection of human central nervous system tumors: use of migration stimulating activity of the cerebrospinal fluid. *Surg. For.* **34**, 532–534.

Chodak G. W., Hospelhom V., Judge S. M., Mayforth, R., Koeppen, H. and Sasse, J. (1988). Increased levels of fibroblast growth factor like activity in urine from patients with bladder or kidney cancer. *Cancer Res.* **48**, 2083–2088.

Courty, J., Dauchel, M. C., Mereau, A., Badet, J. and Barritault, D. (1988). Presence of basic fibroblast growth factor receptors in bovine brain membranes. *J. biol. Chem.* **263**, 11 217–11 220.

Courty, J., Loret, C., Chevallier, B., Moenner, M. and Barritault, D. (1987). Biochemical comparative studies between eye- and brain-derived growth factors. *Biochimie* **69**, 511–516.

Damon, D. H., D'Amore, P. A. and Wagner, J. A. (1988). Sulfated glycosaminoglycans modify growth factor-induced neurite outgrowth in PC12 cells. *J. cell. Physiol.* **135**, 293–300.

Davies, A. M. (1988). Role of neurotrophic factors in development. *Trends Genet.* **4**, 139–143.

Delaunoy, J. P., Langui, D., Ghandour, S., Labourdette, G. and Sensenbrenner, M. (1988). Influence of basic fibroblast growth factor on carbonic anhydrase expression by rat glial cells in primary culture. *Int. J. Dev. Neurosci.* **6**, 129–136.

Delli-Bovi P., Curatola A. M., Kern F. G., Greco, A., Ittmann, M. and Basilico, C. (1987). An oncogene isolated by transfection of Kaposi's sarcoma DNA encodes a growth factor that is a member of the FGF family. *Cell* **50**, 729–737.

Deuel, T. F. (1987). Polypeptide growth factors: roles in normal and abnormal growth. *A. Rev. Cell Biol.* **3**, 443–492.

Dickson, C. and Peters, G. (1987). Potential oncogene product related to growth factors. *Nature* **326**, 833.

DiMario, J., Buffinger, N., Yamada, S. and Strohmann, R. C. (1989). Fibroblast growth factor in the extracellular matrix of dystrophic (mdx) mouse muscle. *Science* **244**, 688–690.

Doherty, P., Seaton, P., Flanigan, T. P. and Walsh, F. S. (1988). Factors controlling the expression of the NGF receptor in PC12 cells. *Neurosci. Lett.* **92**, 222–227.

Dreyer, D., Lagrange, A., Grothe, C. and Unsicker, K. (1989). Basic fibroblast growth factor prevents ontogenetic neuron death *in vivo*. *Neurosci. Lett.* **99**, 35–38.

Eccleston, P. A., Gunton, D. J. and Silberberg, D. H. (1985). Requirements for brain cell attachment, survival and growth in serum-free medium: effects of extracellular matrix, epidermal growth factor and fibroblast growth factor. *Dev. Neurosci.* **7**, 308–322.

Eccleston, P. A. and Silberberg, D. H. (1985). Fibroblast growth factor is a mitogen for oligodendrocytes *in vitro*. *Dev. Brain Res.* **21**, 315–318.

Esch, F., Baird, A., Ling, N., Ueno, N., Hill, F., Denoroy, L., Klepper, R., Gospodarowicz, D., Boehlen, P. and Guillemin, R. (1985). Primary structure of bovine pituitary basic fibroblast growth factor (FGF) and comparison with the amino-terminal sequence of bovine brain acidic FGF. *Proc. natn. Acad. Sci. U.S.A.* **82**, 6507–6511.

Ferrara, N., Ousley, F. and Gospodarowicz, D. (1988). Bovine brain astrocytes express basic fibroblast growth factor, a neurotropic and angiogenic mitogen. *Brain Res.* **462**, 223–232.

Ferrari, G., Minozzi, M. C., Toffano, G., Leon, A. and Skaper, S. D. (1989). Basic fibroblast growth factor promotes the survival and development of mesencephalic neurons in culture. *Devl Biol.* **133**, 140–147.

Florkiewicz, R. Z. and Sommer, A. (1989). Human basic fibroblast growth factor gene encodes four polypeptides: three initiate translation from non-AUG codons. *Proc. natn. Acad. Sci. U.S.A.* **86**, 3978–3981.

Foster, M. B. (1988). Peptide growth factors: a clinical precis. *Semin. Reproduct. Endocrinol.* **6**, 29–34.

Gavaret, J. M., Matricon, C., Pomerance, M., Jacquemin, C., Toru-Delbauffe, D. and Pierre, M. (1989). Activation of S6 kinase in astroglial cells by FGFa and FGFb. *Dev. Brain Res.* **45**, 77–82.

Gensburger, C., Labourdette, G. and Sensenbrenner, M. (1987). Brain basic fibroblast growth factor stimulates the proliferation of rat neuronal precursor cells *in vitro*. *FEBS Lett.* **217**, 1–5.

Gimenez-Gallego, G., Conn, G., Hatcher, V. B. and Thomas, K. A. (1986). Human brain-derived acidic and basic fibroblast growth factors: amino terminal sequences and specific mitogenic activities. *Biochem. biophys. Res. Commun.* **135**, 541–548.

Giulian, D., Vaca, K. and Johnson, B. (1988). Secreted peptides as regulators of neuron–glia and glia–glia interactions in the developing nervous system. *J. Neurosci. Res.* **21**, 487–500.

Glenner, G. G. (1988). Alzheimer's disease: its proteins and genes. *Cell* **52**, 307–308.

Gonzales, A. M., Buscaglia, M., Ong, M. and Baird, A. (1990). Distribution of basic fibroblast growth factor in the 18 day rat fetus: localization in the basement membranes of diverse tissues. *J. Cell Biol.* **110**, 753–765.

Gonzales, A. M., Emoto, N., Walicke, P., Shimasaki, S. and Baird, A. (1989). The distribution of basic fibroblast growth factor mRNA in the adult rat brain. *Abstr. Soc. Neurosci.* **15**, 710.

Gospodarowicz, D. (1975). Purification of a fibroblast growth factor from bovine pituitary. *J. biol. Chem.* **250**, 2515–2520.

Gospodarowicz, D., Baird, A., Cheng, B. J., Lui, G. M., Esch, F. and Bohlen, P. (1986*b*). Isolation of fibroblast growth factor from bovine adrenal gland: physicochemical and biological characterization. *Endocrinology* **118**, 82–90.

Gospodarowicz, D., Cheng, J., Lui, G. M., Baird, A. and Boehlen, P. (1984). Isolation of brain fibroblast growth factor by heparin-Sepharose affinity chromatography: identity with pituitary fibroblast growth factor. *Proc. natn. Acad. Sci. U.S.A.* **81**, 6963–6967.

Gospodarowicz, D., Neufeld, G. and Schweigerer, L. (1986*a*). Molecular and biological characterization of fibroblast growth factor, an angiogenic factor which also controls the proliferation and differentiation of mesoderm and neuroectoderm derived cells. *Cell Differ.* **19**, 1–17.

Goustin, A. S., Leof, E. B., Shipley, G. D. and Moses, H. L. (1986). Growth factors and cancer. *Cancer Res.* **46**, 1015–1029.

Grothe, C., Otto, D. and Unsicker, K. (1989). Basic fibroblast growth factor promotes *in vitro* survival and cholinergic development of rat septal neurons: comparison with the effects of nerve growth factor. *Neuroscience* **31**, 649–661.

Grothe, C. and Unsicker, K. (1989). Immunocytochemical localization of basic fibroblast growth factor in bovine adrenal gland, ovary and pituitary. *J. Histochem. Cytochem.* **37**, 1877–1883.

Grothe, C. and Unsicker, K. (1990). Immunocytochemical mapping of basic fibroblast growth factor in the developing and adult rat adrenal gland. *Histochem. J.* **94**, 141–147.

Grothe, C., Zachmann, K., Unsicker, K. and Westermann, R. (1990). High molecular weight forms of basic fibroblast growth factor recognized by a new anti-bFGF antibody. *FEBS Lett.* **260**, 35–38.

Hefti, F., Hartikka, J. and Knusel, B. (1989). Function of neurotrophic factors in the adult and aging brain and their possible use in the treatment of neurodegenerative diseases. *Neurobiol. Aging* **10**, 515–533.

HENDRY, I. A. AND BELFORD, D. A. (1990). Lack of retrograde axonal transport of fibroblast growth factors in peripheral neurons. *Int. J. Dev. Neurosci.*, Abstr. 8th Biennial Meeting ISDN, p93.

HENDRY, I. A., BELFORD, D. A., CROUCH, M. F. AND HILL, C. E. (1990). Parasympathetic neurotrophic factors. *Int. J. Dev. Neurosci.*, Abstr. 8th Biennial Meeting ISDN, p43.

HEYMANN, D., BOEHLEN, P., GAUTSCHI, P., GROTHE, C. AND UNSICKER, K. (1988). Evidence for basic fibroblast growth factor activity in neuroblastoma cells. In *Sense Organs, Proceedings of the 16th Goettingen Neurobiology Conference* (ed. N. Elsner and F. G. Barth), p. 342. Georg Thieme Verlag: Stuttgart and New York.

HO, P. L., JAKES, R., NORTHROP, F. D. AND GAMBARINI, A. G. (1988). Pituitary fibroblast growth factors: immunocharacterization of an acidic component and N-terminal sequence analysis of a truncated basic component. *Biochem. Int.* **17**, 973–980.

HOHN, A., LEIBROCK, J., BAILEY, K. AND BARDE, Y. A. (1990). Identification and characterization of a novel member of the nerve growth factor/brain-derived neurotrophic factor family. *Nature* **344**, 339–341.

HUANG, S. S., TSAI, C. C., ADAMS, S. P. AND HUANG, J. S. (1987). Neuron localization and neuroblastoma cell expression of brain-derived growth factor. *Biochem. biophys. Res. Commun.* **144**, 81–87.

IMAMURA, T., TOKITA, Y. AND MITSUI, Y. (1988). Purification of basic FGF receptors from rat brain. *Biochem. biophys. Res. Commun.* **155**, 583–590.

INGBER, D. E. AND FOLKMAN, J. (1989). Mechanochemical switching between growth and differentiation during fibroblast growth factor-stimulated angiogenesis *in vitro*: role of extracellular matrix. *J. Cell Biol.* **109**, 317–330.

JANET, T., GROTHE, C., PETTMANN, B., UNSICKER, K. AND SENSENBRENNER, M. (1988). Immunocytochemical demonstration of fibroblast growth factor in cultured chick and rat neurons. *J. Neurosci. Res.* **19**, 195–201.

JANET, T., MIEHE, M., PETTMANN, B., LABOURDETTE, G. AND SENSENBRENNER, M. (1987). Ultrastructural localization of fibroblast growth factor in neurons of rat brain. *Neurosci. Lett.* **80**, 153–157.

JAYE M., HOWK R., BURGESS W., RICCA, G. A., CHIN, I. M., RAVERA, M. W., O'BRIEN, S. J., MODI, W. S., MACIAG, T. AND DROHAN, W. N. (1986). Human endothelial cell growth factor: cloning, nucleotide sequence, and chromosome localization. *Science* **233**, 541–545.

JEANNY, J. C., FAYEIN, N., MOENNER, M., CHEVALLIER, B., BARRITAULT, D. AND COURTOIS, Y. (1987). Specific fixation of bovine brain and retinal acidic and basic fibroblast growth factors to mouse embryonic eye basement membranes. *Expl Cell Res.* **171**, 63–75.

KLAGSBRUN, M., SMITH, S., SULLIVAN, R., SHING, Y., DAVIDSON, S., SMITH, J. A. AND SASSE, J. (1987). Multiple forms of basic fibroblast growth factor: amino terminal cleavages by tumor cell- and brain cell-derived acid proteinases. *Proc. natn. Acad. Sci. U.S.A.* **84**, 1839–1843.

KNISS, D. A. AND BURRY, R. W. (1988). Serum and fibroblast growth factor stimulate quiecent astrocytes to re-enter the cell cycle. *Brain Res.* **439**, 281–288.

KNOECHEL, W. AND TIEDEMANN, H. (1989). Embryonic inducers, growth factors, transcription factors and oncogenes. *Cell Differ. Devl.* **26**, 163–171.

LATZKOVITS, L., TORDAY, C., LABOURDETTE, G., PETTMANN, B. AND SENSENBRENNER, M. (1988). Sodium and potassium uptake in primary cultures of proliferating astroglial cells induced by short-term exposure to an astroglial growth factor. *Neurochem. Res.* **13**, 837–848.

LEDOUX, D., MEREAU, A., DAUCHEL, M. C., BARRITAULT, D. AND COURTY, J. (1989). Distribution of basic fibroblast growth factor binding sites in various tissue membrane preparations from adult guinea pig. *Biochem. biophys. Res. Commun.* **159**, 290–296.

LEHWALDER, D., JEFFREY, P. L. AND UNSICKER, K. (1989). Survival of purified embryonic chick retinal ganglion cells in the presence of neurotrophic factors. *J. Neurosci. Res.* **24**, 329–337.

LEIBROCK, J., LOTTSPEICH, F., HOHN, A., HOFER, M., HENGERER, B., MASIAKOWSKI, P., THOENEN, H. AND BARDE, Y. A. (1989). Molecular cloning and expression of brain-derived neurotrophic factor. *Nature* **341**, 149–152.

LEVI-MONTALCINI, R. (1987). The nerve growth factor: thirty-five years later. *EMBO J.* **6**, 1145–1154.

LOBB, R. R. (1988). Clinical applications of heparin-binding growth factors. *Eur. J. clin. Invest.* **18**, 321–336.

LOPEZ-POUSA, S., FERRIER, I., VICK, J. M. AND DOMENECH-MATEU, J. (1981). Angiogenic activity in the CSF in human malignancies. *Experientia* **37**, 413–414.

LORET, C., LAENG, P., SENSENBRENNER, M. AND LABOURDETTE, G. (1989). Acidic and basic fibroblast growth factors similarly regulate the rate of biosynthesis of rat astroblast proteins. *FEBS Lett.* **257**, 324–328.

LORET, C., SENSENBRENNER, M. AND LABOURDETTE, G. (1988). Maturation-related gene expression of rat astroblasts *in vitro* studied by two-dimensional polyacrylamide gel electrophoresis. *Cell Differ. Devl.* **25**, 37–46.

LUEDECKE, G. AND UNSICKER, K. (1990*a*). Basic FGF and NGF as mitogens for human neuroblastoma cells. *Clin. Chem. Enzym. Comms.* **2**, 293–298.

LUEDECKE, G. AND UNSICKER, K. (1990*b*). Mitogenic effects of neurotrophic factors on human IMR32 neuroblastoma cells. *Cancer* **65**, 2270–2278.

MACHIDA, C. M., RODLAND, K. D., MATRISIAN, L., MAGUN, B. E. AND CIMENT, G. (1989). NGF induction of the gene encoding the protease transin accompanies neuronal differentiation in PC12 cells. *Neuron* **2**, 1587–1596.

MAIER, J. A. M., RUSNATI, M., RAGNOTTI, G. AND PRESTA, M. (1990). Characterization of a M_r 20,000 basic fibroblast growth factor-like protein secreted by normal and transformed fetal bovine aortic endothelial cells. *Expl Cell Res.* **186**, 354–361.

MAISONPIERRE, P. C., BELLUSCIO, L., SQUINTO, S., IP, N. Y., FURTH, M. E., LINDSAY, R. M. AND YANCOPOULOS, G. D. (1990). Neurotrophin-3: a neurotrophic factor related to NGF and BDNF. *Science* **247**, 1446–1451.

MARICS, I., ADELAIDE, J., RAYBAUD, F., MATTEI, M. G., COULIER, F., PLANCHE, J., DE LAPEYRIERE, O. AND BIRNBAUM, D. (1989). Characterization of the HST-related FGF.6 gene, a new member of the fibroblast growth factor gene family. *Oncogene* **4**, 335–340.

MASCARELLI, F., RAULAIS, D. AND COUTOIS, Y. (1989). Fibroblast growth factor phosphorylation and receptors in rod outer segments. *EMBO J.* **8**, 2265–2273.

MATTSON, M. P., MURRAIN, M., GUTHRIE, P. B. AND KATER, S. B. (1989). Fibroblast growth factor and glutamate: opposing roles in the generation and degeneration of hippocampal neuroarchitecture. *J. Neurosci.* **9**, 3728–3740.

MCMANAMAN, J., CRAWFORD, F., CLARK, R., RICHKER, J. AND FULLER, F. (1989). Multiple neurotrophic factors from skeletal muscle: demonstration of effects of basic fibroblast growth factor and comparisons with the 22kD choline acetyltransferase development factor. *J. Neurochem.* **53**, 1763–1771.

MEREAU, A., PIERI, I., GAMBY, C., COURTY, J. AND BARRITAULT, D. (1989). Purification of basic fibroblast growth factor receptors from bovine brain. *Biochimie* **71**, 865–871.

MORRISON, R. S., SHARMA, A., DE VELLIS, J. AND BRADSHAW, R. A. (1986). Basic fibroblast growth factor supports the survival of cerebral cortical neurons in primary culture. *Proc. natn. Acad. Sci. U.S.A.* **83**, 7537–7541.

MOSCATELLI, D. (1988). Metabolism of receptor-bound and matrix-bound basic fibroblast growth factor by bovine capillary endothelial cells. *J. Cell Biol.* **107**, 753–759.

MOSCATELLI, D., JOSEPH-SILVERSTEIN, J., MANEJIAS, R. AND RIFKIN, D. B. (1987). M_r 25,000 heparin-binding protein from guinea pig brain is a high molecular weight form of basic fibroblast growth factor. *Proc. natn. Acad. Sci. U.S.A.* **84**, 5778–5782.

MUESCH, A., HARTMANN, E., ROHDE, K., RUBARTELLI, A., SITIA, R. AND RAPOPORT, T. A. (1990). *Trends Biochem. Sci.* **15**, 86–88.

MUNAIM, S. I., KLAGSBRUN, M. AND TOOLE, B. P. (1988). Developmental changes in fibroblast growth factor in the chicken embryo limb bud. *Proc. natn. Acad. Sci. U.S.A.* **85**, 8091–8093.

NEUFELD, G., GOSPODAROWICZ, D., DODGE, L. AND FUJII, D. K. (1987). Heparin modulation of the neurotropic effects of acidic and basic fibroblast growth factors and nerve growth factor on PC12 cells. *J. cell. Physiol.* **131**, 131–140.

NOJI, S., MATSUO, T., KOYAMA, E., YAMAAI, T., NOHNO, T., MATSUO, N. AND TANIGUCHI, S. (1990). Expression pattern of acidic and basic fibroblast growth factor genes in adult rat eyes. *Biochem. biophys. Res. Commun.* **168**, 343–349.

OKUMURA, N., TAKIMOTO, K., OKADA, M. AND NAKAGAWA, H. (1989). C6 glioma cells produce basic fibroblast growth factor that can stimulate their own proliferation. *J. Biochem., Tokyo* **106**, 904–909.

OLSON, L. (1988). Grafting in the mammalian central nervous system: basic science with clinical promise. *Disc. Neurosci.* V-4, 13–73.

OLWIN, B. B. AND HAUSCHKA, S. D. (1990). Fibroblast growth factor receptor levels decrease during chick embryogenesis. *J. Cell Biol.* **110**, 503–509.

OPPENHEIM, R. W. (1989). The neurotrophic theory and naturally occuring motoneuron death. *Trends Neuosci.* **12**, 252–255.

OTTO, D., FROTSCHER, M. AND UNSICKER, K. (1989). Basic fibroblast growth factor and nerve growth factor administered in gel foam rescue medial septal neurons after fimbria fornix transection. *J. Neurosci. Res.* **22**, 83–91.

OTTO, D. AND UNSICKER, K. (1990). Basic FGF reverses chemical and morphological deficits in the nigrostriatal system of MPTP treated mice. *J. Neurosci.* **10**, 1912–1921.

OTTO, D., UNSICKER, K. AND GROTHE, C. (1987). Pharmacological effects of nerve growth factor and fibroblast growth factor applied to the transectioned sciatic nerve on neuron death in adult rat dorsal root ganglia. *Neurosci. Lett.* **83**, 156–160.

PATERNO, G. D., GILLESPIE, L. L., DIXON, M. S., SLACK, J. M. W. AND HEATH, J. K. (1989). Mesoderm-inducing properties of INT-2 and kFGF: two oncogene-encoded growth factors related to FGF. *Development* **106**, 79–83.

PAULUS, W., GROTHE, C., SENSENBRENNER, M., JANET, T., BAUR, I., GRAF, M. AND ROGGENDORF, W. (1990). Localization of basic fibroblast growth factor, a mitogen and angiogenic factor, in human brain tumors. *Acta Neuropath.* **79**, 418–423.

PERRAUD, F., BESNARD, F., LABOURDETTE, G. AND SENSENBRENNER, M. (1988*a*). Proliferation of rat astrocytes, but not of oligodendrocytes, is stimulated *in vitro* by protease inhibitors. *Int. J. Dev. Neurosci.* **6**, 261–266.

PERRAUD, F., BESNARD, F., PETTMANN, B., SENSENBRENNER, M. AND LABOURDETTE, G. (1988*b*). Effects of acidic and basic fibroblast growth factors (aFGF and bFGF) on the proliferation and the glutamine synthetase expression of rat astroblasts in culture. *Glia* **1**, 124–131.

PERRAUD, F., BESNARD, F., SENSENBRENNER, M. AND LABOURDETTE, G. (1987). Thrombin is a potent mitogen for rat astroblasts but not for oligodendroblasts and neuroblasts in primary culture. *Int. J. Dev. Neurosci.* **5**, 181–188.

PERRAUD, F., LABOURDETTE, G., ECLANCHER, F. AND SENSENBRENNER, M. (1990). Primary cultures of astrocytes from different brain areas of newborn rats and effects of basic fibroblast growth factor. *Devl Neurosci.* **12**, 11–21.

PETTMANN, B., GENSBURGER, C., WEIBEL, M., PERRAUD, F., SENSENBRENNER, M. AND LABOURDETTE, G. (1987). Isolation of two astroglial growth factors from bovine brain: comparison with other growth factors; cellular localization. *NATO ASI Series* **H2**, 451–478.

PETTMANN, B., LABOURDETTE, G., WEIBEL, M. AND SENSENBRENNER, M. (1985). Brain-derived astroglial growth factors. *Funkt. biol. Med.* **4**, 243–248.

PETTMANN, B., LABOURDETTE, G., WEIBEL, M. AND SENSENBRENNER, M. (1986). The brain fibroblast growth factor (FGF) is localized in neurons. *Neurosci. Lett.* **68**, 175–180.

PRATS, H., KAGHAD, M., PRATS, A. C., KLAGSBRUN, M., LELIAS, J. M., LIAUZUN, P., CHALON, P., TAUBER, J. P., AMALRIC, F., SMITH, J. A. AND CAPUT, D. (1989). High molecular mass forms of basic fibroblast growth factor are initiated by alternative CUG codons. *Proc. natn. Acad. Sci. U.S.A.* **86**, 1836–1840.

PRESTA, M., FOIANI, M., RUSNATI, M., JOSEPH-SILVERSTEIN, J., MAIER, J. A. M. AND RAGNOTTI, G. (1988*a*). High molecular weight immunoreactive basic fibroblast growth factor-like proteins in rat pituitary and brain. *Neurosci. Lett.* **90**, 308–313.

PRESTA, M., RUSNATI, M., MAIER, J. A. M. AND RAGNOTTI, G. (1988*b*). Purification of basic fibroblast growth factor from rat brain: identification of a M_r 22,000 immunoreactive form. *Biochem. biophys. Res. Commun.* **155**, 1161–1172.

QUON, D., CATALANO, R. AND CORDELL, B. (1990). Fibroblast growth factor induces β-amyloid precursor mRNA in glial but not neuronal cultured cells. *Biochem. biophys. Res. Commun.* **167**, 96–102.

RIFKIN, D. B. AND MOSCATELLI, D. (1989). Recent developments in the cell biology of basic fibroblast growth factor. *J. Cell Biol.* **109**, 1–6.

ROGELJ, S., WEINBERG, R. A., FANNING, P. AND KLAGSBRUN, M. (1988). Basic fibroblast growth factor fused to a signal peptide transforms cells. *Nature* **331**, 173–175.

Rogelj, S., Weinberg, R. A., Fanning, P. and Klagsbrun, M. (1989). Characteriztion of tumors produced by signal peptide-basic fibroblast growth factor-transformed cells. *J. cell. Biochem.* **39**, 13–23.

Rogister, B., Leprince, P., Pettmann, B., Labourdette, G., Sensenbrenner, M. and Moonen, G. (1988). Brain basic fibroblast growth factor stimulates the release of plasminogen activators by newborn rat cultured astroglial cells. *Neurosci. Lett.* **91**, 321–326.

Rubin J. S., Osada H., Finch P. W., Taylor, W. G., Rudikoff, S. and Aaronson, S. A. (1989). Purification and characterization of a newly identified growth factor specific for endothelial cells. *Proc. natn. Acad. Sci. U.S.A.* **86**, 802–806.

Rydel, R. E. and Greene, L. A. (1987). Acidic and basic fibroblast growth factors promote stable neurite outgrowth and neuronal differentiation in cultures of PC12 cells. *J. Neurosci.* **7**, 3639–3653.

Sato, Y., Murphy, P. R., Sato, R. and Friesen, H. G. (1989). Fibroblast growth factor release by bovine endothelial cells and human astrocytoma cells in culture is density dependent. *Molec. Endocri.* **3**, 744–748.

Schubert, D., Jin, L. W., Saitoh, T. and Cole, G. (1989). The regulation of amyloid β protein precursor secretion and its modulatory role in cell adhesion. *Neuron* **3**, 689–694.

Schubert, D., Ling, N. and Baird, A. (1987). Multiple influences of a heparin-binding growth factor on neuronal development. *J. Cell Biol.* **104**, 635–643.

Schweigerer L., Neufeld G., Friedman J., Abraham, J. A., Fidder, J. C. and Godspodarowicz, D. (1987*b*). Capillary endothelial cells express basic fibroblast growth factor, a mitogen that promotes their own growth. *Nature* **325**, 257–259.

Senior, R. M., Huang, S. S., Griffin, G. L. and Huang, J. S. (1986). Brain-derived growth factor is a chemoattractant for fibroblasts and astroglial cells. *Biochem. biophys. Res. Commun.* **141**, 67–72.

Seno, M., Sasada, R., Iwane, M., Sudo, K., Kurokawa, T., Ito, K. and Igarashi, K. (1988). Stabilizing basic fibroblast growth factor using protein engineering. *Biochem. biophys. Res. Commun.* **151**, 701–708.

Sensenbrenner, M., Pettmann, B., Labourdette, G. and Weibel, M. (1985). Properties of a brain growth factor promoting proliferation and maturation of rat astroglial cells in culture. In *Hormones and Cell Regulation* (ed. J. E. Dumont, B. Hamprecht and J. Nunez), vol. 9, pp. 345–360. Elsevier: Amsterdam.

Sievers, J., Hausmann, B., Unsicker, K. and Berry, M. (1987). Fibroblast growth factors promote the survival of adult rat retinal ganglion cells after transection of the optic nerve. *Neurosci. Lett.* **76**, 157–162.

Sigmund, O., Naor, Z., Anderson, D. J. and Stein, R. (1990). Effect of nerve growth factor and fibroblast growth factor on SCG10 and c-fos expression and neurite outgrowth in protein kinase C-depleted PC12 cells. *J. biol. Chem.* **265**, 2257–2261.

Slack, J. M. W., Darlington, B. G., Heath, J. K. and Godsave, S. F. (1987). Mesoderm induction in early *Xenopus* embryos by heparin-binding growth factors. *Nature* **326**, 197–200.

Smith, E. P., Russell, W. E., French, F. S. and Wilson, E. M. (1989). A form of basic fibroblast growth factor is secreted into the adluminal fluid of the rat coagulating gland. *Prostate* **14**, 353–365.

Stemple, D. L., Mahanthappa, N. K. and Anderson, D. J. (1988). Basic FGF induces neuronal differentiation, cell division, and NGF dependence in chromaffin cells: a sequence of events in sympathetic development. *Neuron* **1**, 517–525.

Taira M., Yoshida T., Miyagawa K., Sakamoto, H., Terada, H. and Sugimura, T. (1987). cDNA sequence of human transforming gene hst and identification of the coding sequence required for transforming activity. *Proc. natn. Acad. Sci. U.S.A.* **84**, 2980–2984.

Thanos, S. and von Boxberg, Y. (1989). Factors influencing regeneration of retinal ganglion cell axons in adult mammals. *Metabol. Brain Disease* **4**, 67–72.

Thomas, K. A. (1987). Fibroblast growth factors. *FASEB J.* **1**, 343–440.

Thomas, K. A. (1988). Transforming potential of fibroblast growth factor genes. *Trends biochem. Sci.* **13**, 327–328.

Togari, A., Dickens, G., Kuzuya, H. and Guroff, G. (1985). The effect of fibroblast growth factor on PC12 cells. *J. Neurosci.* **5**, 307–316.

Ueno, N., Baird, A., Esch, F., Ling, N. and Guillemin, R. (1986). Isolation of an amino terminal extended form of basic fibroblast growth factor. *Biochem. biophys. Res. Commun.* **138**, 580–588.

Unsicker, K., Reichert-Preibsch, H., Schmidt, R., Pettmann, B., Labourdette, G. and Sensenbrenner, M. (1987). Astroglial and fibroblast growth factors have neuronotrophic functions for cultured peripheral and central nervous system neurons. *Proc. natn. Acad. Sci. U.S.A.* **84**, 5459–5463.

Unsicker, K., Seidl, K. and Hofmann, H. D. (1989). The neuro-endocrine ambiguity of sympathoadrenal cells. *Int. J. Dev. Neurosci.* **7**, 413–417.

Vaca, K., Stewart, S. S. and Appel, S. H. (1989). Identification of basic fibroblast growth factor as a cholinergic growth factor from human muscle. *J. Neurosci. Res.* **23**, 55–63.

van Zoelen, E. J. J., Ward-van Oostwaard, T. M. J., Nieuwland, R., van der Burg, B., van den Eijnden-van Raaij, A. J. M., Mummery, C. L. and de Laat, S. W. (1989). Identification and characterization of polypeptide growth factors secreted by murine embryonal carcinoma cells. *Devl Biol.* **133**, 272–283.

Veomett, G., Kuszynski, C., Kazakoff, P. and Rizzino, A. (1989). Cell density regulates the number of cell surface receptors for fibroblast growth factor. *Biochem. biophys. Res. Commun.* **159**, 694–700.

Vlodavski, I., Folkman, J., Sullivan, R., Fridman, R., Ishai-Michaeli, R., Sasse, J. and Klagsbrun, M. (1987). Endothelial cell-derived basic fibroblast growth factor: synthesis and deposition into subendothelial extracellular matrix. *Proc. natn. Acad. Sci. U.S.A.* **84**, 2292–2296.

Wagner, J. A. and D'Amore, P. (1986). Neurite outgrowth induced by an endothelial cell mitogen isolated from retina. *J. Cell Biol.* **103**, 1363–1367.

Walicke, P., Cowan, W. M., Ueno, N., Baird, A. and Guillemin, R. (1986). Fibroblast growth factor promotes survival of dissociated hippocampal neurons and enhances neurite extension. *Proc. natn. Acad. Sci. U.S.A.* **83**, 3012–3016.

Walicke, P. A. (1988). Basic and acidic fibroblast growth factors have trophic effects on neurons from multiple CNS regions. *J. Neurosci.* **8**, 2618–2627.

Walicke, P. A. (1989). Novel neurotrophic factors, receptors, and oncogenes. *A. Rev. Neurosci.* **12**, 103–126.

Watters, D. J. and Hendry, I. A. (1987). Purification of a ciliary neurotrophic factor from bovine heart. *J. Neurochem.* **49**, 705–713.

Weibel, M., Fages, C., Belakebi, M., Tardy, M. and Nunez, J. (1987). Astriglial growth factor-2 (AGF2) increases alpha-tubulin in satroglial cells cultured in a defined medium. *Neurochem. Int.* **11**, 223–228.

Weibel, M., Pettmann, B., Labourdette, G., Miehe, M., Bock, E. and Sensenbrenner, M. (1985). Morphological and biochemical maturation of rat astroglial cells grown in a chemically defined medium: influence of astroglial growth factor. *Int. J. Dev. Neurosci.* **3**, 617–630.

Werner, S., Hofschneider, P. H., Stuerzl, M., Dick, I. and Roth, W. K. (1989). Cytochemical and molecular properties of simian virus 40 transformed Kaposi's sarcoma-derived cells: evidence for the secretion of a member of the fibroblast growth factor family. *J. cell. Physiol.* **141**, 490–502.

Westermann, R., Johannsen, M., Unsicker, K. and Grothe, C. (1990). Basic fibroblast growth factor (bFGF) immunoreactivity is present in chromaffin granules. *J. Neurochem.* **55**, 285–292.

Westermann, R. and Unsicker, K. (1990). Basic fibroblast growth factor (bFGF) and rat C6 glioma cells: regulation of expression, absence of release, and response to exogenous bFGF. *Glia*, in press.

Westphal, M., Brunken, M., Rohde, E. and Hermann, H.-D. (1988). Growth factors in cultured human glioma cells: differential effects of FGF, EGF and PDGF. *Cancer Lett.* **38**, 283–296.

Westphal, M. and Herrmann, H. D. (1989). Growth factor biology and oncogene activation in human gliomas and their implications for specific therapeutic concepts. *Neurosurg.* **25**, 681–694.

Wolburg, H., Neuhaus, J., Pettmann, B., Labourdette, G. and Sensenbrenner, M. (1986).

Decrease in the density of orhtogonal arrays of particles in membranes of cultured rat astroglial cells by the brain fibroblast growth factor. *Neurosci. Lett.* **72**, 25–30.

YONG, V. W., KIM, S. U., KIM, M. W. AND SHIN, D. H. (1988*b*). Growth factors for human glial cells in culture. *Glia* **1**, 113–123.

YONG, V. W., KIM, S. U. AND PLEASURE, D. E. (1988*a*). Growth factors for fetal and adult human astrocytes in culture. *Brain Res.* **444**, 59–66.

ZHAN X., BATES B., HU X. AND GOLDFARB, M. (1988). The human FGF-5 oncogene encodes a novel protein related to fibroblast growth factors. *Molec. cell. Biol.* **8**, 3487–3495.

J. Cell Sci. Suppl. 13, 119–130 (1990)
Printed in Great Britain © The Company of Biologists Limited 1990

Growth factors as inducing agents in early *Xenopus* development

J. M. W. SLACK
Imperial Cancer Research Fund Developmental Biology Unit, Department of Zoology, University of Oxford, South Parks Road, Oxford OX1 3PS, UK

Summary

Factors from two growth factor families have been identified as having mesoderm inducing activity. These include activin and TGFβ2 from the TGFβ superfamily, and all members of the fibroblast growth factor (FGF) family.

When isolated ectoderm explants are treated with any of these factors, a proportion of their cells are caused to differentiate into mesodermal tissue types instead of epidermis. There are several differences in the biological activities which can broadly be summarized by saying that activin yields dorsal type inductions and FGF ventral type inductions.

Both bFGF and an FGF receptor have been detected in *Xenopus* blastulae, but it has not been shown that bFGF is normally secreted from vegetal cells. Various TGFβ-like mRNAs have also been detected and it is expected that an activin-like molecule will prove to be responsible for induction of the dorsal mesoderm *in vivo*.

Introduction

Embryonic inducing factors have been shadowy substances existing at the margins of respectable biochemistry for many decades. In the last few years a combination of better protein purification methods with the characterization of many hormones and cytokines in other branches of cell biology have made it clear that inducing factors are not a special class of substances but are factors already known to be responsible for other biological activities in later life.

This paper will deal with mesoderm induction in *Xenopus*. This is a process which has now been intensively studied biologically and the moderate, although not complete, degree of understanding we have at the biological level has been crucial for the work on the inducing factors. I shall deal not only with the work of my own laboratory, which has concentrated on the role of the fibroblast growth factors (FGFs), but also with results obtained by others working with factors from the TGFβ superfamily.

Biology of mesoderm induction

Amphibian embryos have long been favoured for experimental embryology because their large size and accessibility makes them much more favourable for micromanipulation than mammals. Several species have been used in the past

Key words: *Xenopus*, mesoderm, induction, fibroblast growth factors, activins.

but *Xenopus laevis*, the African clawed frog, is now the world standard organism for this type of work. Although the situation is now improving, the account given in many textbooks is rather misleading, describing neural induction as 'primary embryonic induction' and confusing several processes which are logically distinct, such as formation of the mesoderm and determination of anteroposterior levels within the mesoderm.

Our current understanding of mesoderm induction is based on work by Nieuwkoop and of Nakamura in the early seventies (Nieuwkoop, 1969; Nakamura *et al.* 1971), which has been refined and extended but not fundamentally challenged in recent years by ourselves and others (Dale *et al.* 1985; Gurdon *et al.* 1985; Jones and Woodland, 1987). The results of these studies are very briefly as follows.

The mesoderm arises during the blastula stages, when the embryo consists of a hollow ball of cells. It is formed as a ring of cells around the equator called the marginal zone. At least half and perhaps all of the mesoderm is induced from the animal hemisphere as a result of inductive signals from the vegetal hemisphere. It also cannot be excluded that part of the mesoderm is formed by inheritance of a cytoplasmic determinant in the fertilized egg. The whole animal hemisphere is competent to become mesoderm but less than half of it becomes induced to do so *in vivo*, this being the ring of tissue abutting the vegetal hemisphere. The vegetal tissue is signalling and the animal tissue is competent to respond during the blastula stages but both signalling and competence decline shortly after gastrulation has commenced. Most of these facts have been established by variations on the combination experiment first introduced by Nieuwkoop and shown in Fig. 1, in which tissue from the animal pole region (an animal cap) is combined *in vitro* with an explant from the vegetal hemisphere.

The mesoderm will later become several tissues: in the trunk region notochord, somite, kidney, lateral plate and blood islands appear in a dorsal to ventral sequence. It is thought that the initial mesodermal induction creates perhaps only two zones, a small zone about 60° in circumference called the organizer and a larger zone comprising the remaining 300° of circumference, which initially has a ventral type of specification. The organizer region has several important properties which are quite difficult to disentangle mechanistically: it is the first region to involute during gastrulation and undergoes the most profound extension movements; it becomes the dorsal midline structure of the mesoderm of the entire body from head to tail; it dorsalizes the surrounding mesoderm to form the somites and kidney; it induces the overlying ectoderm to form the central nervous system. The ventral mesoderm undergoes later and less pronounced extension movements: its cells tend to move toward the dorsal midline during gastrulation and those which end up near it become dorsalized to form the somites. Several microsurgical experiments have shown that the specificity for the induction of organizer *versus* ventral type mesoderm lies with the signalling and not with the responding tissue. So the signal is complex, consisting at least of one substance at two concentrations or perhaps of more than one substance. The

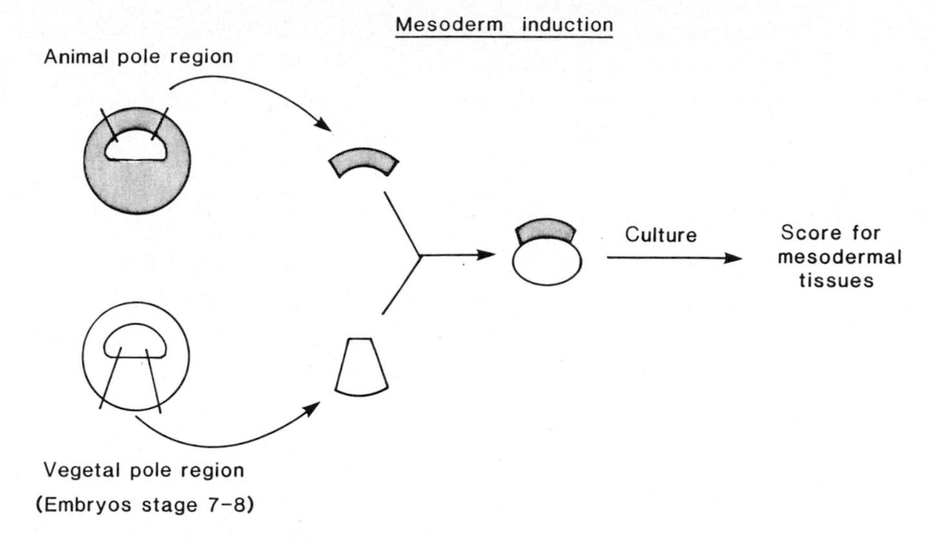

Fig. 1. Basic protocol for experiments on mesoderm induction. An animal cap from a *Xenopus* blastula is combined with part or all of the vegetal region. After a culture period of 1–3 days the combination is assayed for mesoderm formation by molecular or histological methods.

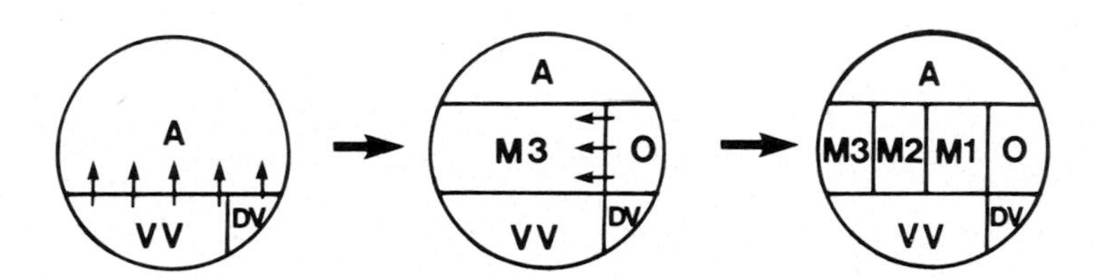

Fig. 2. The three signal model for mesoderm induction. It is proposed that distinct signals are responsible for inducing the organizer (O) and ventral mesodermal (M3) regions, possibly activin A and bFGF respectively. Specification of other territories within the mesoderm occurs in response to a later signal emitted from the organizer. A, animal; VV, ventrovegetal; DV, dorsovegetal.

most generally accepted model of the inductive signals responsible for the formation and dorsoventral specification of the mesoderm is called the 'three signal model' (Dale and Slack, 1987) and is shown schematically in Fig. 2.

Mesoderm induction is not a cell contact-mediated process. This has been shown by transfilter experiments in which the vegetal and animal tissues are placed on opposite sides of a nucleopore membrane with pores too small to admit cellular processes (Grunz and Tacke, 1986). The inducing signal(s) can cross the liquid gap quite effectively. It has also been shown that the signals do not require the presence of functional gap junctions (Warner and Gurdon, 1987) and that the range of the effect is about 80 microns which is a few cell diameters at the embryonic stages in question (Gurdon, 1989).

These experiments tell us quite a lot about what to expect of the inductive signals. We are looking for one or more substances which are capable of causing mesodermal differentiation in isolated ectoderm. They should not be intracellular but capable of secretion by the vegetal cells. They should be diffusible, but not so diffusible that they would spread all over the embryo.

The candidate mesoderm inducing factors

Those who do not work with amphibian embryos should note that progress in this area has not been made by treating intact embryos with cytokines or by injecting cytokines into them. Such experiments are very difficult to interpret because of uneven access, a multiplicity of effects of different parts of the embryo and multiple secondary events which may involve alterations in cell movements as well as cell differentiation. The advantage of *Xenopus* compared to the mouse or the chick lies in the availability of a test tissue which can be used for bioassays *in vitro*. This is the 'animal cap' (Fig. 1) which forms 100 % epidermis when cultured *in vitro* but will respond to mesoderm inducing factors (MIFs) in solution by forming some mesoderm as well. Explants in which some mesoderm is present will elongate during the first day of culture and form characteristic vesicles on the second or third day (Fig. 3). It is therefore possible to score an explant as induced or uninduced simply by visual inspection down the dissecting microscope. This is

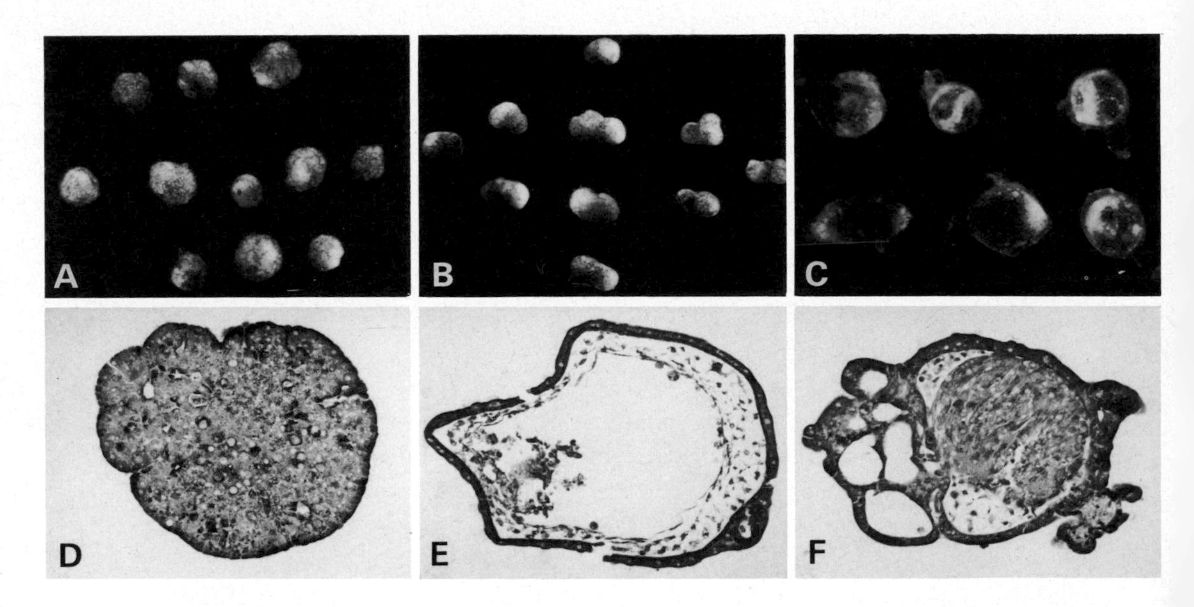

Fig. 3. Mesoderm induction by bFGF. (A) Untreated animal caps after overnight culture have rounded up into balls. (B) Caps treated with bFGF have elongated. (C) After 3 days culture, treated explants have become fluid filled vesicles. (D) Section of untreated explant after 3 days, showing 100 % epidermal differentiation. (E) Explant treated with 4 units ml^{-1} *Xenopus* bFGF, showing formation of mesenchyme and mesothelium. (F) Explant treated with 32 units ml^{-1}, showing formation of a large muscle mass.

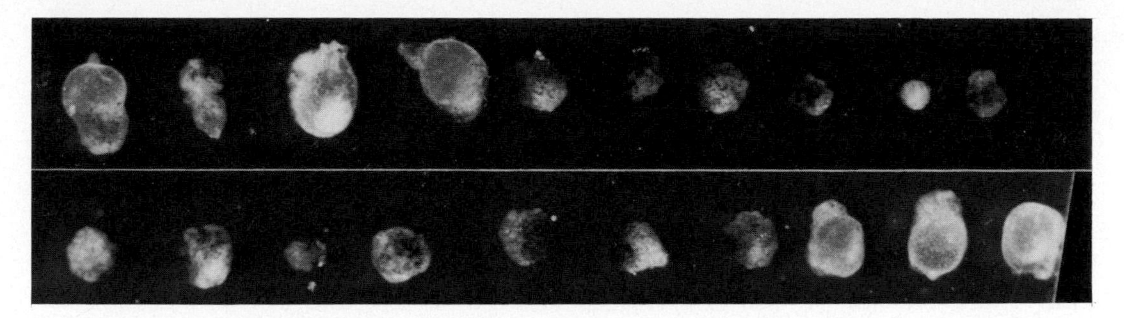

Fig. 4. Animal cap titrations. The top row are explants treated with serial twofold dilutions of *Xenopus* bFGF starting at $10\,\text{ng}\,\text{ml}^{-1}$. The first four are positive. The bottom row shows explants which were all treated with $10\,\text{ng}\,\text{ml}^{-1}$ bFGF but also with twofold dilutions of a neutralizing IgG. The last three are positive, meaning that $3\,\mu\text{g}\,\text{ml}^{-1}$ of IgG can just neutralize $10\,\text{ng}\,\text{ml}^{-1}$ of bFGF.

made into a quantitative assay by testing serial dilutions and observing the threshold concentration above which induction occurs, which is defined as $1\,\text{unit}\,\text{ml}^{-1}$ (Godsave *et al.* 1988; Fig. 4). A competent operator can dissect about 100 animal caps in one afternoon which is enough for 10–15 titrations, so this bioassay is simple enough to be used for protein purification.

The factors shown to be active are, at the time of writing, activin A (XTC-MIF), TGFβ-2, and all members of the fibroblast growth factor family (aFGF, bFGF, kFGF, FGF-5, ECDGF and int-2). XTC-MIF was purified from a *Xenopus* cell line by Smith *et al.* (1988), activin A was found to be active by Asashima *et al.* (1990) and it has recently been shown that XTC-MIF is the *Xenopus* homologue of mammalian activin A (Smith *et al.* 1990). TGFβ-2 was shown to be active by Rosa *et al.* (1988). TGFβ-1 and -3 are not active on their own, although as we shall see they have synergistic effects with the FGFs. bFGF, ECDGF and aFGF were shown to be active by Slack *et al.* (1987), and were later joined by kFGF, FGF-5 and int-2, although int-2 has much lower specific activity than the others (Paterno *et al.* 1989).

All these MIFs were already known through work in other areas of cell biology. Activin is a hormone secreted by the ovary which promotes release of follicle stimulating hormone from the pituitary (Ling *et al.* 1988). There are two gene products called β_A and β_B which can assemble into homo and heterodimers. They are quite similar in sequence and belong to the TGFβ-superfamily. Activin A is the homodimer of β_A and it is probable, although not yet known for sure, that the other forms will also have mesoderm inducing activity. The TGFβs are well known cytokines which have a variety of biological activities in different systems, sometimes promoting and sometimes inhibiting cell division, sometimes promoting and sometimes inhibiting differentiation (Massagué, 1987). The FGFs are mitogens, particularly active on capillary endothelial cells, and characterised (except int-2) by tight binding to heparin (Gospodarowicz *et al.* 1987). The prototype members of the family, a and bFGF, were identified by protein purification, while the others considered here, kFGF, FGF-5 and int-2, were initially identified as oncogenes and later shown to code for mitogenic proteins.

Biological activity of MIFs

Detailed studies have been carried out for activin A and for a and bFGF (Slack *et al.* 1988; Green *et al.* 1990), and less information is available about the rest. However, it seems that there are a number of differences in biological properties between activin A on the one hand and all the other factors, including TGFβ-2, on the other.

The dose–response curves show that activin A has a specific activity of about 5×10^{6} units mg^{-1}, meaning that it is active above 0.2 ng ml^{-1}. The FGFs have specific activities about 5× lower, at about 1×10^{6} units mg^{-1}. There is little difference between the effectiveness of bFGF from bovine and *Xenopus* sources. For intact explants the dose response curves are very steep, and this is the basis of the serial dilution assay mentioned above. But for isolated cells, at least with the FGFs the dose response is quite extended, indicating that individual cells differ in their response thresholds (S. F. Godsave and J. M. W. Slack, unpublished observations).

Different doses also elicit the formation of different mesodermal tissues. For activin A, low concentrations induce ventral patterns of differentiation consisting of mesenchyme and mesothelium, higher concentrations induce muscle (Fig. 5), and the highest concentrations induce notochord and endoderm-like tissue. The full range of responses is spanned by about a 10× change in concentration above

Fig. 5. A clone of muscle cells arising from a single, isolated animal cap cell treated with murine activin A. Stained with a muscle-specific antibody (12/101).

the threshold. For the FGFs low concentrations also induce ventral structures, but then there is an extended concentration range going right up to toxic levels (micrograms ml^{-1}) in which there is a progressively greater proportion of muscle induced. When cultures of isolated cells are induced (for method, see Godsave and Slack, 1989), 100% can be converted to muscle, suggesting that the mixtures of cell types typically found in induced explants arise because only the exposed blastocoelic surface is accessible to added factors. But it is very rare to see notochord in FGF inductions at whatever dose. It should also be noted that neither type of factor induces blood cells or pronephros although these are induced by vegetal tissue, either in combination or in transfilter experiments.

The time of competence of the animal cap can be assessed by short term treatment of tissue isolated from different stage embryos. This has shown that competence to respond to activin A extends into the early gastrula stages, while competence to respond to the FGFs is lost just before the onset of gastrulation.

These four differences, the specific activity, the steepness of the dose–response curve, the formation of notochord, and the stages of competence could all be interpreted as saying that the FGFs are just weak inducers, evoking the same response as activin A but less effectively. But the last difference argues against this. The FGFs are more effective than activin A at turning on a homeobox-containing gene called *xhox-3*, which is normally expressed in the posteroventral regions of the mesoderm (Ruiz i Altaba and Melton, 1989*a*). This makes the two groups of factor seem to differ in a qualitative way with activin A having an organizer-inducing capability and the FGFs having a ventral mesoderm-inducing capability.

Modifications of biological activity

It was early discovered that TGFβ-1 could enhance the effect of a given dose of FGF and provoke a degree of muscle formation equivalent to a higher dose (Kimelman and Kirschner, 1987). We have found that the same is also true for TGFβ-3. However, the quality of the induction is not altered from FGF-like to activin-like, and in particular the induction of notochord remains rare, as is found for FGF alone.

A similar enhancement of FGF effect is seen with lithium ion. Lithium has been shown to cause hyperdorsalization of whole embryos, which is interpreted as an expansion in the size of the organizer territory (Kao *et al.* 1986; Kao and Elinson, 1988). It has no effect on isolated animal explants but will enhance the effect of FGF treatment and will also greatly increase the amount of muscle formed by explants from the ventral marginal zone, which normally form predominantly blood cells and mesenchyme (Slack *et al.* 1988).

A negative effect on bFGF is shown by heparin. This binds tightly to all forms of FGF but its inhibitory action is most apparent on bFGF. Inhibition is seen in the microgram ml^{-1} range meaning that a molar excess of >1000 is required (Slack *et al.* 1987). This suggests that those complexes of heparin and bFGF which are presumably formed at lower heparin concentrations are still active and that inhibition may result from the occupation not of the high affinity heparin sites on bFGF but of the low affinity sites. It is often thought that FGFs *in vivo* are immobilized on cell surface heparin sulphate, so it is important that immobilization is not necessarily incompatible with biological activity.

Factors and receptors present in the embryo

A number of laboratories are energetically cloning likely factors from *Xenopus* cDNA libraries. Several TGFβ-like clones have been obtained but it is not yet known which of them produce biologically active proteins or which produce proteins at the appropriate embryonic stages. A TGFβ-like mRNA which is present in the early embryo is called *vg-1* and was discovered by differential screening of an oocyte library with animal and vegetal cDNA (Weeks and Melton, 1987). The *vg-1* message becomes localized to the vegetal hemisphere during oogenesis. Protein is synthesized before and after fertilization and because the message is inherited by the vegetal cells it is in these cells that the zygotic synthesis occurs (Tannahill and Melton, 1989). However, it has been difficult to demonstrate biological activity on the part of the protein, and although it may have synergistic activity with other factors it probably does not have mesoderm inducing activity in its own right. The present consensus is that *vg-1* probably is involved in mesoderm induction in some capacity, but it is probably not itself an inducing factor.

mRNA for bFGF was found in embryos by Kimelman *et al.* (1987, 1988). The levels are low and decrease from the oocyte to the early embryo. There is also an excess of an 'anti-message' which is derived from the other DNA strand and codes for an entirely different protein (Volk *et al.* 1989). Although the actual protein coding sequences do not overlap, the untranslated regions of each mRNA overlap the coding sequences of the other, so it seems unlikely that both mRNAs could be transcribed from the same genes at the same time. There are also low levels of heparin-binding inducing proteins present in early embryos (Kimelman *et al.* 1988; Slack and Isaacs, 1989). Two protein bands have been identified, of about 20 and $14\times10^3\,M_r$. The $20\times10^3\,M_r$ form is certainly bFGF and the $14\times10^3\,M_r$ form reacts with two antibodies prepared against bFGF, but since these particular antibodies may cross react with other types of FGF the identification of the $14\times10^3\,M_r$ species is not unambiguous. The purification of these proteins from *Xenopus* embryos so far represents the only recovery of MIFs from the embryos themselves.

It has already been mentioned that the mesoderm inducing signals can cross a liquid gap in a transfilter apparatus. Using a modification of the transfilter method, in which the tissues are held about 100 microns apart, I have been able to show that both dorsal and ventral vegetal tissues can provoke inductions transfilter, and must therefore be secreting active factors. However, all attempts to condition even tiny volumes of medium by vegetal calls have failed, so the levels must be very low or the factors very unstable. It is reproducibly found that heparin in solution can inhibit the transmission of these transfilter signals, from whole or ventral half vegetal pieces (Slack *et al.* 1987). Heparin also inhibits the action of bFGF *in vitro* and so this may seem a good piece of evidence for the secretion of bFGF by vegetal cells. However, neutralizing antibodies to bFGF do not inhibit the transmission of the transfilter signals, and the specificity of heparin for bFGF, although reasonable, is unlikely to be absolute.

A receptor for FGF (probably binding all forms) has been detected in early *Xenopus* embryos by chemical cross-linking (Gillespie *et al.* 1989). It is a protein of about $130 \times 10^3 M_r$ and seems similar in its properties to the FGF receptors identified on mammalian cells. The levels detectable on the surfaces of animal cap cells rise and fall in a very similar way to the competence of explants to respond to FGFs, reaching a peak in the midblastula and falling by a factor of 10 by the beginning of gastrulation. Binding studies on dissected explants suggest that there are not significant differences in receptor density in different parts of the embryo, so it is present both in the region that normally responds (the marginal zone) and in the remainder of the animal hemisphere that does not respond *in vivo* but is capable of doing so *in vitro*. This is consistent with the inference from the embryological studies that the range of the response is controlled by the distribution of the signal, rather than by differential competence. While bFGF will compete efficiently with aFGF for binding to this receptor, TGFβ-2 will not. In mammalian cells, TGFβ receptors are distinct from those for the FGFs, and in view of the various differences in the responses to activin A and bFGF it also seems likely that in *Xenopus* embryos each group of factors will have its own receptor.

Later events of mesoderm induction

As in so many other systems, there is evidence that the immediate events following binding of the factors involve protein phosphorylation of tyrosine. The mammalian FGF receptor is a tyrosine kinase, and we have data to show that the same is true for the *Xenopus* receptor (L. L. Gillespie, unpublished observations). Injection of synthetic mRNA for the polyoma middle T protein has a mesoderm inducing effect (Whitman and Melton, 1989). This binds to and activates the cellular protein pp60-c-src which is a tyrosine kinase. From a technical standpoint it should be noted once again that these experiments were not based on observation of effects on whole embryos. They used the more discriminating method of injecting the mRNA into zygotes, raising the embryos to blastulae and then explanting animal pole tissue and looking for auto-induction of the explants. As far as other signal transduction pathways are concerned it seems that cyclic AMP and cyclic GMP are not involved, nor is protein kinase C activated, although it does seem to be involved in neural induction at somewhat later stages (Otte *et al.* 1989).

Several genes are turned on as a consequence of exposure of animal caps to

MIFs (Rosa, 1989). One, turned on by activin A but not by FGFs, is a homeobox gene called *mix-1* which is normally expressed in the endoderm. Whether parts of the endoderm are formed *in vivo* by induction is not known, but it does seem to happen *in vitro*. Another homeobox-containing gene which is preferentially turned on by FGFs is *xhox-3* (Ruiz i Altaba and Melton, 1989*a*). This is normally expressed in the posteroventral parts of the mesoderm. Ectopic over-expression in whole embryos, provoked by injection of excess synthetic mRNA into the zygotes, causes suppression of the head (Ruiz i Altaba and Melton, 1989*b*) and it has been argued by these authors that *xhox-3* is a coding factor for the posterior part of the body. But there may of course be other events intervening between FGF treatment and *xhox-3* elevation. There are certainly other events before the activation of muscle specific markers, such as α-actin, which have been much used for studies on mesoderm induction.

How well do we now understand the mechanism of mesoderm induction?

This field has received a disproportionate share of publicity in recent years and there have been almost more mini-reviews than original papers. Outsiders might be forgiven for thinking that it was all over. But actually it has only just begun. What we are dealing with is not the activation of one gene in a bottle of fairly homogeneous tissue culture cells. Regional specification in the early embryo is a complex, three dimensional problem, and at any given embryonic stage there are several different processes going on in the same cells. If we look at the much more profound revolution of understanding that has occurred in *Drosophila* we find that events are indeed complex and each process requires a number of components. Sometimes the same phenotype is obtained by mutating components to inactivity at different stages of a causal chain. Therefore it should not be surprising that similar effects can be obtained by different treatments at different stages, for example dorsalization provoked by lithium in cleavage stages, and by an organizer graft in the early gastrula. In *Drosophila* we also find that specification of territories along the two principal body axes occurs more or less independently. In *Xenopus* it is hard to imagine that one or two substances control the whole of early development, as is implied in some accounts. It is much more likely that each identifiable biological process has its own set of factors, responses and genes, and indeed that many more components are at work than would be strictly necessary if the maximum combinatorial economy was sought. We are presently still very short of sufficiently discriminating markers to dissect the individual processes, for example when we see more muscle we do not know if we are looking at a dorsalization, a posteriorization or simply more responding cells in the population.

The strongest candidate for a mesoderm inducing factor *in vivo* is probably still bFGF. It is present at the relevant embryonic stages, it is there in sufficient quantity for biological activity, and its receptor rises and falls during the blastula stages in a manner suggesting some role for the ligand. But we still do not know its localization in the embryo; there is no evidence that it is secreted by vegetal cells; and its biological activity only covers a partial range of the responses seen *in vivo*. Since there is some serious doubt about whether bFGF can be secreted at all, due to its lack of a signal sequence, it is possible that it plays some intracellular role in the responding tissue rather than being an inducing factor *per se*. Everyone in the field expects activin A to be the organizer-inducing signal, but it remains to be shown that it is present in the embryo and that it is secreted by vegetal cells.

What we have done in the last three years is not so much to provide definitive answers as to bring together the fields of experimental embryology and cytokine research. This is proving to be a very fruitful relationship for both sides. For the embryologists it provides explanations of previously mysterious phenomena, crudely summarised by the slogans: 'inducing factors are growth factors; competence means receptors for growth factors; determination means activation of particular signal transduction pathways'. It also provides us with a basket of relevant molecular probes which should eventually turn out to be the key to unlock the embryonic box. For cell biologists working with cytokines, the discovery of their role in early development opens new possibilities for research on their *in vivo* function, always the weak link in a field previously dominated by research using tissue culture cells.

References

ASASHIMA, M., NAKANO, H., SHIMADA, K., KINOSHITA, K., ISHII, K., SHIBAI, H. AND UENO, N. (1990). Mesoderm induction in early amphibian embryos by activin A (erythroid differentiation factor). *Wilhelm Roux Arch. Devl Biol.* **198**, 330–335.

DALE, L. AND SLACK, J. M. W. (1987). Regional specification within the mesoderm of early embryos of *Xenopus laevis*. *Development* **100**, 279–295.

DALE, L., SMITH, J. C. AND SLACK, J. M. W. (1985). Mesoderm induction in *Xenopus laevis*. A quantitative study using a cell lineage label and tissue specific antibodies. *J. Embryol. exp. Morph.* **89**, 289–313.

GILLESPIE, L. L., PATERNO, G. D. AND SLACK, J. M. W. (1989). Analysis of competence: Receptors for fibroblast growth factor in early *Xenopus* embryos. *Development* **106**, 203–208.

GODSAVE, S. F., ISAACS, H. AND SLACK, J. M. W. (1988). Mesoderm inducing factors: a small class of molecules. *Development* **102**, 555–566.

GODSAVE, S. F. AND SLACK, J. M. W. (1989). Clonal analysis of mesoderm induction. *Devl Biol.* **134**, 486–490.

GOSPODAROWICZ, D., FERRARA, N., SCHWEIGERER, L. AND NEUFELD, G. (1987). Structural characterization and biological functions of fibroblast growth factor. *Endocrine Rev.* **8**, 95–114.

GREEN, J. B. A., HOWES, G., SYMES, K., COOKE, J. AND SMITH, J. C. (1990). The biological effects of XTC-MIF: quantitative comparison with *Xenopus* bFGF. *Development* **108**, 173–183.

GRUNZ, H. AND TACKE, L. (1986). The inducing capacity of the presumptive endoderm of *Xenopus laevis* studied by transfilter experiments. *Wilhelm Roux Arch. Devl Biol.* **195**, 467–473.

GURDON, J. B. (1989). The localization of an inductive response. *Development* **105**, 27–33.

GURDON, J. B., FAIRMAN, S., MOHUN, T. J. AND BRENNAN, S. (1985). The activation of muscle specific action genes in *Xenopus* development by an induction between animal and vegetal cells of a blastula. *Cell* **41**, 913–922.

JONES, E. A. AND WOODLAND, H. R. (1987). The development of animal cap cells in *Xenopus*: a measure of the start of animal cap competence to form mesoderm. *Development* **101**, 557–563.

KAO, K. R. AND ELINSON, R. P. (1988). The entire mesodermal mantle behaves as Spemann's organizer in dorsoanterior enhanced *Xenopus laevis* embryos. *Devl Biol.* **127**, 64–77.

KAO, K. R., MASUI, Y. AND ELINSON, R. P. (1986). Lithium induced respecification of pattern in *Xenopus laevis* embryos. *Nature* **322**, 371–373.

KIMELMAN, D., ABRAHAM, J. A., HAAPARANTA, T., PALISI, T. M. AND KIRSCHNER, M. W. (1988). The presence of fibroblast growth factor in the frog egg: its role as a natural mesoderm inducer. *Science* **242**, 1053–1056.

KIMELMAN, D. AND KIRSCHNER, M. (1987). Synergistic induction of mesoderm by FGF and TGF-β and the identification of an mRNA coding for FGF in the early *Xenopus* embryo. *Cell* **51**, 869–877.

LING, N., UENO, N., YING, S. Y., ESCH, F., SHIMASAKI, S., HOTTA, M., CUEVAS, P. AND GUILLEMIN, R. (1988). Inhibins and activins. *Vitamins and Hormones* **44**, 1–46.

MASSAGUÉ, J. (1987). The TGF-β family of growth and differentiation factors. *Cell* **49**, 437–438.

NAKAMURA, O., TAKASAKI, H. AND ISHIHARA, M. (1971). Formation of the organizer by combinations of presumptive ectoderm and endoderm. *Proc. Jap. Acad.* **47**, 313–318.

NIEUWKOOP, P. D. (1969). The formation of the mesoderm in urodelean amphibians I. Induction by the endoderm. *Wilhelm Roux Arch. EntwMech. Org.* **162**, 341–373.

OTTE, A. P., VAN RUN, P., HEIDEVALD, M., VAN DRIEL, R. AND DURSTON, A. J. (1989) Neural induction is mediated by cross talk between the protein kinase C and cyclic AMP pathways. *Cell* **58**, 641–648.

PATERNO, G. D., GILLESPIE, L. L., DIXON, M. S., SLACK, J. M. W. AND HEATH, J. K. (1989). Mesoderm inducing properties of int-2 and kFGF: two oncogene encoded growth factors related to FGF. *Development* **106**, 79–83.

ROSA, F., ROBERTS, A. B., DANIELPOUR, D., DART, L. L., SPORN, M. B. AND DAWID, I. B. (1988). Mesoderm induction in amphibians: The role of TGF-β2-like factors. *Science* **239**, 783–785.

ROSA, F. M. (1989). Mix-1, a homeobox mRNA inducible by mesoderm inducers, is expressed mostly in the presumptive endodermal cells of *Xenopus* embryos. *Cell* **57**, 965–974.

RUIZ I ALTABA, A. AND MELTON, D. A. (1989*a*). Interaction between peptide growth factors and homeobox genes in the establishment of anteroposterior polarity in frog embryos. *Nature* **341**, 33–38.

RUIZ I ALTABA, A. AND MELTON, D. A. (1989*b*). Involvement of the *Xenopus* gene xhox-3 in pattern formation along the anteroposterior axis. *Cell* **57**, 317–326.

SLACK, J. M. W., DARLINGTON, B. G., HEATH, J. K. AND GODSAVE, S. F. (1987). Mesoderm induction in early *Xenopus* embryos by heparin- binding growth factors. *Nature* **326**, 197–200.

SLACK, J. M. W. AND ISAACS, H. V. (1989). Presence of basic fibroblast growth factor in the early *Xenopus* embryo. *Development* **105**, 147–154.

SLACK, J. M. W., ISAACS, H. V. AND DARLINGTON, B. G. (1988). Inductive effects of fibroblast growth factor and lithium ion on *Xenopus* blastula ectoderm. *Development* **103**, 581–590.

SMITH, J. C., PRICE, B. M. J., VAN NIMMEN, K. AND HUYLEBROEK, D. (1990). XTC-MIF: a potent *Xenopus* mesoderm inducing factor, is a homologue of activin A. *Nature* **345**, 729–731.

SMITH, J. C., YAQOOB, M. AND SYMES, K. (1988). Purification, partial characterisation and biological effects of the XTC mesoderm-inducing factor. *Development* **103**, 591–600.

TANNAHILL, D. AND MELTON, D. A. (1989). Localized synthesis of the vg-1 protein during early *Xenopus* development. *Development* **106**, 775–785.

VOLK, R., KÖSTER, M., PÖTING, A., HARTMAN, L. AND KNÖCHEL, W. (1989). An antisense transcript from the *Xenopus laevis* bFGF gene coding for an evolutionarily conserved 24 kD protein. *EMBO J.* **8**, 2983–2988.

WARNER, A. E. AND GURDON, J. B. (1987). Functional gap junctions are not required for muscle gene activation by induction in *Xenopus* embryos. *J. Cell Biol.* **104**, 557–564.

WEEKS, D. L. AND MELTON, D. A. (1987). A maternal messenger RNA localised to the vegetal hemisphere in *Xenopus* eggs codes for a growth factor related to TGF-β. *Cell* **51**, 861–867.

WHITMAN, M. AND MELTON, D. (1989). Induction of mesoderm by a viral oncogene in early *Xenopus* embryos. *Science* **244**, 803–806.

J. Cell Sci. Suppl. 13, 131–138 (1990)
Printed in Great Britain © The Company of Biologists Limited 1990

Transforming growth factor-β receptors and binding proteoglycans

FREDERICK T. BOYD*, SELA CHEIFETZ, JANET ANDRES, MARIKKI LAIHO AND JOAN MASSAGUÉ

Cell Biology and Genetics Program, Memorial Sloan-Kettering Cancer Center, Sloan-Kettering Division of the Graduate School of Medical Sciences, Cornell University, New York, New York 10021, USA

Summary

Transforming growth factors-beta (TGFs-β) are representative of a superfamily whose members were first identified as regulators of morphogenesis and differentiation, and subsequently found to be structurally related. Other members of the family include the activins and inhibins, BMPs, MIS, the DPP-C gene product and Vg-1. When assayed by affinity-labelling techniques, TGFs-β bind to three distinct cell surface proteins which are present on most cells. These proteins are all of relatively low abundance but bind TGFs-β with affinities consistent with the biological potency of the factors. The Type I and Type II binding proteins are glycoproteins with estimated molecular weights of 53 and $73\times10^3 M_r$, respectively. They both bind TGF-β1 significantly better than TGF-β2. The Type I receptor has been identified as the receptor which mediates many of the responses of TGFs-β, based on somatic cell genetic studies of epithelial cell mutants unresponsive to TGFs-β. Betaglycan is the third binding protein present on many, but not all, cell types and is a large proteoglycan ($\sim280\times10^3 M_r$) with $100–120\times10^3 M_r$ core proteins. A soluble form of this molecule is present in conditioned media of many cell lines and may be derived from the cell surface-associated molecule by cleavage of a small membrane anchor. Betaglycan binds TGF-β1 and TGF-β2 with similar affinity and this binding is to the core proteins, not the glycosaminoglycan side chains. This molecule may have a function in the localization and delivery or the clearance of activated TGFs-β. The molecular basis of TGF-β signalling is still largely unknown, but it is possible that one or more of these cell surface molecules signals *via* a novel mechanism, as the TGFs-β are biologically quite distinct from other factors that act *via* well-characterized signalling systems.

Introduction

Transforming growth factors-beta (TGFs-β) are a family of polypeptide hormones which probably act over relatively confined physiological spaces *in vivo*. They are concentrated in platelets and released in an inactive form upon platelet degranulation (Pircher *et al.* 1986). Presumably they are activated in a local area by proteases or other specific activating conditions (Lyons *et al.* 1988). In addition, *in situ* hybridization and immunohistochemical studies have demonstrated that the factors are synthesized in many *in vivo* sites (Heine *et al.* 1987; Lehnert and Adhurst, 1988; Pelton *et al.* 1989). Our laboratory has attempted to identify and

*Present address: Laboratory of Medicine and Pathology, University of Minnesota, Minneapolis, Minnesota 55455, USA.

Key words: Transforming growth factor-beta (TGF-β), receptors, binding proteoglycans.

characterize those cell surface proteins with which TGFs-β can interact and which presumably mediate the biological signalling of the factor into the cell.

This is a somewhat complicated task for a number of reasons. As indicated above, TGF-β is a family of five closely related factors (TGF-β1 to TGF-β5) within a superfamily of several other factors implicated in essential elements of development. These other factors include the decapentaplegic gene product of *Drosophila*, the *Vg-1* gene product and mesoderm inducing factor in *Xenopus*, Mullerian Inhibiting Substance, the Bone Morphogenic Proteins, and the activins and inhibins (Massagué, 1990). This suggests that there may be a family of related receptors with differential affinities for different members of the superfamily. Another complicating element is that the known effects of TGF-β are varied and in some instances diametrically opposed in different cell systems. It is intriguing that TGF-β has a potent growth inhibitory effect on some epithelial cell lines, but it is also mitogenic in some culture systems, such as osteoblasts (Centrella *et al.* 1987) and some fibroblast lines (Leof *et al.* 1986). It is an inhibitor of differentiation in myogenic and adipogenic model systems *in vitro* (Massagué *et al.* 1986; Ignotz and Massagué, 1985), but stimulates differentiation of chondroblasts (Seyedin *et al.* 1985). And finally, experimentally, we have found that there is a relatively low level of specific cell-surface TGF-β binding and that this activity is distributed among three distinct cell surface binding proteins.

Distribution of specific TGF-β-binding proteins

In the course of investigating the activities of TGF-β over the last six years, our laboratory has characterized the TGF-β binding patterns of over one hundred different cell lines, primary cells and tissues (Fig. 1A) (Massagué, 1990). It is striking that the general pattern is so similar in most cells analyzed. Briefly, the experimental approach is to bind iodinated TGF-β to cells, crosslink the ligand to cell surface proteins to which it is associated with a bifunctional crosslinking reagent such as disuccinimidyl suberate, solubilize the cell membranes with detergent and separate the labeled proteins on SDS–polyacrylamide gels with subsequent autoradiography. This method commonly identifies three proteins

Fig. 1. Distribution of TGF-β receptors and binding proteoglycans in various cell types. A. Summary of TGF-β receptors in various cell types. All cell lines were screened for the presence of TGF-β receptors (I, II, III) using the affinity-labelling protocol outlined in the text. The presence of a receptor type in any cell line is signified with an open circle (○). The majority of cell lines screened express all three protein species, some lines express only the Type II and Type I receptors, a few lines express only the Type I receptor. No cell line which responds to TGF-β with established assays lacks the Type I receptor. Cell lines which lack any receptor type are signified with a closed circle (●) under the column relating to that receptor type. B. Receptor profiles from representative cell lines. Mouse BALB/c-3T3 cells, 3T3-L1 cells, rat NRK cells, chick embryo fibroblasts (CEF) and mink lung epithelial cells (Mv1Lu) were screened by affinity labelling with 50 pM ^{125}I-TGF-β in the presence of no (a), 50 (b), 100 (c), 200 (d), 700 (e), or 3000 (f) pM native TGF-β. These experiments show the affinity of these proteins for TGF-β and the general nature of the affinity labeled species (Cheifetz *et al.* 1986).

A

Human Normal	I	II	III
GM316 skin fibroblasts	○	○	○
GM370 skin fibroblasts	○	○	○
GM5877skin fibroblasts	○	○	○
GM1262 conjunctive cells	○	○	●
Primary lung fibroblast	○	○	○
Primary skin fibroblasts	○	○	○
WI-38 Embryo lung fibroblasts	○	○	○
IMR-90 Embryo lung fibroblasts	○	○	○
Retina, primary	○	○	○
E1MC3T3 Osteoblasts	○	○	○
MC3T3 Osteoblasts	○	○	○
HS14 Muscle	○	○	○

Human Transformed			
A431 Epithelioid carcinoma	○	○	○
T24 Bladder carcinoma	○	○	○
MDA-MB231 Breast carcinoma	○	○	○
MCF-7/2 Breast carcinoma	○	●	○
MCF-7/4 Breast carcinoma	○	●	○
MCF-7/5 Breast carcinoma	○	●	○
MCF-7/1 Breast carcinoma	○	●	●
SCC25 Squamous cell carcinoma	○	●	○
A875 Melanoma	○	○	○
A549 Lung adenocarcinoma	○	○	○
B16F1 Melanoma	○	○	○
B6F10 Melanoma	○	○	○
HS 224 Rhabdomyosarcoma	○	○	○
MG-63 Osteosarcoma	○	○	○
OS-1P Osteosarcoma	○	○	○
7922 Osteosarcoma	○	○	○
A1684 Chondrosarcoma	○	○	○
HT-1080 Fibrosarcoma	○	○	○
MNNG-HOS Osteogenic sarcoma	○	○	○
VA-13 fibroblast SV-40	○	○	○
Weri Retinoblastoma	●	●	○
RB7 Retinoblastoma	●	●	●
Y79 Retinoblastoma	●	●	●
RB27 Retinoblastoma	●	●	●
RB13 Retinoblastoma	●	●	●
RB20 Retinoblastoma	●	●	●
RB22 Retinoblastoma	●	●	●
RB24 Retinoblastoma	●	●	●
HL-60 Monomyelocytic leuk. (phage)	○	●	●
U937 Monocytic leukemia	○	●	○
IMR32 Neuroblastoma	○	●	○
SK-N-SH Neuroblastoma	○	○	○
HS4 Schwanoma	○	○	○
HS0929 Wilm's tumor	○	○	○
AD-12 Retina, adenovirus	○	○	○
GM01972A Progeria	○	○	○
TE671 Medulloblastoma	○	○	○
ADL-D Glioblastoma	○	○	○

Rat Normal			
FR Skin fibroblasts	○	○	○
NRK Kidney fibroblasts	○	○	○
H9 Heart myoblast	○	○	○
L8 Skeletal muscle myoblasts	○	○	●
A7r5 Aorta smooth muscle myoblast	○	○	○
BRL-3A Newborn rat liver	○	○	○
Primary calvaria osteoblasts	○	○	○
Primary hepatocytes	○	○	○
Primary cardiocytes	○	○	○
Primary pituitary	○	○	○
Fetal tissue	○	○	○
Placenta	○	○	○

Rat Transformed			
FeSV-Fisher rat embryo fibroblast	○	○	○
tsKiMSV-NRK Kidney epithelial	○	○	○
C6 Glioma	○	○	○
H35 Hepatoma	○	○	●
PC12 Pheochromocytoma	●	●	●
L6 Skeletal muscle myoblast	○	○	●
L6E9 Skeletal Muscle Myoblast	○	○	●
L6E9 myotubes	○	○	●

Mouse Normal			
Swiss 3T3 embryo fibroblasts	○	○	○
Swiss 3T3/NR-6 embryo fibroblasts	○	○	○
Balb/c 3T3 embryo fibroblasts	○	○	○
3T3-L1 preadipocyte	○	○	○
3T3-L1 adipocyte	○	○	○
MK Keratinocyte	○	○	○
AKR-2B Fibroblast	○	○	○
B6SUtA Hematopoietic progenitors	○	●	●
B6SUt-CL27 hematopoietic progenitors	○	●	●
32D-C13 Hematopoietic progenitors	○	●	●

Mouse Transformed			
3T3 MSV-transformed fibroblasts	○	○	○
N2A Neuroblastoma	●	●	○
EL Strain E, Ascites carcinoma	○	○	○
C1271 Mammary tumor	○	○	○
MB66-MCA 10T1/2 Fibroblast	○	○	○

Avian			
Chick embryo fibroblasts	○	○	○

Monkey			
BSC-1 Kidney epithelial	○	○	○
COS Kidney fibroblast-like	○	○	○

Mink			
Mv1Lu (CCL64) Lung epithelial	○	○	○

Bovine			
FBHE Fetal heart endothelial	○	○	●

Dog			
MDCK kidney epithelial	○	○	●

Hamster			
DDT1 Myosarcoma	○	○	○
CCL-39 Lung Fibroblast	○	○	○
BHK kidney	○	●	●
CHO-K1 Ovary	○	○	○
CHO pgsA-745	○	○	○
CHO ldlD	○	○	○

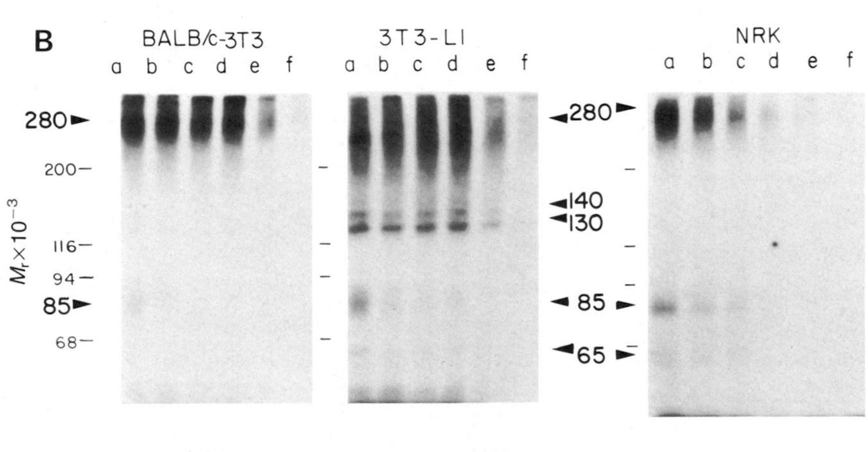

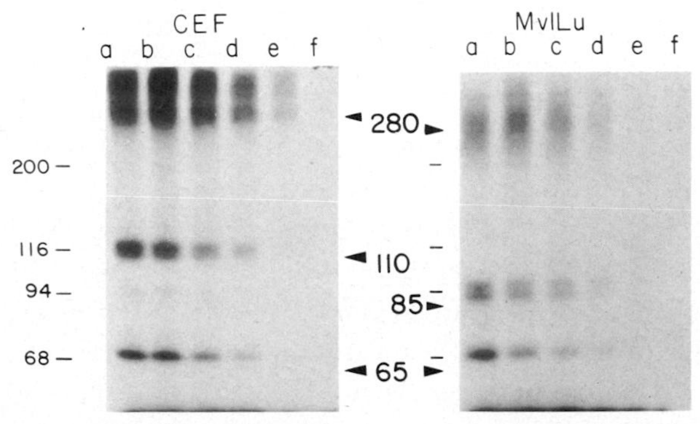

labeled specifically (Fig. 1B). The Type I protein is an affinity-labeled species of approximately 65×10^3 M_r. The Type II species is approximately 85×10^3 M_r, and the Type III species, termed betaglycan, is a broad band typically centered around 280×10^3 M_r. All of these apparent molecular weights include an associated monomer of TGF-β of 12.5×10^3 M_r, so the presumed size of the binding proteins is correspondingly 12.5×10^3 M_r smaller than the apparent molecular weight on SDS gels. Each of these proteins have high affinities for TGFs-β, with K_d values in the range of 5–500 pM. They bind TGF-β1, TGF-β2 and TGF-β3, but not more distantly related members of the TGF-β superfamily. The Type I binding protein is ubiquitous, with every cell type that responds to TGFs-β having the Type I protein. There are several examples of hematopoietic progenitor cell lines which respond to TGF-β1, TGF-β1.2 and TGF-β2 differentially with an order of potencies that parallels the order of affinities of the factors for the Type I protein, the only TGF-β binding protein detectable on these cell lines (Ohta *et al.* 1987; Cheifetz *et al.* 1988). Human and bovine vascular endothelial cells possess both the Type I and Type II proteins and respond differentially to TGF-β1 or TGF-β2. In addition, L6E9 myoblasts also have only the Type I and Type II proteins, yet respond equivalently to TGF-β1 and TGF-β2. The most common pattern of TGF-β cell surface binding proteins is the presence of the Type I protein, Type II protein and betaglycan together. These data are suggestive, though certainly not definitive, that the Type I protein is sufficient to confer TGF-β responsiveness to cells. However, the other TGF-β binding proteins may modulate the availability or activities of the TGFs-β or mediate particular responses in particular cell types.

Somatic cell mutants defective in TGF-β responsiveness

More direct evidence supporting specific roles of these proteins in TGF-β signalling has come from studies of somatic cell mutants in the Mv1Lu epithelial cell line which are non-responsive to TGFs-β. The parental line is exquisitely sensitive to TGF-β and is virtually completely growth inhibited by 5 pM TGF-β1; it also responds to TGF-β with increased expression of extracellular matrix components such as fibronectin and plasminogen activator inhibitor (PAI-1). In addition, the parental line has all three types of putative TGF-β receptor proteins. Ethyl methane sulfonate (EMS)-mutagenized Mv1Lu cells were selected which were able to grow in the presence of 100 pM TGF-β1. The mutant clones were completely resistant to the growth inhibitory effects of TGF-β1 and TGF-β2 and have also lost all other responses to TGFs-β assayed for. In addition, several of the clones were defective in expression of the Type I binding protein (Fig. 2). Clones of Mv1Lu cells obtained from a non-mutagenized population and analyzed for TGF-β binding proteins always have all three binding proteins present. On this basis, we suggest that the Type I protein is the receptor which mediates epithelial cell responsiveness to TGF-β (Boyd and Massagué, 1989). In addition to these receptor defective mutants, termed R mutants, mutants were isolated which have normal binding protein patterns, yet are deficient in all TGF-β responses. These are

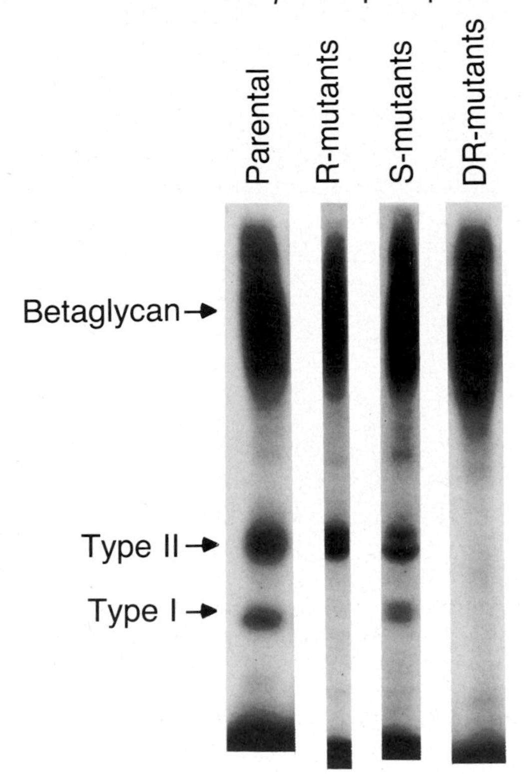

Fig. 2. TGF-β receptor profiles of TGF-β resistant mutants of mink lung epithelial cells. Mutants of Mv1Lu cells which are not responsive to TGF-β with respect to growth inhibition were generated as described in the text. Representative clones of several mutant phenotypes were affinity labeled with 100 pM ^{125}I-TGF-β. Three distinct types of mutants have been isolated as characterized by TGF-β binding profiles. When compared with the parental Mv1Lu cells, it is apparent that two of the mutant types have distinct deficits in TGF-β binding. The R-mutants are lacking the Type I receptor. In addition, the DR-mutants are lacking both the Type I and the Type II receptors. In contrast, the S-mutants have an apparently normal receptor profile.

called signalling, or S, mutants. When complementation analysis was performed, none of the mutant hybrids were complementary, suggesting that all the R and S mutants isolated are mutants in the same gene, presumably the TGF-β receptor gene. It is also apparent from this analysis that all the mutants isolated were recessive mutations. All mutant–parental fusions were fully responsive to TGF-β with normal receptor profiles.

These studies have been extended and selection of Mv1Lu mutants has been performed with lower doses (25 pM) of TGF-β2. In addition to the R and S mutant classes, a third class of mutants defective in the expression of both the Type I and Type II proteins has been isolated (DR mutants) (M. Laiho and others, unpublished observations). Complementation analysis of these mutants is not complete, so the genetic basis of these mutations is still unknown. Several explanations for this phenotype are possible. DR mutants may be the result of mutations of two distinct loci encoding for the Type I and Type II proteins, although the frequency of isolation of these mutants argues strongly against this. Another possibility is that the two genes are linked and a single large deletion can knock out expression of both genes. A third possibility is that the two proteins are associated intracellularly and some mutations in one or the other of the proteins prevent the expression of both. These possibilities are the subject of active investigation in the laboratory. We are also attempting to complement these mutants by cDNA and genomic transfection to clone the genes responsible for these mutations.

Betaglycan

While mutant analysis has revealed the functional significance of the Type I and Type II proteins, the fact remains that in the majority of cells we have screened, the major component of cell surface TGF-β binding activity is associated with a large proteoglycan species with apparent molecular weight of $200-400\times10^3\,M_r$. This molecule is a complex mixed chondroitin/heparan sulfate proteoglycan (Segarini and Seyedin, 1988; Cheifetz *et al.* 1988) with multiple deglycosylated core proteins of $100-120\times10^3\,M_r$. Unlike other growth factors which associate with proteoglycans *via* relatively non-specific binding to the glycosaminoglycan chains, TGFs-β bind to betaglycan *via* the core proteins. The core proteins are expressed and bind TGFs-β in metabolic mutants which do not synthesize glycosaminoglycan side chains (Cheifetz and Massagué, 1989). There is also a soluble form of betaglycan, found in the media of tissue culture cells which express betaglycan, which is capable of binding TGFs-β (Andres *et al.* 1989). This form is slightly smaller than the membrane-associated form and is incapable of being incorporated into phospholipid vesicles. The soluble form of betaglycan also associates with the extracellular matrix. It is intriguing that TGF-β, which is a well-characterized modulator of the extracellular matrix, binds specifically to a TGF-β binding proteoglycan which associates with the extracellular matrix and is the major species of TGF-β binding activity associated with cells. This may be a mechanism which can be modulated by TGFs-β, by which TGFs-β are sequestered in the intercellular space.

A model of TGF-β binding protein function

The definitive identification of the TGF-β receptor awaits cloning of the gene and reconstitution of a TGF-β-responsive phenotype to TGF-β-receptor mutants. However, the features of TGF-β-binding outlined above suggests the following model. The identification of cell lines which possess only the Type I protein and the common defect in Type I binding activity in TGF-β-response mutants suggests that the Type I protein is the signalling receptor for TGF-β. The identification of

TGF-β-response mutants defective in both the Type I and Type II proteins suggests that the Type II protein may be part of a higher order complex with the Type I protein that associates and is a functional entity in some cell types. The dimeric structures of TGFs-β are reminiscent of the structure of other dimeric growth factors which bind to dimeric receptors. Finally, betaglycan may be an extracellular storage site or mechanism of inactivation of the ligand. The physiology of the TGFs-β suggests that specificity of action may come about as a result of acute regulation of availability of the active ligand. The fact that betaglycan is a proteoglycan and is present in membrane associated form as well as soluble and matrix associated forms suggests the possibility of acute regulation of the molecule, which might make it well suited to a role in clearance or storage of the ligands.

References

ANDRES, J. L., STANLEY, K., CHEIFETZ, S. AND MASSAGUÉ, J. (1989). Membrane-anchored and soluble forms of betaglycan, a polymorphic proteoglycan that binds transforming growth factor. *J. Cell Biol.* **109**, 3137–3145.

BOYD, F. T. AND MASSAGUÉ, J. (1989). Transforming growth factor-beta-inhibition of epithelial cell proliferation linked to the expression of a 53 kD membrane receptor. *J. biol. Chem.* **264**, 2272–78.

CENTRELLA, M., MCCARTHY, T. L. AND CANALIS, E. (1987). Transforming growth factor-beta is a bifunctional regulator of replication and collagen synthesis in osteoblast-enriched cell cultures from fetal rat bone. *J. biol. Chem.* **262**, 2869–2874.

CHEIFETZ, S. AND MASSAGUÉ, J. (1989). The TGF-β-receptor proteoglycan. Cell surface expression and ligand binding in the absence of glycosaminoglycan chains. *J. biol. Chem.* **264**, 12 025–12 028.

CHEIFETZ, S., ANDRES, J. L. AND MASSAGUÉ, J. (1988). The transforming growth factor-β-receptor type III is a membrane proteoglycan. Domain structure of the receptor. *J. biol. Chem.* **263**, 16 984–16 991.

HEINE, U. I., MUNOZ, E. F., FLANDERS, K. C., ELLINGSWORTH, L. R., LAM, H. Y. P., THOMPSON, N. L., ROBERTS, A. B. AND SPORN, M. B. (1987). Role of transforming growth factor-β in the development of the mouse embryo. *J. Cell Biol.* **105**, 2861–2867.

IGNOTZ, R. A. AND MASSAGUÉ, J. (1985). Type-β transforming growth factor controls the adipogenic differentiation of 3T3 fibroblasts. *Proc. natn. Acad. Sci. U.S.A.* **82**, 8530–8534.

LEHNERT, S. A. AND ADHURST, R. J. (1988). Embryonic expression pattern of TGF-β-type 1 RNA suggests both paracrine and autocrine mechanisms of action. *Development* **104**, 263–273.

LEOF, E. B., PROPER, J. A., GOUSTIN, A. S., SHIPLEY, G. D., DICORLETO, P. E. AND MOSES, H. L. (1986). Induction of c-sis mRNA and activity similar to platelet-derived growth factor-β by transforming growth factor-β: a proposed model for indirect mitogenesis involving autocrine activity. *Proc. natn. Acad. Sci. U.S.A.* **83**, 2453–2457.

LYONS, R. M., KESKI-OJA, J. AND MOSES, H. L. (1988). Proteolytic activation of latent transforming growth factor-β from fibroblast-conditioned medium. *J. Cell. Biol.* **106**, 1659–1665.

MASSAGUÉ, J. (1990). The transforming growth factor-β-family. *A. Rev. Cell Biol.* **6**, In press.

MASSAGUÉ, J., CHEIFETZ, S., ENDO, T. AND NADAL-GINARD, B. (1986). Type-β transforming growth factor is an inhibitor of myogenic differentiation. *Proc. natn. Acad. Sci. U.S.A.* **83**, 8206–8210.

MILLER, D. A., LEE, A., MATSUI, Y., CHEN, E. Y., MOSES, H. L. AND DERYNCK, R. (1989). Complementary DNA cloning of the murine transforming growth factor-β3 (TGF-β3) precursor and the comparative expression of the TGF-β3 and TGF-β1 messenger RNA in murine embryos and adult tissues. *Molec. Endocrinol.* **3**, 1926–1934.

OHTA, M., GREENBERGER, J. S., ANKLESARIA, P., BASSOLS, A. AND MASSAGUÉ, J. (1987). Two forms of transforming growth factor-β distinguished by multipotential haematopoietic progenitor cells. *Nature* **329**, 539–541.

Pelton, R. W., Nomura, S., Moses, H. L. and Hogan, B. L. M. (1989). Expression of transforming growth factor-β2 RNA during murine embryogenesis. *Development* **106**, 759–768.

Pircher, R., Julien, P. and Lawrence, D. A. (1986). Transforming growth factor-β is stored in human blood platelets as a latent high molecular weight complex. *Biochem. biophys. Res. Comm.* **136**, 30–37.

Segarini, P. R. and Seyedin, S. M. (1988). The high molecular weight receptor to transforming growth factor-β contains glycosaminoglycan chains. *J. biol. Chem.* **263**, 8366–8370.

Seyedin, S. M., Thomas, T. C., Thompson, A. Y., Rosen, D. M. and Piez, K. A. (1985). Purification and characterization of two cartilage-inducing factors from bovine demineralized bone. *Proc. natn. Acad. Sci. U.S.A.* **82**, 2267–2271.

Sporn, M. B., Roberts, A. B., Wakefield, L. M. and de Crombrugge, B. (1987). Some recent advances in the chemistry and biology of transforming growth factor-β. *J. Cell Biol.* **105**, 1039–1045.

J. Cell Sci. Suppl. 13, 139–148 (1990)
Printed in Great Britain © The Company of Biologists Limited 1990

Regulation of transforming growth factor-β subtypes by members of the steroid hormone superfamily

LALAGE WAKEFIELD[1,*], SEONG-JIN KIM[1], ADAM GLICK[1], THOMAS WINOKUR[1], ANTHONY COLLETTA[2] AND MICHAEL SPORN[1]

[1]*Laboratory of Chemoprevention, National Cancer Institute, Bethesda, Maryland 20892, USA*
[2]*Department of Clinical Biochemistry, Addenbrookes Hospital, Cambridge CB2 2QR, UK*

Summary

Transforming growth factor-βs (TGF-βs) are potent regulators of cell growth and differentiation. Expression of the closely related TGF-β subtypes *in vivo* is differentially regulated both temporally and spatially. Members of the steroid hormone superfamily may play an important role in this gene- and tissue-specific regulation. We have shown that anti-estrogens induce the production of TGF-β1 in mammary carcinoma cells and fetal fibroblasts, whereas retinoic acid specifically induces TGF-β2 in primary epidermal keratinocytes. The induction of TGF-β2 by retinoids is accompanied by an increase in TGF-β2 mRNAs, but little change in transcription rates, suggesting an effect of retinoids on message stability or processing. In contrast, TGF-β1 mRNA levels are unchanged by anti-estrogen treatment, suggesting these compounds may regulate the translatability of the TGF-β1 message or some post-translational processing event. We have identified a stable stem–loop structure in the 5′ untranslated region (UTR) of the TGF-β1 mRNA that inhibits translation of a heterologous reporter gene, and we are investigating the possibility that anti-estrogens may regulate the activity of this element, and hence the translatability of the TGF-β1 message. A significant fraction (25–90 %) of the TGF-β induced by retinoids and anti-estrogens is in the biologically active rather than the latent form. We have shown that active TGF-β has a much shorter *in vivo* half-life than latent TGF-β, suggesting that the TGF-β induced by retinoids and steroids may act locally at the site of production. Since many tumor cells retain sensitivity to the growth inhibitory effects of active TGF-β, the use of members of the steroid hormone superfamily for inducing this potent growth inhibitor locally at the tumor site may have therapeutic potential.

Introduction

Members of the transforming growth factor-β superfamily have emerged as key regulators of many aspects of cell growth, differentiation and function (for review, see Roberts and Sporn, 1990). Five structurally highly related TGF-β subtypes have been identified to date, and multiple subtypes appear to be expressed in all species examined. Thus TGF-βs 1, 2 and 3 are expressed in human; 1, 2, 3 and 4 in chick; and 2 and 5 in frog. *In situ* and immunohistochemical studies suggest that these isoforms are differentially regulated, both temporally and spatially (Heine *et al.* 1987; Pelton *et al.* 1989; Miller *et al.* 1989). While the different TGF-β

* Author for correspondence.

Key words: transforming growth factor-β, steroid hormones, translational control, growth regulation, chemoprevention.

subtypes are equipotent in the majority of biological assays *in vitro* (reviewed by Roberts and Sporn, 1990), there is some indication that in more complex systems, involving interactions between different cell types, the TGF-β subtypes differ in their activities. For example, only TGF-β2 and TGF-β3 are active in inducing mesoderm formation in *Xenopus* embryos (Roberts *et al.* 1990). The significance of multiple TGF-β subtypes may therefore be twofold: (1) different regulatory elements controlling expression of the various TGF-β subtypes may allow differential regulation of the same biological activity in an organ- and time-specific manner, and (2) subtype switching in a given tissue may result in expression of a different biological activity in the context of that tissue. The recent cloning and analysis of the promoters for the TGF-β1, β2 and β3 genes indicate that these are very different in structure, suggesting one potential mechanism for the differential regulation of TGF-β subtypes (Kim *et al.* 1989; Noma *et al.*; Lafyatis *et al.* unpublished).

We have examined the possible role of members of the steroid hormone superfamily in the regulation of TGF-β subtypes. The goal here has been (1) to identify endogenous regulatory molecules responsible for the selective expression of different TGF-β subtypes observed *in vivo*, and (2) to identify analogs or antagonists of these agents which may allow us to manipulate the endogenous levels of various TGF-β subtypes *in vivo*, in a target-specific manner.

Results and discussion

Several members of the steroid hormone superfamily have been shown to regulate the production of TGF-β subtypes

The data in Table 1 summarize work from our laboratories and others showing an induction of TGF-β family members by steroids and related compounds in a

Table 1. *Induction of TGF-β subtypes by members of the steroid hormone superfamily*

Agent	Target cells	TGF-β subtype	Reference
Estrogen	Osteosarcoma cells	1*	Komm *et al.* (1988)
Anti-estrogens	Breast carcinoma cells (ER+)	1 and 2	Knabbe *et al.* (1987) Colletta *et al.* (1990*a*)
Anti-estrogens	Fetal fibroblasts (ER−)	1	Colletta *et al.* (1990*b*)
Tamoxifen (anti-estrogen)	Prostatic adenocarcinoma	2	Ikeda *et al.* (1987)
Testosterone	Ventral prostate	1*	Kyprianou and Isaacs (1989)
Gestodene (synthetic progestin)	Breast carcinoma cells	1	Colletta *et al.* (1990*a*)
1,25-dihydroxy-vitamin D_3	Calvarial cells	n.d.	Pfeilschifter and Mundy (1987)
Retinoic acid	Keratinocytes	2	Glick *et al.* (1989)

ER, estrogen receptor; n.d., not done.
*Demonstrated only at mRNA level.

variety of different target cell types. In addition to these examples of positive regulation, it should be noted that negative regulatory effects have been observed in other systems. For example, estrogen appears to depress TGF-β2 and TGF-β3 mRNAs in human breast carcinoma cells (Arrick, Korc and Derynck, 1990), and to cause a decrease in TGF-β protein (subtype not identified) in the same cells (Knabbe *et al.* 1987), and progestins decrease the level of TGF-β1 mRNA in the T47D human breast carcinoma line (Murphy and Dotzlaw, 1989). It is apparent that members of the steroid hormone superfamily can regulate the production of distinct TGF-β subtypes *in vitro*, in a target-specific manner. The inhibitory effects of the anti-estrogens and gestodene on breast carcinoma cells, and of retinoids on keratinocytes, are partially reversed by neutralizing antibodies to TGF-β (Knabbe *et al.* 1987; Colletta *et al.* 1990*a*,*b*; Glick *et al.* 1989). This suggests that the induction of TGF-β subtypes has functional significance in the mechanism of action of steroids and retinoids.

The regulation of TGF-β family members by steroids/retinoids is not simply an *in vitro* curiosity, since the induction of TGF-β2 in keratinocytes by retinoids has now been confirmed *in vivo* (Glick *et al.* 1989). Retinoic acid was applied to the shaved skin of adult Balb/c mice, and treated and control skins were processed for immunohistochemistry 48 h later. Using subtype-specific anti-TGF-β antibodies, a dramatic increase in immunoreactive TGF-β2 was demonstrated in the follicular and interfollicular epithelium of retinoid-treated skins. Little staining for TGF-β1 was observed in either treated or untreated skins. Thus retinoic acid can also function as subtype-specific inducer of TGF-β in the more complex *in vivo* situation. Retinoids have been implicated in the control of proliferation and normal differentiation in a variety of epithelia (Sporn and Roberts, 1984). Since TGF-βs are strongly growth inhibitory for many epithelial cells (reviewed by Roberts and Sporn, 1990), inducing terminal differentiation in some instances, this indicates that the observed actions of retinoids *in vivo* may be due in part to their ability to modulate local levels of TGF-β2 in target epithelia.

Regulation of TGF-β subtypes by retinoids and steroids is frequently at the post-transcriptional level

Retinoic acid treatment of mouse keratinocytes results in a greater than 20-fold increase in the steady-state levels of the four major transcripts coding for TGF-β2 (Glick *et al.* 1989). However, nuclear run-on experiments indicate that there is no increase in the rate of transcription of the TGF-β2 mRNAs on retinoic acid treatment, so the observed increase in transcript levels must be due to post-transcriptional effects (Glick *et al.* 1989). Possible mechanisms include effects of retinoids on message processing, transport or stability. Post-transcriptional effects of retinoids on other mRNAs have been observed, indicating this may be a major pathway by which retinoids control gene expression (Smits *et al.* 1987).

In contrast, the induction of TGF-β1 in breast carcinoma cells in response to anti-estrogens or gestodene, and in fibroblasts in response to anti-estrogens, is accompanied by little or no change in TGF-β1 mRNA levels (Colletta *et al.* 1990*a*; Colletta *et al.* 1990*b*). For example, treatment with 500 nM gestodene causes a greater than 90-fold induction of TGF-β1 protein, with less than a 3-fold increase in corresponding mRNA levels (Colletta *et al.* 1990*a*). This suggests that the anti-estrogens and gestodene may be affecting the efficiency of translation of the TGF-β1 mRNA, or enhancing some subsequent post-translational step, such as post-translational processing or secretion. Comparably large effects on TGF-β1 protein levels without effects on the mRNA are observed in other systems. For example, activation of lymphocytes with phytohaemagglutinin is accompanied by a large increase in TGF-β mRNA within 6 h, but an increase in secreted TGF-β protein is not observed for a further 2 days (Kehrl *et al.* 1986). Similarly, activation of monocytes with lipopolysaccharide results in increased TGF-β secretion with no change in mRNA levels (Assoian *et al.* 1987). Important translational or post-translational regulatory effects on the TGF-β1 message are therefore observed in a variety of systems. Experimentally, the presence of TGF-β1 mRNA should not be assumed necessarily to indicate the presence of the cognate protein.

In common with a number of cytokines and oncogenes, the TGF-β1 mRNA has a long (>800 bp) 5′ untranslated region (UTR) (Kim *et al.* 1989). Computer-aided analysis of possible secondary structure formation in the TGF-β1 mRNA indicates the potential presence of two highly stable ($\Delta G \approx -16.84$ kJ) (1 kJ=0.239 kcal) stem–loop structures in the 5′ UTR, shown diagrammatically in Fig. 1 (S.-J. Kim, unpublished). The ferritin mRNA has a very similar stem–loop structure in the 5′ UTR, and this stem–loop has been shown to be the site of binding of a regulatory protein that acts as a repressor of ferritin mRNA translation under conditions of low availability of iron (Klausner and Harford, 1989). By analogy, we propose that the putative stem–loops in the 5′ UTR of the TGF-β1 mRNA may be involved in regulating the translatability of the TGF-β1 message in a similar fashion. Preliminary experiments in which cDNA fragments of the 5′ UTR were inserted upstream from the coding sequence of a human growth hormone reporter gene, indicate that the downstream stem–loop structure has an inhibitory effect on the translation of the reporter gene, in certain cell backgrounds (S.-J. Kim, unpublished). Further work is underway to define more closely the structures and specific sequence elements that are responsible for translational regulation of this message.

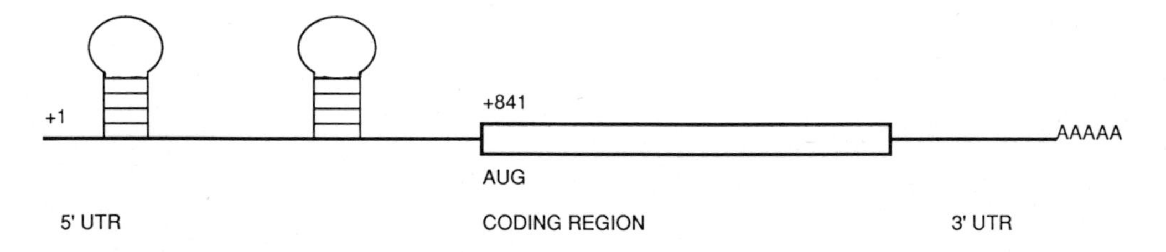

Fig. 1. Diagrammatic representation of the TGF-β1 mRNA, showing location of potential stem–loop structures in the 5′ untranslated region.

A significant fraction of the TGF-β induced by steroids or retinoids is in the biologically active form

The TGF-βs differ from the majority of growth regulatory factors in that they are generally synthesized and secreted in a biologically latent form, which must be activated before TGF-β can exert its biological effects on target cells (Lawrence *et al.* 1984; Wakefield *et al.* 1987). For TGF-β1, the latent form is a non-covalent complex in which the homodimeric active TGF-β1 is non-covalently associated with a dimer of the remainder of its precursor 'pro' region, and this in turn is disulfide-bonded to a third, structurally unrelated, protein of $135\,M_r \times 10^{-3}$ (Wakefield *et al.* 1988; Miyazono *et al.* 1988). The nature of the activation mechanism *in vivo* is unclear, but may involve proteases, and in some instances be dependent on cell–cell interactions (Antonelli-Orlidge *et al.* 1989; Sato and Rifkin, 1989).

For most cells in culture, the TGF-β secreted is more than 95% latent (Wakefield *et al.* 1987). However, for cells induced to secrete TGF-β in response to retinoids and steroids, a significant fraction of the secreted TGF-β is in the biologically active form (see Table 2). The active fraction ranges from 25% to nearly 100% of the total TGF-β secreted. Using latent TGF-β1 generated by combining iodinated active TGF-β1 with the TGF-β1 precursor pro region from recombinant sources, we have demonstrated that the latent form of TGF-β1 has a greatly extended plasma half-life *in vivo* when compared with active TGF-β1 (see Table 3) (Wakefield *et al.* 1990). This is consistent with data from our laboratory and others suggesting that active TGF-β1 may be rapidly cleared from the extracellular fluid and plasma as a complex with alpha-2-macroglobulin (O'Connor-McCourt and Wakefield, 1987; Coffey *et al.* 1987). The latent form of TGF-β does not bind to alpha-2-macroglobulin and cannot be cleared by this route (Wakefield *et al.* 1988). We have therefore proposed that whereas latent TGF-β may be able to exert a long-range, endocrine type of action, active TGF-β probably acts very locally to its site of production, in an autocrine/paracrine fashion (see Fig. 2).

Table 2. *The fraction of TGF-β that is induced in the active, as opposed to biologically latent, form in response to members of the steroid hormone superfamily*

Agent	Target cells	% TGF-β in active form
Retinoic acid	Keratinocytes	~25
Anti-estrogens	Breast carcinoma cells (ER+)	>25
Anti-estrogens	Fetal fibroblasts (ER−)	>70
Gestodene (synthetic progestin)	Breast carcinoma cells	~100

ER, estrogen receptor.

The observation that steroids and retinoids frequently induce TGF-β in its active form is important for two reasons. First, as indicated above, it means that the TGF-β induced by these agents is likely to have a very local action. Second, any cell in the vicinity that possesses TGF-β receptors is a potential target for the

Table 3. *Comparison of the plasma half-lives of active and latent forms of TGF-β1*

Form of TGF-β1	Plasma half-life (min)
Active TGF-β1	2.7±0.4
Latent TGF-β1	108.6±8.2

Latent TGF-β1 was formed by combining ^{125}I-TGF-β1 with the dimeric TGF-β1 precursor pro region, corresponding to residues 30–278 of the TGF-β1 precursor, purified from recombinant sources. The labelled material was injected into the femoral vein of an anesthetized, heparinized rat and plasma samples were drawn at timed intervals from a cannula in the contralateral iliac artery for half-life determinations.

Values represent the mean±standard deviation of 3 determinations.

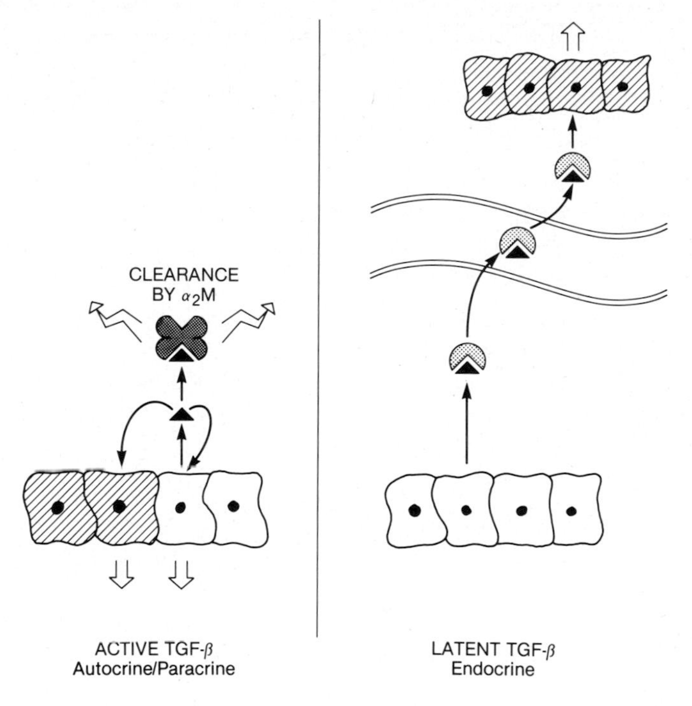

Fig. 2. Active and latent TGF-β1 may have a different range of action. Active TGF-β1 has a very short plasma half-life. Active TGF-β1 secreted by cells is likely to bind rapidly to the ubiquitous cell surface binding proteins for TGF-β1, or to be cleared from the vicinity, possibly as a complex with alpha-2-macroglobulin (α_2M). This is likely to restrict the site of action of active TGF-β to target cells close to the site of production. By contrast, latent TGF-β does not bind to alpha-2-macroglobulin, and has a much longer plasma half-life than active TGF-β. It may therefore be carried by the circulation to more distant targets, and exert a more long-range endocrine type of action.

induced TGF-β. This contrasts with the situations where latent TGF-β is induced, since cells that are capable of activating latent TGF-β have yet to be identified.

Implications for chemoprevention or therapy of epithelial malignancies

TGF-βs are potent inhibitors of the growth of normal epithelial cells (for review, see Roberts and Sporn, 1990). One might expect therefore that loss of response to this endogenous growth inhibitor could contribute to the genesis of epithelial tumors. Indeed, there is experimental evidence in the rat tracheal epithelial system suggesting that neoplastic progression following carcinogen treatment is accompanied by increased resistance to growth inhibition by TGF-β (Hubbs, Hahn and Thomassen, 1989). Similarly, less aggressive, well-differentiated human colon carcinoma cells are growth-inhibited by TGF-β, whereas more agressive, poorly differentiated colon tumor cells are not (Hoosein *et al.* 1989). However, a recent survey of the literature indicated that of the 37 human carcinoma cell lines analysed for a response to TGF-β *in vitro*, more than half (20) of these retained the ability to be inhibited by TGF-β (Wakefield and Sporn, 1990). This suggests that there may be a window of opportunity during neoplastic progression, when raising local TGF-β concentrations in the vicinity of the developing tumor might restore a measure of growth control and slow down the progression of the affected cells to full-blown malignancy. Since resistance to TGF-β appears to increase with malignant progression, and because tumors become more heterogeneous as they progress, it would obviously be most effective to intervene as early as possible, preferably during the preneoplastic phase, to prevent tumor development.

The pleiotropic effects of TGF-βs on multiple cell types suggests that systemic elevation of TGF-β levels in a clinical situation should probably be avoided. Ideally, therefore, one would like to develop pharmacological agents that could cause a highly localized induction of TGF-β in specific target tissues. Analogs and antagonists of members of the steroid hormone superfamily seem particularly promising in this regard; only cells expressing the appropriate ligand receptor will be targets for the steroid action. Furthermore, as noted above, the TGF-β secreted in response to these agents is largely in the biologically active form, which means that its action will be local, and that even tumor cells that lack or have lost the ability to activate the latent form will be growth-inhibited.

From the point of view of achieving very specific tissue targeting, one molecule of potential interest is the synthetic progestin, gestodene, which is a component of an oral contraceptive preparation widely used in Europe. Although gestodene binds to the classical progestin receptor, an additional novel binding site for this agent has been demonstrated in malignant breast tissue and cells lines (Iqbal *et al.* 1986; Colletta *et al.* 1989). This binding site does not appear to be present in normal cells or cells of other malignancies (Iqbal *et al.* 1986). We have shown that, acting through the novel gestodene binding site, gestodene will induce TGF-β1 in breast cancer cell lines and inhibit their growth (Colletta *et al.* 1990*a*). Other progestins that only bind to the classical progesterone receptor have no effect.

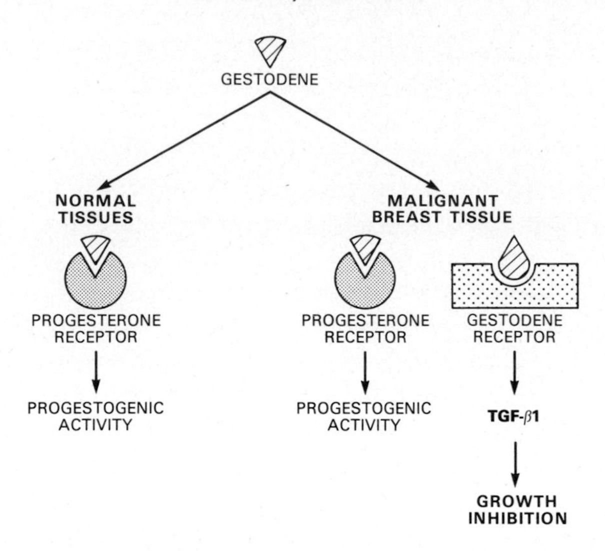

Fig. 3. Possible roles of the synthetic progestin, gestodene, in normal and malignant tissues. In normal tissues, gestodene binds to the classical progesterone receptor and exerts a progestagenic effect. In malignant breast cells, which express a novel binding site for gestodene, gestodene induces the production of TGF-β1, which may then inhibit the growth of the malignant cell in an autocrine fashion. For more details, see text.

Here, therefore, we may have a prototype of a novel form of chemopreventive agent. In the normal breast, and other tissues, gestodene would bind to the progestin receptor and exert its normal progestogenic activity. However, in malignant breast tissue that expresses the unique gestodene binding site, gestodene may bind to this site and induce TGF-β production, thereby slowing growth of the malignant cells (Fig. 3). It is not yet known at what stage in malignant progression the gestodene binding site is first expressed, but obviously the earlier this occurs, the greater the chance of effective prevention. While it remains to be shown that gestodene will actually prevent development of breast cancers *in vivo*, the work nevertheless indicates that agents may be found that can cause induction of TGF-β subtypes in very restricted target tissues, and suggests that the use of pharmacological compounds to induce local production of endogenous growth inhibitors in malignant or premalignant cells represents an important new approach to the problem of prevention and treatment of epithelial malignancies.

Conclusions

Members of the steroid hormone superfamily have been shown to regulate the production of TGF-β subtypes in a variety of target cell types, and experiments with neutralizing antibodies indicate that the TGF-βs may be local mediators of some of the biological activities of steroids/retinoids. Regulation of TGF-β subtypes by these agents appears to involve predominantly post-transcriptional mechanisms, and much of the TGF-β secreted is in the biologically active, not the more common latent, form. Since active TGF-β is much more rapidly cleared from the circulation than the latent form, this suggests that the TGF-β induced by steroids/retinoids may have a very local action. Since TGF-βs are highly potent inhibitors of epithelial cell growth, the targetted local induction of this peptide family could be exploited in the development of novel chemopreventive or chemotherapeutic strategies for the management of epithelial malignancies.

References

Antonelli-Orlidge, A., Saunders, K. B., Smith, S. R. and D'Amore, P. A. (1989). An activated form of transforming growth factor β is produced by cocultures of endothelial cells and pericytes. *Proc. natn. Acad. Sci. U.S.A.* **86**, 4544–4548.

Arrick, B. A., Korc, M. and Derynck, R. (1990). Differential regulation of expression of three transforming growth factor-β species in human breast cancer cell lines by estradiol. *Cancer Res.* **50**, 299–303.

Assoian, R. K., Fleurdelys, B. E., Stevenson, H. C., Miller, P. J., Madtes, D. K., Raines, E. W., Ross, R. and Sporn, M. B. (1987). Expression and secretion of type β transforming growth factor by activated human macrophages. *Proc. natn. Acad. Sci. U.S.A.* **84**, 6020–6024.

Coffey, R. J. Jr, Kost, L. J., Lyons, R. M., Moses, H. L. and Larusso, N. F. (1987). Hepatic processing of transforming growth factor β in the rat. *J. clin. Invest.* **80**, 750–757.

Colletta, A. A., Howell, F. V. and Baum, M. (1989). A novel binding site for a synthetic progestagen in breast cancer cells. *J. Steroid Biochem.* **33**, 1055–1061.

Colletta, A. A., Wakefield, L. M., Howell, F. V., Danielpour, D., Baum, M. and Sporn, M. B. (1990*a*). The growth inhibition of human breast cancer cells by a novel synthetic progestin is partly mediated by the induction of transforming growth factors β. *J. Clin. Invest.*, in press.

Colletta, A. A., Wakefield, L. M., Howell, F. V., Van Roozendaal, K. E. P., Danielpour, D., Ebbs, S. R., Sporn, M. B. and Baum, M. (1990*b*). Antioestrogens induce the secretion of active transforming growth factor β from human fibroblasts. *Br. J. Cancer*, in press.

Glick, A. B., Flanders, K. C., Danielpour, D., Yuspa, S. H. and Sporn, M. B. (1989). Retinoic acid induces transforming growth factor-β2 in cultured keratinocytes and mouse epidermis. *Cell Reg.* **1**, 87–97.

Heine, U. I., Munoz, E. F., Flanders, K. C., Ellingsworth, L. R., Lam, H.-Y. P., Thompson, N. L., Roberts, A. B. and Sporn, M. B. (1987). Role of transforming growth factor-β in the development of the mouse embryo. *J. Cell Biol.* **105**, 2861–2876.

Hoosein, N. M., McKnight, M. K., Levine, A. E., Mulder, K. M., Childress, K. E., Brattain, D. E. and Brattain, M. G. (1989). Differential sensitivity of subclasses of human colon carcinoma cell lines to the growth inhibitory effects of transforming growth factor-β1. *Expl Cell. Res.* **181**, 442–453.

Hubbs, A. F., Hahn, F. F. and Thomassen, D. G. (1989). Increased resistance to transforming growth factor beta accompanies neoplastic progression of rat tracheal epithelial cells. *Carcinogenesis* **10**, 1599–1605.

Iqbal, M. J., Colletta, A. A., Houmayoun-Valyani, S. D. and Baum, M. (1986). Differences in oestrogen receptors in malignant and normal breast tissue as identified by the binding of a new synthetic progestogen. *Br. J. Cancer* **54**, 447–452.

Kehrl, J. H., Wakefield, L. M., Roberts, A. B., Jakowlew, A., Alvarez-Mon, M., Derynck, R., Sporn, M. B. and Fauci, A. S. (1986). Production of transforming growth factor-β by human T lymphocytes and its potential role in the regulation of T cell growth. *J. exp. Med.* **163**, 1037–1050.

Kim, S.-J., Glick, A., Sporn, M. B. and Roberts, A. B. (1989). Characterization of the promoter region of the human transforming growth factor-β1 gene. *J. biol. Chem.* **264**, 402–408.

Klausner, R. D. and Harford, J. B. (1989). Cis-trans models for post-transcriptional gene regulation. *Science* **246**, 870–872.

Knabbe, C., Lippman, M. E., Wakefield, L. M., Flanders, K. C., Kasid, A., Derynck, R. and Dickson, R. B. (1987). Evidence that transforming growth factor-β is a hormonally regulated negative growth factor in human breast cancer cells. *Cell* **48**, 417–428.

Komm, B. S., Terpening, C. M., Benz, D. J., Graeme, K. A., Gallegos, A., Korc, M., Greene, G. L., O'Malley, B. W. and Haussler, M. R. (1988). Estrogen binding, receptor mRNA, and biologic response in osteoblast-like osteosarcoma cells. *Science* **241**, 81–84.

Kyprianou, N. and Isaacs, J. T. (1989). Expression of transforming growth factor-β in the rat ventral prostate during castration-induced programmed cell death. *Molec. Endocr.* **3**, 1515–1522.

Lawrence, D. A., Pircher, R., Kryceve-Martinerie, C. and Jullien, P. (1984). Normal embryo fibroblasts release transforming growth factors in a latent form. *J. cell. Physiol.* **121**, 184–188.

Miller, D. A., Lee, A., Matsui, Y., Chen, E. Y., Moses, H. L. and Derynck, R. (1989). Complementary DNA cloning of the murine transforming growth factor-β3 (TGF-β3) precursor and the comparative expression of TGF-β3 and TGF-β1 messenger RNA in murine embryos and adult tissues. *Molec. Endocr.* **3**, 1926–1934.

Miyazono, K., Hellman, U., Wernstedt, C. and Heldin, C-H. (1988). Latent high molecular weight complex of transforming growth factor β1: purification from human platelets and structural characterization. *J. biol. Chem.* **263**, 6407–6415.

Murphy, L. C. and Dotzlaw, H. (1989). Regulation of transforming growth factor-alpha and transforming growth factor-β messenger ribonucleic acid abundance in T-47D, human breast cancer cells. *Molec. Endocr.* **3**, 611–617.

O'Connor-McCourt, M. D. and Wakefield, L. (1987). Latent transforming growth factor-β in serum: a specific complex with alpha-2-macroglobulin. *J. biol. Chem.* **262**, 14 090–14 099.

Pelton, R. W., Nomura, S., Moses, H. L. and Hogan, B. L. M. (1989). Expression of transforming growth factor β2 RNA during murine embryogenesis. *Development* **106**, 759–767.

Pfeilschifter, J. and Mundy, G. R. (1987). Modulation of type β transforming growth factor activity in bone cultures by osteotropic hormones. *Proc. natn. Acad. Sci. U.S.A.* **84**, 2024–2028.

Roberts, A. B., Kondaiah, P., Rosa, F., Watanabe, S., Good, P., Roche, N. S., Rebbert, M. L., Dawid, I. B. and Sporn, M. B. (1990). Mesoderm induction in *Xenopus laevis* distinguishes between the various TGF-β isoforms. *Growth Factors*, in press.

Roberts, A. B. and Sporn, M. B. (1990). The transforming growth factor-βs. In *Handbook of Experimental Pharmacology* (ed. M. B. Sporn and A. B. Roberts), vol. 95, pp. 419–472. Springer-Verlag, Heidelberg.

Sato, Y. and Rifkin, D. B. (1989). Inhibition of endothelial cell movement by pericytes and smooth muscle cells: activation of a latent transforming growth factor-β1-like molecule by plasmin during co-culture. *J. Cell Biol.* **109**, 309–315.

Smits, H. L., Floyd, E. E. and Jetten, A. M. (1987). Molecular cloning of sequences regulated during squamous differentiation of tracheal epithelial cells and controlled by retinoic acid. *Molec. cell. Biol.* **7**, 4017–4023.

Sporn, M. B. and Roberts, A. B. (1984). In *The Retinoids* (ed. M. B. Sporn, A. B. Roberts and D. S. Goodman), vol. 1. New York: Academic Press.

Wakefield, L. M., Smith, D. M., Flanders, K. C. and Sporn, M. B. (1988). Latent transforming growth factor-β from human platelets: a high molecular weight complex containing precursor sequences. *J. biol. Chem.* **263**, 7646–7654.

Wakefield, L. M., Smith, D. M., Masui, T., Harris, C. C. and Sporn, M. B. (1987). Distribution and modulation of the cellular receptor for transforming growth factor-β. *J. Cell Biol.* **105**, 965–975.

Wakefield, L. M. and Sporn, M. B. (1990). Suppression of carcinogenesis: a role for TGF-β and related molecules in prevention of cancer. In *Tumor Suppressor Genes* (ed. G. Klein), pp. 217–243. Marcel Dekker, Inc., New York.

Wakefield, L. M., Winokur, T. S., Hollands, R. S., Christopherson, K., Levinson, A. D. and Sporn, M. B. (1990). Recombinant latent transforming growth factor-β1 has a longer plasma half-life in rats than active transforming growth factor-β1, and a different tissue distribution. *J. Clin. Invest.*, in press.

J. Cell Sci. Suppl. 13, 149–156 (1990)

Printed in Great Britain © The Company of Biologists Limited 1990

Growth factors influencing bone development

J. M. WOZNEY, V. ROSEN, M. BYRNE, A. J. CELESTE, I. MOUTSATSOS AND E. A. WANG

Genetics Institute, Inc., 87 Cambridge Park Drive, Cambridge, MA 02140, USA

Summary

We have approached the study of growth factors affecting cartilage and bone development by investigating those factors present in bone which are able to initiate new cartilage and bone formation *in vivo*. This has led to the identification and molecular cloning of seven novel human factors which we have named BMP-1 through BMP-7. Six of these molecules are related to each other, and are also distantly related to TGF-β. The presence of one of these molecules, recombinant human BMP-2 (rhBMP-2) is sufficient to produce the complex developmental system of cartilage and bone formation when implanted subcutaneously in a rat assay system. In this model, administration of rhBMP-2 ultimately results in the formation of a piece of trabecular bone, which is filled with mature bone marrow. While our studies demonstrate that rhBMP-2 by itself has the ability to induce cartilage and bone formation *in vivo*, we find other BMP molecules present along with BMP-2 in our highly purified nonrecombinant bone-inductive material. These results suggest that the bone inductive capacity of bone-derived proteins may reside in the combinatorial or synergistic activities of this set of BMP-2 related molecules.

Introduction

The growth and maintenance of bone tissue are complex processes influenced by many systemic hormones and locally produced growth factors. Parathyroid hormone, calcitonin, and the vitamin D metabolites are known effectors of bone remodeling. Growth factors such as TGF-β1, TGF-β2, bFGF, aFGF, IGF-I and IGF-II all are present in significant quantities in bone matrix and have been demonstrated to have various effects on the growth properties and functions of bone cells *in vitro* (Hauschka *et al.* 1988; Canalis *et al.* 1988; Seyedin *et al.* 1985). In spite of these studies, the knowledge of which factors initiate and control the formation of cartilage and bone, both in the adult and during embryogenesis, is minimal. For example, morphological investigations of bone development during embryogenesis have indicated that bone can form in two distinct sequences: through direct condensation of mesenchyme (intramembranous bone formation) and through the formation and removal of a cartilage model (endochondral bone formation). The end result of each of these pathways is a functional bone with bone marrow. Whether or not the cell types and initiating signalling systems are the same for both of these pathways is unknown.

In the adult, protein extracts of bone are able to induce bone formation through the endochondral process. This activity, often referred to as bone morphogenetic protein, or BMP, initiates and possibly maintains a sequence of events *in vivo*

Key words: BMP, bone, cartilage, TGF-β, mRNA.

which culminates in bone formation (Urist *et al.* 1973). The cellular events caused by BMP have been described by histological examination of BMP implanted subcutaneously in a rat assay system (Reddi, 1981). These include chemotactic events (infiltration of the implant site with cartilage and bone cell precursors), proliferative events, differentiation of precursor cells into chondrocytes, induction of vascularization, maturation of the chondrocytes and differentiation of cells into osteoblasts. This complex process results in a new piece of bone tissue, complete with osteoblasts, osteocytes, osteoclasts, and bone marrow elements. Through our investigations, we have discovered growth factors responsible for this series of events, and have begun examining which of the many specific events in this process they induce.

Identification and molecular cloning of growth factors responsible for BMP activity

The extensive (over 300 000-fold) purification of BMP activity from bovine bone has been described previously (Wang *et al.* 1988). Biochemical analysis of the proteins present in the most active bone-derived BMP indicated the presence of several polypeptides of approximately $30\times10^3\,M_r$, the majority of which are found as polypeptides of 16 and $18\times10^3\,M_r$ after reduction. Further purification of these $30\times10^3\,M_r$ species resulted in substantial loss of activity, and as disulfide reduction destroys BMP activity, the individual polypeptides could not be assayed. Trypsin digestion of the $30\times10^3\,M_r$ nonreduced material, as well as the 16 and $18\times10^3\,M_r$ reduced components, produced a set of short peptides whose amino acid sequences were determined. These sequences were then used to design and synthesize oligonucleotide probes, which were used to screen bovine genomic or cDNA libraries. Recombinant clones were identified which accounted for the various sequences derived from bovine BMP. These clones were then used to screen human cDNA libraries to obtain recombinant clones encoding the complete human equivalents of the bovine proteins. Using this strategy, we have obtained the sequences of seven novel human proteins, which we call BMP-1 through BMP-7 (Wozney *et al.* 1988; Wozney, 1989; Celeste *et al.* unpublished).

Six of these seven proteins (BMP-2 through BMP-7) are closely related to each other and are distantly related to TGF-β. A schematic representation of their structural features is given in Fig. 1. Like all molecules in the TGF-β family, the BMPs are synthesized as precursor proteins. Their primary translation products, as derived from the cDNA sequences, contain hydrophobic secretory leader sequences as well as substantial propeptides. The mature portion of each molecule, as determined by homology to other members of the TGF-β family, constitutes the carboxy-terminal portion of the precursor peptide. Furthermore, all of the peptide sequences found in bovine BMP have their counterparts in these regions of the molecules. Each of the molecules contains seven cysteine residues in their carboxy-terminal portions, in positions corresponding to those present in all members of the TGF-β superfamily. Unlike TGF-β, all the mature BMP

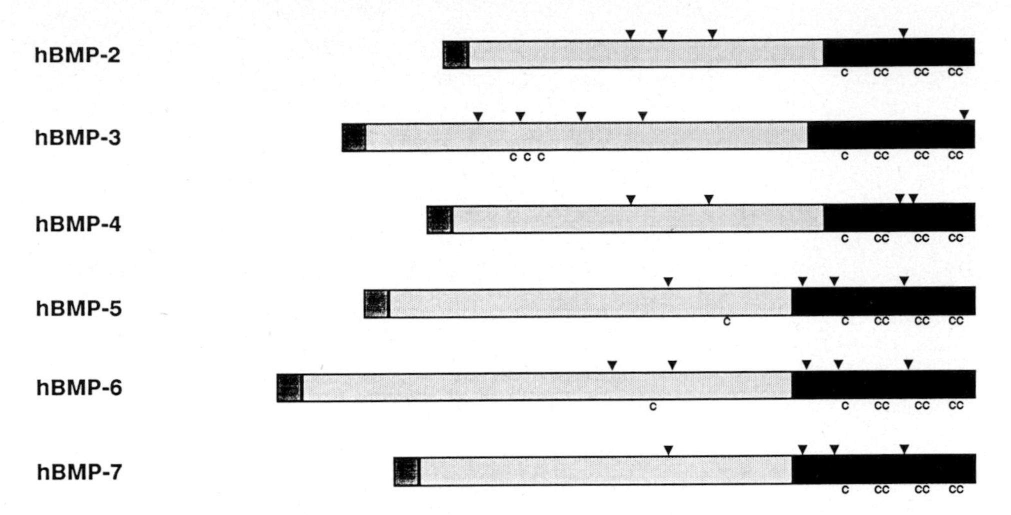

Fig. 1. Schematic representation of the human TGF-β-like BMPs. The amino acid sequence of each derived from the full-length cDNA is represented by a bar. Each precursor peptide contains a hydrophobic secretory leader sequence, a substantial propeptide sequence (light shading), and the mature region (filled shading) at the carboxy terminus. Potential N-linked glycosylation sites (Asn-X-Ser/Thr) are represented by triangles (▼) above the bars. Cysteine residues (exclusive of the leader sequences) are indicated by (c) below the bars. The molecule originally designated BMP-2A (Wozney *et al.* 1988) is now called BMP-2; BMP-2B has been renamed BMP-4.

proteins contain potential N-linked glycosylation sites. Indeed, the BMP proteins found in bovine bone do contain $2\text{–}3\times10^3\,M_r$ of carbohydrate per chain, as evidenced by reduction of their molecular weights with *N*-glycanase (Wang *et al.* 1988). The propeptide portions also contain potential N-linked glycosylation sites, similar to TGF-β. No Arg-Gly-Asp potential cell recognition sequences are present as they are in the TGF-β1 and TGF-β3 propeptide sequences. Three of the BMP family members, BMP-2, BMP-4, and BMP-7 do not contain cysteine residues in their propeptides. The presence of cysteine residues required for dimerization of the propeptide have been implicated in the formation of a latent (inactive) complex between TGF-β and its precursor (Brunner *et al.* 1989). This suggests that these three BMPs are secreted in active forms.

Interrelationships of the BMP proteins and homologies with other TGF-β superfamily members

Examination of the amino acid sequences of the human BMP proteins demonstrates that they are closely related to each other, and that they form a subgroup of the TGF-β superfamily. When the mature regions of the BMPs are aligned, it is evident that, in addition to the seven cysteine residues present in all members of the TGF-β superfamily, substantial regions of sequence identity exist. Analysis of the amino acid identities between all pairs of the BMPs indicates that these molecules can be divided into three subfamilies. BMP-2 and BMP-4 are

A

	Within group		To BMP-2	
	Pro	Mature	Pro	Mature
BMP-2/4	57%	86%	-	-
BMP-5/6/7	59%	78%	27%	56%
BMP-3	-	-	21%	45%

B

	TGF-β1,2,3	Inhibin β_A	Vg1	dpp
BMP-2/4	37%	44%	57%	75%
BMP-5/6/7	35%	42%	53%	52%
BMP-3	34%	37%	49%	43%

Fig. 2. Inter-relationships of the BMPs and relationship to TGF-β superfamily members. (A) Amino acid identities between members of a subgroup of the BMPs to each other in either the propeptide (PRO) or mature regions. (B) Amino acid identities to other members of the TGF-β family. Numbers are an average of percent amino acid identities of pairwise comparisons aligned using the UWGCG sequence analysis system.

quite closely related (86 % amino acid identity in this region, see Fig. 2A) and form one group. BMP-5, BMP-6, and BMP-7 are also closely related to one another (an average of 78 %), and form a second group. BMP-3, while more closely related to the other BMP proteins than to other TGF-β family members, is the most different and by itself forms the third group. The relationships of these groups of BMP proteins to representative members of the TGF-β family are given in Fig. 2B, comparing the sequences from the regions containing the seven cysteine residues. While distantly related to TGF-β1, TGF-β2, and TGF-β3 (35 % sequence identity), they are more closely related to the inhibin β family of molecules (41 %). All of the BMP proteins are even more closely related to the developmentally implicated proteins Vg1 from *Xenopus* (Weeks and Melton, 1987) and dpp from *Drosophila* (Padgett *et al.* 1987). From the substantial homology between dpp and the BMP-2 and BMP-4 subfamily (75 %), as well as substantial regions of sequence identity in the propeptide regions, we infer that either (or both) are the human equivalents of the dpp gene product. Human BMP-6 shows 91 % sequence identity across the entire precursor molecule with Vgr-1, a polypeptide defined by the derived amino acid sequence of a cDNA isolated from a mouse embryo cDNA library by cross-hybridization with Vg1 (Lyons *et al.* 1989). For this reason, we presume that Vgr-1 is the murine homolog of BMP-6.

Distribution of mRNAs for the BMP proteins

As part of an effort to understand the roles of the BMP proteins in cartilage and bone formation, as well as their possible growth factor activities in other tissues, we have examined the tissue distribution of mRNAs for BMP-1 through BMP-4 in various bovine tissues. These values are compared with the levels of mRNAs in primary subperiosteal bone cell cultures (a generous gift from Dr Marian Young, N.I.H.) in Table 1. Using the sensitive technique of mRNA protection, low levels of BMP-1, BMP-2, and BMP-4 mRNA can be found in all tissues. With BMP-1, substantially higher levels of mRNA are present in cerebellum and liver; with approximately 3-fold higher levels in bone cells than in cerebellum. BMP-2 mRNA was present at about 40-fold higher levels in bone cells than in any other tissue examined. The relative mRNA distribution of BMP-4 is distinct from that of BMP-2, though the two molecules are extremely closely related. The levels of BMP-4 mRNA found in bone cells are only about 2-fold higher than in lung and kidney. The tissue distribution of BMP-3 mRNA is the most restricted. BMP-3 mRNA is undetectable in most tissues, with some small amount present in cerebrum and significantly higher levels in lung. Interestingly, the mRNAs for all four of these proteins can be found in brain. Many growth factors, including aFGF (Gimenez-Gallego *et al.* 1985), bFGF (Esch *et al.* 1985; Emoto *et al.* 1989), inhibin α, β_A, and β_B (Sawchenko *et al.* 1988) have been found in brain, though their roles there are unclear.

Though low levels of BMP-1, BMP-2, and BMP-4 mRNAs are found in all tissues examined, it is difficult to interpret this result. The technique used to detect mRNA in these studies is quite sensitive, and several reports indicate there

Table 1. *Relative distribution of the mRNAs for BMPs*

	BMP-1	BMP-2	BMP-3	BMP-4
Cerebrum	5.4	4.9	(1.0)	(1.0)
Cerebellum	140	4.5	0	7.8
Heart	(1.0)	(1.0)	0	1.2
Lung	7.6	3.7	20	21
Kidney	5.7	3.8	0	16
Spleen	6.9	14	0	5.2
Liver	43	8.0	0	4.4
Bone cell (subperiosteal)	420	539	0	48

mRNA levels were quantitated by RNase protection (Zinn *et al.* 1983). Quantitation was by scanning of the autoradiographs, except for BMP-3 which was estimated visually. Numbers in each group are normalized to the lowest detectable signal being (1.0); numbers can therefore only be compared within a column.

is a basal expression of any mRNA in any tissue (Chelly *et al.* 1988; Chelly *et al.* 1989; Sarkar and Sommer, 1989). Furthermore, the presence of small amounts of mRNA does not necessarily indicate that it is translated into protein. With these caveats in mind, it is interesting that the highest levels of BMP-1 and BMP-2 expression are found in the subperiosteal bone cells, consistent with a role for these molecules in bone formation. The levels of BMP-1 mRNA are 3-fold higher in these bone cells than in the next most highly expressing tissue, cerebellum. Almost 40-fold higher levels of BMP-2 mRNA are found in bone cells than in any other tissue examined. While BMP-3 mRNA was undetectable in these bone cells, it should be pointed out that these cells represent a distinct subset of cells of the osteoblastic lineage and other bone cell populations may synthesize BMP-3 mRNA.

In vivo activity of BMP-2

The clear definition of the *in vivo* and *in vitro* activities of any growth factor require its purification to absolute biochemical homogeneity and/or expression in a recombinant system. The derivation of mammalian cell lines which stably express BMP-2 has allowed us to examine its activity in the rat ectopic bone formation assay used to identify BMP activity (Wang *et al.* 1990). In this system, recombinant human BMP-2 (rhBMP-2) is reconstituted with a bone-derived collagenous matrix. Histological examination of the implants removed at various times after implantation of moderate doses (2–20 μg) of rhBMP-2 show that rhBMP-2 alone can induce cartilage and bone formation in a subcutaneous site in the rat. The cellular events observed are similar if not identical to those observed with crude or purified nonrecombinant bone-derived BMP. Undifferentiated mesenchymal cells infiltrate the implant site, proliferate, and differentiate into chondroblasts by day 5. These cartilage-forming cells mature, hypertrophy, and mineralize; vascularization of the implant area can be seen at about this time. Osteoblasts appear in the implant site and bone matrix (osteoid) is deposited while the cartilage is removed. The bone matrix is then mineralized, and by day 14 only the newly formed trabecular bone tissue remains, containing osteoblasts, osteoclasts, and bone marrow elements. The normal bone remodeling sequence continues, with osteoblasts depositing bone and osteoclasts resorbing bone. Interestingly, the dose of rhBMP-2 administered affects the time at which these events occur. This process can be accelerated such that bone is observable at day 14 with 0.5 μg, at day 7 with 12 μg, and at day 5 with 115 μg of rhBMP-2. When the latter amount of rhBMP-2 is administered, large amounts of both cartilage and bone formation are observable at day 5. This may suggest that BMP-2 acts on both cartilage and bone induction directly, and that the sequence of cartilage induction followed by bone formation observed with lower amounts of BMP in this assay system is not truly a cascade. While reconstitution of rhBMP-2 with collagenous bone matrix decreases the amount of BMP needed to see cartilage and bone induction, it is not necessary for BMP activity. Implantation of rhBMP-2 without

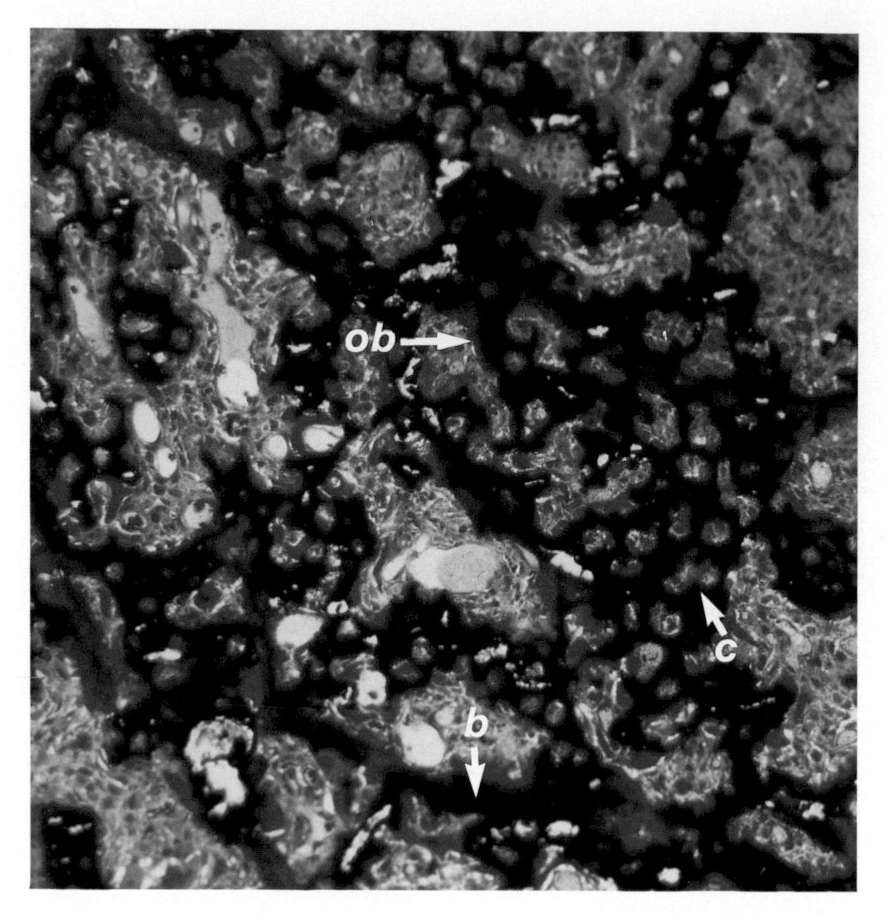

Fig. 3. Induction of cartilage and bone by rhBMP-2 without addition of carrier matrix. BMP-2 was implanted subcutaneously in a rat for 9 days. Von Kossa stain. c, calcified cartilage; b, bone; ob, osteoblasts.

any matrix carrier induces both cartilage and bone formation, as seen in Fig. 3. Therefore our working hypothesis is that the matrix contributes slow release and/or immobilization characteristics to the BMP-2, but does not contribute any additional factors necessary for bone induction. The dissection of which particular events in this complex system of *in vivo* cartilage and bone induction are affected by BMP-2 will require further studies on the *in vivo* and *in vitro* activities of BMP-2.

Bone-derived BMP activity

While rhBMP-2 appears to have the complete cartilage and bone inductive activities of BMP derived from bone, there are several lines of evidence to suggest that BMP-2 is not the sole growth factor responsible for this activity. In our work, we have found that the most active bone-derived material contains a mixture of molecules, BMP-1 through BMP-7. Other researchers have also purified bone-inductive substances from bovine bone (Luyten *et al.* 1989). The reported amino acid sequence of this material, called osteogenin, is the same as that of BMP-3. This may indicate that BMP-3 also has bone-inductive activity by itself, as does BMP-2, but one cannot exclude the possibility that other BMP molecules are also contained in this nonrecombinant material. In addition, the amount of rhBMP-2 necessary to produce *in vivo* bone induction is an order of magnitude higher than that of highly purified bone-derived BMP. Though it is possible that the recombinant BMP-2 made in our mammalian cell expression system is somehow different from bone-derived BMP-2 and therefore is less active, these facts suggest that natural BMP activity is some combination of the activities of multiple BMP molecules or represents synergistic activity between them. The fact that mRNA for the BMP proteins can be found in many non-bony tissues also supports the argument that multiple factors may normally have to work together to reach a critical concentration to induce bone formation in the animal. As a further complication, since all the TGF-β-like BMPs form dimers, it is possible that heterodimers of these molecules exist. By analogy to other members of the family, these heterodimers could have increased or even opposite activities to the homodimers themselves. Again, the possibility of heterodimer formation is supported by the finding that many tissues express several of the BMP mRNAs. Production of the rest of the BMP proteins through recombinant means will allow further clarification of their roles in the cartilage and bone induction process.

References

Brunner, A. M., Marquardt, H., Malacko, A. R., Lioubin, M. N. and Purchio, A. F. (1989). Site-directed mutagenesis of cysteine residues in the pro region of the transforming growth factor β1 precursor. *J. biol. Chem.* **264**, 13 660–13 664.

Canalis, E., McCarthy, T. and Centrella, M. (1988). Isolation of growth factors from adult bovine bone. *Calc. Tiss. Int.* **43**, 346–351.

Chelly, J., Concordet, J.-P., Kaplan, J.-C. and Kahn, A. (1989). Illegitimate transcription: transcription of any gene in any cell type. *Proc. natn. Acad. Sci. U.S.A.* **86**, 2617–2621.

Chelly, J., Kaplan, J. C., Maire, P., Gautron, S. and Kahn, A. (1988). Transcription of the dystrophin gene in human muscle and non-muscle tissues. *Nature* **333**, 858–860.

Emoto, N., Gonzalez, A.-M., Walicke, P. A., Wada, E., Simmons, D. M., Shimasaki, S. and Baird, A. (1989). Basic fibroblast growth factor (FGF) in the central nervous system: identification of specific loci of basic FGF expression in the rat brain. *Growth Factors* **2**, 21–29.

Esch, F., Baird, A., Ling, N., Ueno, N., Hill, F., Denoroy, L., Klepper, R., Gospodarowicz, D., Bohlen, P. and Guillemin, R. (1985). Primary structure of bovine pituitary basic fibroblast growth factor (FGF) and comparison with the amino-terminal sequence of bovine brain acidic FGF. *Proc. natn. Acad. Sci. U.S.A.* **82**, 6507–6511.

Gimenez-Gallego, G., Rodkey, J., Bennett, C., Rios-Candelore, M., DiSalvo, J. and Thomas, K. (1985). Brain-derived acidic fibroblast growth factor: complete amino acid sequence and homologies. *Science* **230**, 1385–1388.

Hauschka, P. V., Chen, T. L. and Mavrakos, A. E. (1988). Polypeptide growth factors in bone matrix. In *Cell and Molecular Biology of Vertebrate Hard Tissues* (ed. Evered, D. and Harnett, S.), pp. 207–225. Chichester: Wiley.

Luyten, F. P., Cunningham, N. S., Ma, S., Muthukumaran, N., Hammonds, R. G., Nevins, W. B., Wood, W. I. and Reddi, A. H. (1989). Purification and partial amino acid sequence of osteogenin, a protein initiating bone differentiation. *J. biol. Chem.* **264**, 13 377–13 380.

Lyons, K., Graycar, J. L., Lee, A., Hashmi, S., Lindquist, P. B., Chen, E. Y., Hogan, B. L. M. and Derynck, R. (1989). Vgr-1, a mammalian gene related to *Xenopus* Vg-1, is a member of the transforming growth factor β gene superfamily. *Proc. natn. Acad. Sci. U.S.A.* **86**, 4554–4558.

Padgett, R. W., St Johnston, R. D. and Gelbart, W. M. (1987). A transcript from a *Drosophila* pattern gene predicts a protein homologous to the transforming growth factor-β family. *Nature, Lond.* **325**, 81–84.

Reddi, A. H. (1981). Cell biology and biochemistry of endochondral bone development. *Collagen Rel. Res.* **1**, 209–226.

Sarkar, G. and Sommer, S. S. (1989). Access to a messenger RNA sequence or its protein product is not limited by tissue or species specificity. *Science* **244**, 331–334.

Sawchenko, P. E., Plotsky, P. M., Pfeiffer, S. W., Cunningham, E. T., Jr, Vaughan, J., Rivier, J. and Vale, W. (1988). Inhibin β in central neural pathways involved in the control of oxytocin secretion. *Nature* **334**, 615–617.

Seyedin, S. M., Thomas, T. C., Thompson, A. Y., Rosen, D. M. and Piez, K. A. (1985). Purification and characterization of two cartilage-inducing factors from bovine demineralized bone. *Proc. natn. Acad. Sci. U.S.A.* **82**, 2267–2271.

Urist, M. R., Iwata, H., Ceccotti, P. L., Dorfman, R. L., Boyd, S. D., McDowell, R. M. and Chien, C. (1973). Bone morphogenesis in implants of insoluble bone gelatin. *Proc. natn. Acad. Sci. U.S.A.* **70**, 3511–3515.

Wang, E. A., Rosen, V., Cordes, P., Hewick, R. M., Kriz, M. J., Luxenberg, D. P., Sibley, B. S. and Wozney, J. M. (1988). Purification and characterization of other distinct bone-inducing factors. *Proc. natn. Acad. Sci. U.S.A.* **85**, 9484–9488.

Wang, E. A., Rosen, V., D'Alessandro, J. S., Bauduy, M., Cordes, P., Harada, T., Israel, D., Hewick, R. M., Kerns, K., Lapan, P., Luxenberg, D. P., McQuaid, D., Moutsatsos, I., Nove, J. and Wozney, J. M. (1990). Recombinant human bone morphogenetic protein induces bone formation. *Proc. natn. Acad. Sci. U.S.A.* **87**, 2220–2224.

Weeks, D. L. and Melton, D. A. (1987). A maternal mRNA localized to the vegetal hemisphere in *Xenopus* eggs codes for a growth factor related to TGF-β. *Cell* **51**, 861–867.

Wozney, J. M. (1989). Bone morphogenetic proteins. *Prog. Growth Factor Res.* **1**, 267–280.

Wozney, J. M., Rosen, V., Celeste, A. J., Mitsock, L. M., Whitters, M. J., Kriz, R. W., Hewick, R. M. and Wang, E. A. (1988). Novel regulators of bone formation: molecular clones and activities. *Science* **242**, 1528–1534.

Zinn, K., DiMaio, D. and Maniatis, T. (1983). Identification of two distinct regulatory regions adjacent to the human β-interferon gene. *Cell* **34**, 865–879.

J. Cell Sci. Suppl. 13, 157–168 (1990)

Printed in Great Britain © The Company of Biologists Limited 1990

Mechanisms of positional signalling in the developing eye of *Drosophila* studied by ectopic expression of *sevenless* and *rough*

ERNST HAFEN AND KONRAD BASLER

Zoologisches Institut der Universität Zürich, Winterthurerstrasse 190, CH-8057 Zürich, Switzerland

Summary

In the developing eye of *Drosophila* cell fate is controlled by a cascade of inductive interactions. Little is known about how the specificity of positional signalling is achieved such that directly adjacent progenitor cells reproducibly choose distinct developmental pathways. The determination of the R7 photoreceptor in each ommatidium depends on the presence of the *sevenless* protein which acts as a receptor for positional information on the R7 precursor. The *rough* gene encodes a homeodomain protein that plays an instructive role in the determination of the R3 and R4 photoreceptor cells. The use of ectopic expression of *sevenless* and *rough* has provided insight into the mechanisms of positional signalling and the normal function of *rough*. Ubiquitous expression of *sevenless* does not alter cell fate suggesting that the inducing signal is both spatially and temporally controlled. Conversely, ectopic expression of *rough* in the R7 precursor causes a transformation of R7 cells into R1–6 type cells. This indicates that *rough* acts, similar to other homeobox genes, as a selector gene that determines the fate of single cells.

Introduction

In the developing eye of *Drosophila*, specification of cell fate can be studied at the single cell level. This is facilitated by the remarkably precise arrangement of cells within both the developing and the mature retinal epithelium (Fig. 1A and 1E). The compound eye consists of a repetitive array of about 800 identical unit eyes, or ommatidia, each of which is a precise assembly of a few distinct cell types (Fig. 1C; for review see Ready, 1989). The eight photoreceptor cells (R1 to R8) of the ommatidial unit can be grouped into three functional classes based upon position, spectral sensitivities, and projection pattern of their axons: R1 to R6, R7, and R8. During development they assemble and differentiate in a fixed temporal sequence: R8 is the founder cell in each ommatidium, subsequently R2/R5, R3/R4, and R1/R6 are added pairwise, and finally R7. Later, the non-neuronal elements, the cone cells and the pigment cells, become incorporated (Fig. 1F).

Cells are directed to their individual fates by inductive signals communicated by neighboring cells (Tomlinson and Ready, 1987). Differentiating cells are thought to express cell type specific signals which undetermined cells receive and interpret in order to choose their differentiation pathways. Mutations in genes

Key words: *sevenless*, *rough*, homeobox gene, ectopic expression, positional signalling, cell–cell interactions, cell fate determination, eye development, *Drosophila*.

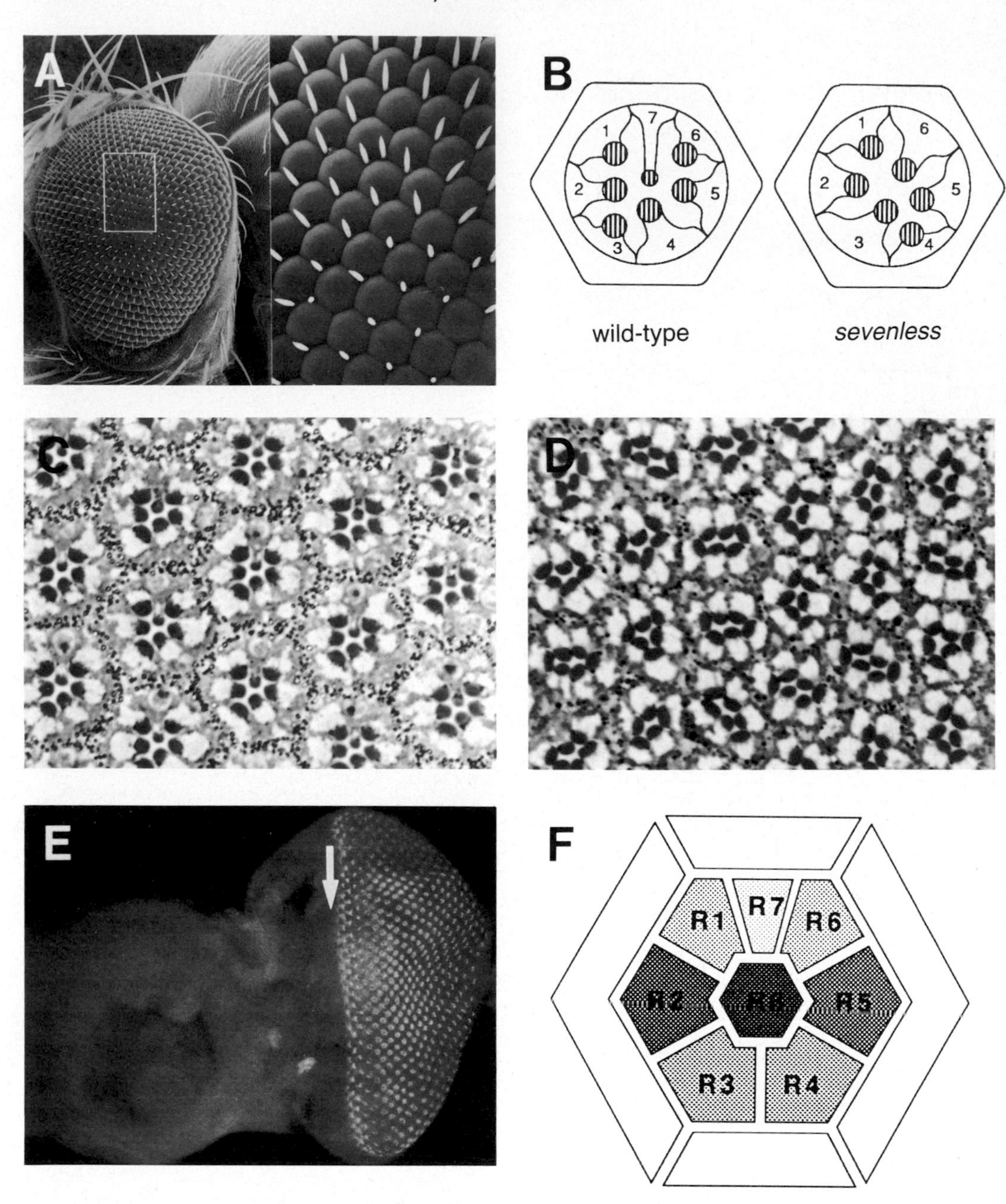

whose products are involved in these inductive pathways can be grouped into two classes. Genes encoding products involved in the signalling pathway will show a non-autonomous mutant phenotype. Conversely, genes encoding products involved in the reception of the inductive signals and the subsequent execution of the developmental program will exhibit a cell autonomous mutant phenotype and are required intrinsically for a cell to differentiate correctly.

Three genes have been identified as important components in the communi-

Fig. 1. Structure and development of the compound eye of *Drosophila*. Anterior is to the left. (A) Scanning electron micrograph of a wild-type eye. The enlargement shows the hexagonal arrangement of the individual ommatidia. (B) Schematic representation of cross-sections through the distal part of a wild–type and a *sevenless* mutant ommatidium. The positions of the photoreceptors R1–R7 are shown. In the *sevenless* mutants the R7 cell is missing. (C) Histological cross-sections through wild-type and (D) *sevenless* mutant eyes. (E) Eye imaginal disc stained with an antiserum against the *sevenless* protein visualizing the assembly of the ommatidia. Each of the stained clusters corresponds to a subset of cells in an ommatidium. Ommatidial assembly begins in a wave (arrow) that moves over the disc epithelium towards the anterior; cells anterior to this wave are unlabeled. The different developmental stages of ommatidial assembly are spatially displayed along the anterior–posterior axis. (F) Schematic representation of the inner 12-cell unit of an ommatidium at the larval stage. The different gray shades indicate the temporal sequence of assembly. R8 is the first cell in the cluster followed by the pairwise addition of photoreceptors R2/R5, R3/R4, R1/R6 and finally by R7.

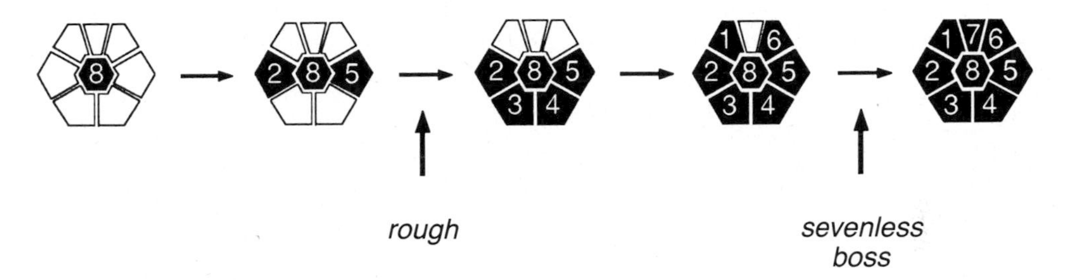

Fig. 2. Mutations interfere with different steps of the assembly sequence of the ommatidial unit. Only the assembly of the 8-photoreceptor cell unit is shown schematically. Mutations in the *rough* gene interfere with the correct development of R3/4. Subsequent steps are disturbed which leads to the roughening of the eye. Mutations in the genes *sevenless* and *boss* interfere with a later step, namely the specification of the R7 photoreceptor cell. In the absence of either *boss* or *sevenless* the R7 precursor does not develop into a photoreceptor cell but into a non-neuronal cone cell.

cation mechanisms (Fig. 2). The *sevenless* mutation prevents the presumptive R7 cell from differentiating correctly and causes the cell to become a lens-secreting cone cell (Harris *et al.* 1976; Tomlinson and Ready, 1986). The *sevenless* gene product is a receptor tyrosine kinase required autonomously in the presumptive R7 cell and is thought to receive signals from the adjacent R8 cell (Basler and Hafen, 1988; Bowtell *et al.* 1988). The *boss* gene is required only in R8 for the R7 cell to develop appropriately (Reinke and Zipursky, 1988). The non-autonomy of *boss* suggests that it is part of the signalling machinery in R8. Similar to *boss*, *rough* is non-autonomously required in cells R2 and R5 for the correct specification of the neighboring R3 and R4 cells (Tomlinson *et al.* 1988; Fig. 2). The *rough* protein appears to act on the signalling side of the inductive pathway. Surprisingly, molecular analysis of *rough* revealed that it encodes a homeodomain protein and not a membrane protein or a secreted factor (Saint *et al.* 1988; Tomlinson *et al.* 1988). It has been postulated that *rough* functions in R2 and R5 as a transcriptional regulator for an R3/4 inducing signal (Tomlinson *et al.* 1988).

Although the characterization of mutations that disrupt ommatidial assembly permits the identification of genes involved in this process, it does not provide information on how these products function and interact such that directly adjacent cells select distinct developmental pathways. To address the question of the specificity of positional signalling we and others have started to examine the effects of ectopic expression of genes coding for sensory and instructive proteins. These analyses have led to important conclusions about the function of these genes and about the multipotency of retinal precursor cells. In this article we will review the results obtained from studying ectopic expression of *sevenless* and *rough*.

Ubiquitous expression of *sevenless* does not alter cell fate

Specification of cell fate by ligand – receptor interactions can be viewed in three different ways. (1) The ligand is ubiquitous but the receptor is expressed only locally. Specificity is controlled by the restricted expression of the receptor (Fig. 3A). (2) The ligand is localized and the receptor is ubiquitously expressed. In this case the specificity is controlled by the local expression of the ligand (Fig. 3C). (3) Both ligand and receptor are partially restricted. Specificity is controlled by the combination and overlap between the expression domains of both ligand and receptor (Fig. 3B).

The *sevenless* gene is expressed in a complex spatial and temporal pattern (Tomlinson *et al.* 1987). During the third instar larval period, it is almost exclusively expressed in the developing eye imaginal disc. Within the eye disc it is expressed transiently only in a subset of the ommatidial precursors (Fig. 1E).

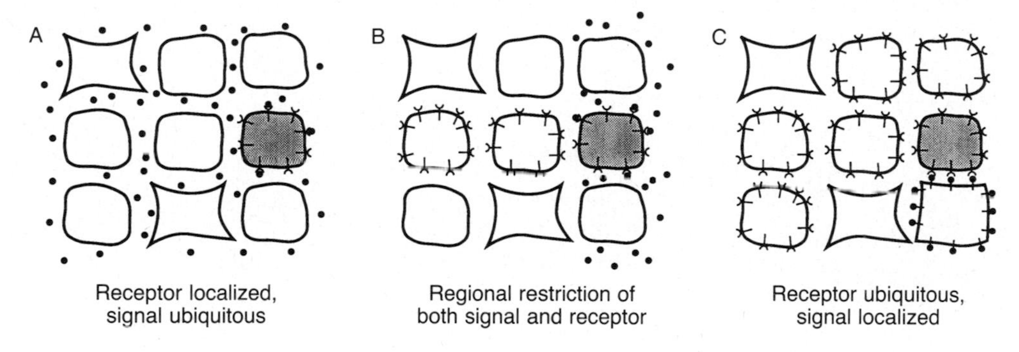

Fig. 3. Models showing how regional specificity can be achieved. Rounded squares represent undetermined cells; already differentiated cells are indicated by rhomboids. (A) Ubiquitous ligand and restricted expression of the receptor. Cell type restricted response is controlled by the expression of the receptor. Prominent examples of this type of regulation are growth factor-mediated responses. (B) Partially restricted expression of signal and receptor. (C) Ubiquitous receptor and localized ligand. All undetermined cells contain receptors but only one contacts the ligand-presenting cell. Position-dependent specification of cell fate is likely to be governed by this mechanism. Y, receptor; ●, diffusible ligand; ♀, membrane bound ligand; stippling, cell that is determined in response to ligand-induced receptor activation.

Based on the expression pattern of *sevenless* it has been postulated that the local specification of R7 cells is controlled according to the model shown in Fig. 3B, by the combination of the partially restricted expression of both ligand and receptor (Tomlinson *et al.* 1987). To test this hypothesis, the *sevenless* gene was placed under the control of an inducible ubiquitous promoter and introduced into the germline of *sevenless* mutant flies (Basler and Hafen, 1989; Basler *et al.* 1989; Bowtell *et al.* 1989*b*). Ubiquitous expression of *sevenless* during development specifies R7 cells in correct positions but does not interfere with the development of other cells where *sevenless* is not normally expressed. This result suggests that the complex spatial and temporal regulation of *sevenless* gene expression does not contribute to the spatially restricted specification of R7 cells. Specificity of R7 selection may therefore be controlled by the local presentation of the *sevenless* ligand rather than by the restricted expression of the receptor (Fig. 3C).

Even if the expression of the ligand is restricted to a single cell, this cell may still contact multiple neighboring cells. For example, in the case of the R7 determination, the R8 cell which most likely produces the ligand for *sevenless*, is in contact with seven cells of which all may express the *sevenless* receptor, yet only the presumptive R7 cell initiates R7 development. It has been postulated that cell fate in the eye is controlled by the combination of different contacts between neighboring cells (Tomlinson and Ready, 1987). Alternatively, specificity may also be achieved if the ligand is not only locally but also temporally restricted, such that at the time of ligand presentation only the presumptive R7 cell is capable of responding since all other cells have selected a different pathway earlier. It is even conceivable that temporal control of ligand presentation alone is sufficient to control specificity. If for example the ligand for *sevenless* is expressed on all photoreceptor cells relatively late during the differentiation, R8, the oldest cells in the cluster, will be the first to express the ligand at a time when the R7 precursor is the only cell that can still respond to the inductive signal. Transient expression of surface proteins during the differentiation of neural cells is well documented (Patel *et al.* 1987). For an undetermined cell it might therefore not only be important what its neighbors are but also how old these neighbors are. The test of which of the two models, combinatorial cell contacts with multiple inducing ligands for a single fate, or the spatially and/or temporally restricted expression of a single inducing signal, is correct has to await the cloning of the gene for such a ligand. By ectopic expression of the ligand it should be possible to distinguish between the two models.

Indiscriminate expression of the homeobox gene *rough* is lethal

The *rough* gene is expressed in a small subpopulation of the ommatidial precursor cells. Persistent expression is observed only in the R2/5 and R3/4 cells (Kimmel *et al.* 1990). To examine the effect of ectopic expression of the *rough* gene we have designed a *rough* minigene under the control of the inducible hsp70 heat shock promoter. Repeated heat-induction at embryonic, larval and pupal stages resulted in lethality of the *hsp-rough* transformants. Under the same conditions, wild-type or *hsp-sev* transformants that contain a heat shock-inducible *sevenless* gene survived. Even single heat shocks at the white prepupal stage caused a substantially lower survival rate. Hatched flies that survived the heat shock have severely reduced eyes bordered anteriorly by a sharp vertical scar. The few ommatidia anterior to this scar were highly irregular whereas ommatidial columns posterior to the scar were normal. Similar results were obtained by Kimmel *et al.* (1990) except that they did not observe lethality associated with the ubiquitous expression of rough. Although we do not know what the reason for this discrepancy is, it is possible that the two hsp-*rough* constructs used, differ in their level of expression upon heat induction.

Ubiquitous expression of *rough* during development appears to interfere with the normal development of cells. Induction of *rough* in adult flies, however, has no effect. The greatly reduced eyes resulting from a single heat shock during the third instar stage indicates that ubiquitous expression of *rough* interferes with the formation of new ommatidial columns in the eye imaginal disc. Heat shock induction of *sevenless* and *Ultrabithorax (Ubx)* carried out in parallel experiments did not produce a similar phenotype. Hence we assume that this effect is specifically caused by *rough*.

Although ectopic expression of *rough* during embryonic development was lethal, we did not observe a detectable alteration in the cuticular pattern of the central or peripheral nervous system of heat-shocked *hsp-rough* transformants. Under the same conditions ubiquitous expression of *Ubx* caused an almost complete transformation of the head and thoracic segments into A1 segments (Gonzàles-Reyes *et al.* 1990). We assume the ectopic expression of *rough* in unrelated cells causes a general disruption of a cellular function rather than inducing an alteration in cell fate.

Localized ectopic expression of *rough* using the *sevenless* enhancer

To limit ectopic expression of *rough* to a defined subset of cells in the developing eye imaginal disc, we used the *sevenless* enhancer sequences that control the *sevenless* expression pattern. In contrast to other tissue specific enhancers which are active only in differentiated cell types, such as the cis-acting sequences that control rhodopsin expression (Mismer *et al.* 1987), *sevenless* is expressed transiently in a subpopulation of ommatidial precursor cells prior to or at the time of their commitment (Tomlinson *et al.* 1987). *sevenless* is expressed strongly in cells R3, R4, R7 and the cone cells, and weakly in R1 and R6. It has been shown previously that a gene-internal fragment of the *sevenless* gene is responsible and sufficient for the temporally and spatially restricted expression of *sevenless* (Basler *et al.* 1989; Bowtell *et al.* 1989*a*). This regulatory element imposes the *sevenless*-specific expression pattern on heterologous promoters such as the hsp70 promoter. We inserted a restriction fragment containing the *sevenless* enhancer upstream of the *hsp-rough* construct, *sev-hsp-rough,* and generated germ-line transformants (Basler *et al.* 1990). A similar analysis was carried out independently by Kimmel *et al.* (1990).

Although the eyes of heterozygous *sev-hsp-rough* transformants appeared normal, histological sections revealed an irregular rhabdomere pattern in 50–80 % of the ommatidia. Wild-type ommatidia contain 8 photoreceptor cells (R1 to R8). The rhabdomeres of R1 to R6 extend through the depth of the retina, and form an asymmetric trapezoid (Fig. 4A, C, and E). The rhabdomere of R7 is smaller in diameter than that of the outer photoreceptors and occupies the central position in the distal two thirds of the retina. This position is occupied proximally by the rhabdomere of R8 (Fig. 4E). Therefore in each plane of cross section, a highly ordered pattern of 7 rhabdomeres is visible (Fig. 4A,C, and E). Apical sections of *sev-hsp-rough* ommatidia also display seven rhabdomeres, but in an abnormal arrangement (Fig. 4B). The few undisturbed ommatidia can be used as an internal reference for the orientation of the rhabdomere pattern. Comparison with adjacent affected ommatidia indicates that photoreceptors R1 to R6 are normal, but that the cell between R1 and R6, corresponding to the R7 cell in wild-type, causes the observed irregularity: its rhabdomere is not in the central position, but is instead between that of R1 and R6 (Fig. 4B and 4E). Furthermore, its rhabdomere diameter is larger than that of a normal R7 cell, and similar to the size of the R1–6 type rhabdomere. In contrast to wild-type R7 cells, these cells extend through the depth of the retina. Therefore in more basal sections we observe 8 rhabdomeres instead of 7 in the ommatidia containing transformed R7 cells: in addition to the small R8 rhabdomere surrounded by the rhabdomeres of R1 to R6, there is an additional rhabdomere present between R1 and R6, corresponding to the transformed R7 cell. Therefore in the *sev-hsp-rough* transformants, a large fraction of R7 cells assume all of the morphological characteristics of the R1–6 type photoreceptor cells. We further demonstrated that in transformed flies homozygous for the *ora* mutation (Stark and Sapp, 1987), which causes specific degeneration of the rhabdomeres of the R1–R6 photoreceptor cells, the transformed R7 cells but not normal R7 cells also degenerate. We therefore conclude the expression of *rough* causes the transformation of the R7 precursor into an R1–6 type photoreceptor.

To find out whether *rough* had to be expressed in the R7 precursor or in a neighboring cell for the transformation of the R7 into an R1–6 type cell we used mitotic recombination to generate eyes in which the *sev-hsp-rough* construct was only present in a subpopulation of cells. These results indicated that expression of *rough* in just the R7 precursor cell is sufficient for the transformation to occur (Basler *et al.* 1990; Kimmel *et al.* 1990). The autonomous requirement of *sev-hsp-rough* for the R7 transformation contrasts the apparent non-autonomy of the *rough* mutant phenotype: In the developing ommatidia, the *rough* protein is required in R2 and R5 for the ommatidia to develop correctly. In its absence from these two cells, R3 and R4 develop aberrantly, but R2 and R5, as judged from morphological criteria, develop normally. Given the role of the *rough* homeodomain protein as putative transcription factor, it has been proposed that *rough*

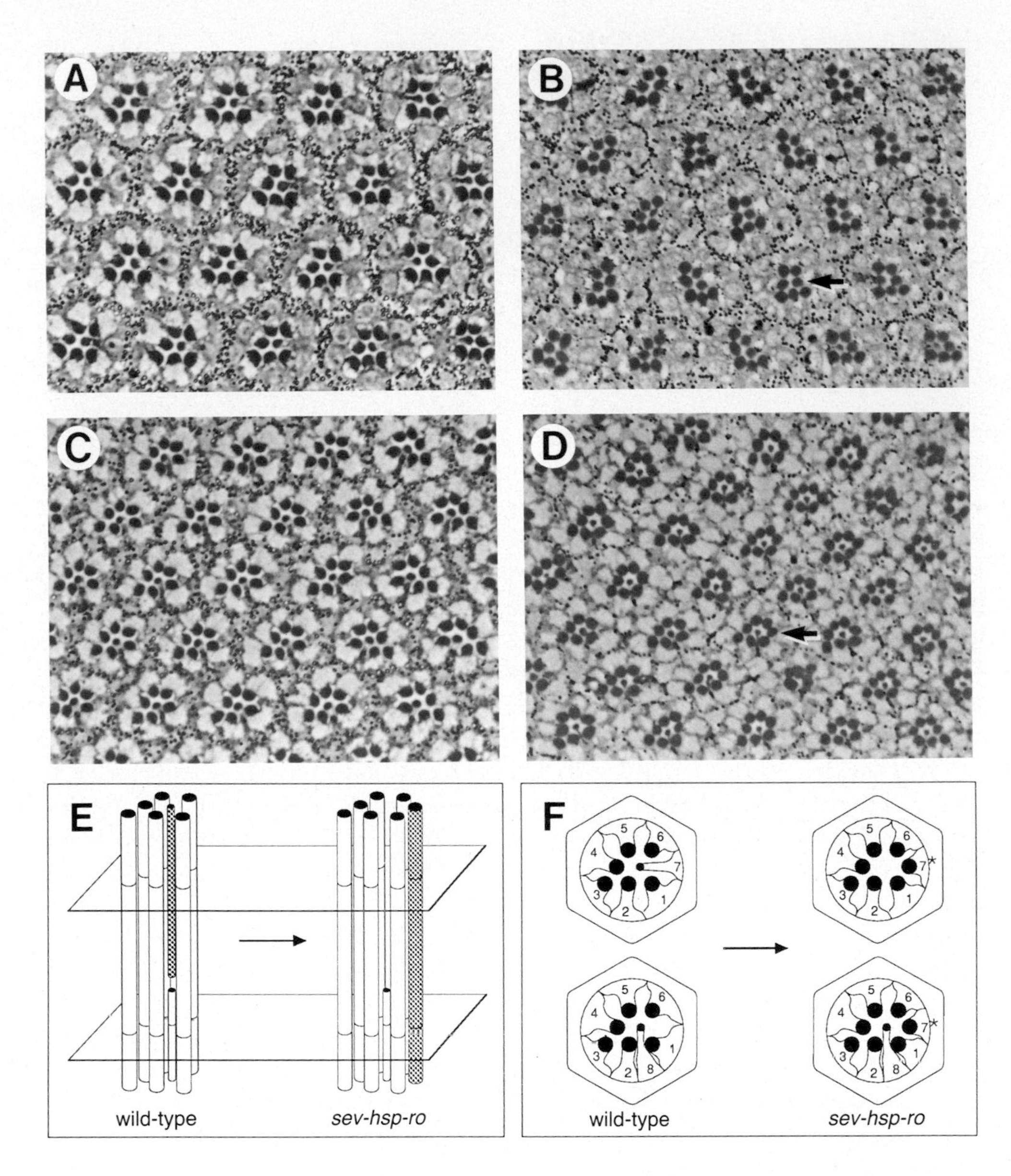

controls in R2 and R5 the expression of a gene encoding the R3/4 inducing signal (Tomlinson *et al.* 1988). One way to reconcile this contradiction is to suggest that in the presumptive R7 cell, the *rough* protein transcriptionally activates the R3/4 inducing signals which, in an autocrine way, trigger the R1–6 developmental pathway in R7. We think, however, that this is unlikely, and propose instead that *rough* serves an autonomous role in controlling the identity of the R2 and R5 cells. Since markers that distinguish between the identities of the outer photoreceptor cells are not available, the failure to obtain R2/5 identity in *rough* mutants could not be detected. According to this model the incorrect development of R3 and R4 in

Fig. 4. Transformation of R7 cells into R1–6 type photoreceptor cells. Histological sections through wild-type (A, C) and *sev-hsp-rough* (B, D) eyes. As schematically illustrated in (E), both apical (A, B) and basal sections (C, D) are shown. Each photoreceptor cell has a microvillar stack of membranes containing rhodopsin, called the rhabdomere, that projects towards the center of the ommatidium. In wild-type, the R7 rhabdomere differs morphologically from the rhabdomeres of R1 to R6 with respect to position, diameter and length: it is located in the center of the trapezoid formed by the larger rhabdomeres of R1 to R6, it is smaller in diameter and it extends to only two thirds the depth of the retina. In many ommatidia of heterozygous *sev-hsp-rough* transformants, the rhabdomere of the R7 cell has all the characteristics of an R1–6 type rhabdomere (arrow in B, D, hatched in E). In contrast to the seven rhabdomeres observed in basal sections of wild-type eyes (C), eight rhabdomeres are visible in altered ommatidia in corresponding sections of *sev-hsp-rough* transformants (D). Anterior is down. Magnification, ×1000.

rough mutants is a mere consequence of the incomplete differentiation of R2 and R5. It appears therefore that *rough,* similar to other homeotic genes, is a selector gene that distinguishes between alternative fates of the cells where it is expressed. Although the progenitors of R2 and R5 enter the photoreceptor cell pathway, in *rough* mutants they fail to assume an R2/5 identity, which is manifested in their failure to induce R3 and R4. In wild-type, *rough* is also expressed in R3 and R4 (Kimmel *et al.* 1990). According to our model the presence of *rough* protein in R3 and R4 also causes these cells to assume an R2/5 identity and thereby distinguishes them from R1 and R6. The failure to assume this identity in *rough* mutant clones, however, cannot be detected because R3 and R4 do not recruit other photoreceptor cells. Similarly, a potential, incorrect signal sent by the transformed R7 cells does not induce a change in cell fate of neighboring cone and pigment cells, presumably because they are not competent to respond to such a signal.

Interestingly, *rough* induced transformation of R7 into an R1–6 type photoreceptor is dependent on the *sevenless* and *boss* gene. When *sev-hsp-rough* is crossed into a *sevenless* or *boss* mutant background we observed a complete absence of cells in the R7 position. Therefore *sevenless* and *boss* are epistatic over *rough* in the R7 precursor. In *sevenless* mutants the R7 precursor appears never to enter the photoreceptor cell pathway. *Rough* expression in the absence of either *sevenless* or *boss* is not sufficient to initiate the photoreceptor cell development. This suggests that specification of photoreceptor cell fate requires more than one function. *Boss* and *sevenless* appear to be involved in the initiation of this pathway in R7, whereas *rough* functions in the specification of photoreceptor cell identity. *Rough* expression in R7 can compete with or even override the step that leads to the specification of the R7 cell fate. The dependence of *rough* function on a prior commitment of the *rough* expressing cell to photoreceptor development is consistent with its normal function in R2 and R5, since these cells apparently initiate photoreceptor cell development in the absence of *rough* (Tomlinson *et al.* 1988). Although *rough* controls photoreceptor cell identity, it is not sufficient to confer neural fate to undetermined cells.

Conclusions

The developing eye as an experimental system to study position-dependent cell fate specification has its limitations in that experimental manipulations such as laser ablation of individual cells and single cell transplantations are not possible. These techniques have been extremely useful in the *Drosophila* embryo (Technau and Campos-Ortega, 1986) and in the grasshopper (Doe *et al.* 1985) for assaying the developmental potential of individual cells in an altered environment. This shortcoming can be overcome by molecular genetic manipulation of the system. Genes controlling developmental decisions or specifying positional information can be fused to heterologous promoters and reintroduced into the germline. In this way positional values in the developing field can be changed without the inherent side effects of experimental manipulation. The specificity of the molecular genetic manipulation depends largely on the genes used and, just as importantly, on the specificity of the control elements that confer ectopic expression. Ectopic expression may be achieved by fusing the structural gene to an inducible promoter such as the hsp70 promoter. Indiscriminate expression of genes, however, is often lethal, and this lethal phenotype is then difficult to relate to the normal function of the gene. Similarly, the results reported here on the effect of ubiquitous expression of *rough* are difficult to compare with the normal function of *rough* in specifying photoreceptor cell identity in postmitotic cells. In contrast, by using the *sevenless* enhancer to drive *rough* expression, we expressed *rough* ectopically in only a small number of related cells, and indeed obtained specific information on the function of *rough*. Although the promoters of other genes have been characterized that are specifically expressed in subsets of ommatidial cells (i.e. rhodopsin, Mismer *et al.* 1987), the *sevenless* enhancer is unique in that it is active early, during the determination process of retinal precursor cells. This approach will become more widely applicable as more stage- and tissue-specific enhancers are identified and characterized by the *enhancer trap* method (O'Kane and Gehring, 1987; Wilson *et al.* 1989; Bier *et al.* 1989).

The dominant phenotype observed by the ectopic expression of *rough* demonstrates that the restricted *rough* expression observed in wild-type is critical for proper differentiation of several cell types in the compound eye. In contrast, the restricted expression pattern of *sevenless*, a putative receptor for an inductive signal, is not important for the correct determination of retinal precursor cells (Basler and Hafen, 1989; Bowtell *et al.* 1989*b*). This reflects the different roles of *rough* and *sevenless* in cell fate specification. While *rough* has an instructive role, *sevenless* acts as a sensory protein in cell fate specification.

Expression of *rough* in the R7 progenitor has demonstrated that the R7 precursor is multipotent. It can assume four distinct fates depending on the state of activity of the *sevenless* and *rough* gene products (Table 1): In wild-type, *sevenless* protein is activated, probably by *boss*, which results in an R7 fate. If *rough* is also present in the cell, it can become an R1–6 type photoreceptor cell. If *sevenless* is not activated, the cell becomes a normal cone cell. We assume that, as in the case of normal cone cells, high levels of *rough* protein abort development of

Table 1. *Potential fates of the R7 precursor cell depending on different combinations of* sevenless *and* rough *activity*

sev^+ ro^-	sev^+ ro^+	sev^- ro^-	sev^- ro^{++}
R7	R1–R6	CC	†

Activation of the *sevenless* protein (sev^+) in the wild-type R7 precursor leads to normal development and is most likely dependent on the *boss* gene product in the neighboring R8 cell. Expression of *rough* in the R7 precursor (ro^+) by *sev-hsp-rough* causes the cell to become an R1–6 type cell. If *sevenless* protein is not activated in *boss* mutants or is not functional in *sevenless* mutants (sev^-) the R7 precursor chooses the cone cell (C-C) fate. The strong *rough* expression in cone cells (ro^{++}) of *sev-hsp-rough* homozygotes causes abortive development (†).

this cell. The observed multipotency indicates that the precision with which cell fate is determined in the eye depends in part on the spatially and temporally restricted expression of positional cues and additionally on the combination of different intracellular signal transducers that respond to the inductive stimulus.

References

Basler, K. and Hafen, E. (1988). Control of photoreceptor cell fate by the *sevenless* protein requires a functional tyrosine kinase domain. *Cell* **54**, 299–311.

Basler, K. and Hafen, E. (1989). Ubiquitous expression of *sevenless*: position-dependent specification of cell fate. *Science* **243**, 931–934.

Basler, K., Siegrist, P. and Hafen, E. (1989). The spatial and temporal expression pattern of *sevenless* is exclusively controlled by gene-internal elements. *EMBO* **8**, 2381–2386.

Basler, K., Yen, D., Tomlinson, A. and Hafen, E. (1990). Reprogramming cell fate in the developing *Drosophila* retina: transfromation of R7 cells by extopic expression of *rough*. *Genes Dev.* **4**, 728–739.

Bier, E., Vaessin, H., Shepherd, S., Lee, K., McCall, K., Barbel, S., Ackerman, L., Carretto, R., Uemura, T., Grell, E., Jan, L. Y. and Jan, Y. N. (1989). Searching for pattern and mutation in the *Drosophila* genome with a P-lacZ vector. *Genes Dev.* **3**, 1273–1287.

Bowtell, D. D. L., Kimmel, B. E., Simon, M. A. and Rubin, G. M. (1989*a*). Regulation of the complex pattern of *sevenless* expression in the developing *Drosophila* eye. *Proc. natn. Acad. Sci. U.S.A.* **86**, 6245–6249.

Bowtell, D. D. L., Simon, M. A. and Rubin, G. M. (1988). Nucleotide sequence and structure of the *sevenless* gene of *Drosophila melanogaster*. *Genes Dev.* **2**, 620–634.

Bowtell, D. D. L., Simon, M. A. and Rubin, G. M. (1989*b*). Ommatidia in the developing *Drosophila* eye require and can respond to *sevenless* for only a restricted period. *Cell* **56**, 931–936.

Doe, C. Q. and Goodman, C. S. (1985). Early events in insect neurogenesis: II. The role of cell interactions and cell lineage in the determination of neuronal precursor cells. *Devl Biol.* **111**, 206–219.

Gibson, G. and Gehring, W. J. (1988). Head and thoracic transformations caused by ectopic expression of Antennapedia during *Drosophila* development. *Development* **102**, 657–675.

Gonzàlez-Reyes, A., Urquia, N., Gehring, W. J., Struhl, G. and Morata, G. (1990). A functional hierarchy of homeotic genes that is not based on cross-regulatory interactions. *Nature* **334**, 78–80.

Hafen, E., Basler, K., Edstroem, J. E. and Rubin, G. M. (1987). *Sevenless*, a cell-specific homeotic gene of *Drosophila*, encodes a putative transmembrane receptor with a tyrosine kinase domain. *Science* **236**, 55–63.

Harris, W. A., Stark, W. S. and Walker, J. A. (1976). Genetic dissection of the photoreceptor system in the compound eye of *Drosophila melanogaster*. *J. Physiol.* **256**, 415–439.

KIMMEL, B., HEBERLEIN, U. AND RUBIN, G. M. (1990). The homeo domain protein *rough* is expressed in a subset of cells in the developing *Drosophila* eye where it can specify photoreceptor cell subtype. *Genes Dev.* **4**, 712–727.

KLEMENZ, R., WEBER, U. AND GEHRING, W. J. (1987). The white gene as a marker in a new P-element vector for gene transfer in *Drosophila. Nucl. Acids Res.* **15**, 3947–3959.

MISMER, D. AND RUBIN, G. M. (1987). Analysis of the promoter of the ninaE opsin gene in *Drosophila melanogaster. Genetics* **116**, 565–578.

O'KANE, C. J. AND GEHRING, W. J. (1987). Detection *in situ* of genomic regulatory elements in *Drosophila. Proc. natn. Acad. Sci. U.S.A.* **84**, 9123–9127.

PATEL, N. H., SNOW, P. M. AND GOODMANN, C. S. (1987). Characterization and cloning of fasciclin III: a glycoprotein expressed on a subset of neurons and axon pathways in *Drosophila. Cell* **48**, 975–988

READY, D. F. (1989). A multifaceted approach to neural development. *Trends Neurosci.* **12**, 102–110.

REINKE, R. AND ZIPURSKY, S. L. (1988). Cell–cell interaction in the *Drosophila* retina: the bride of *sevenless* gene is required in photoreceptor cell R8 for R7 cell development. *Cell* **55**, 321–330.

SAINT, R., KALIONIS, B., LOCKETT, T. J. AND ELIZUR, A. (1988). Pattern formation in the developing eye of *Drosophila melanogaster* is regulated by the homeobox gene, *rough. Nature* **334**, 151–154.

STARK, W. S. AND SAPP, R. (1987). Ultrastructure of the retina of *Drosophila melanogaster*: the mutant *ora* (outer rhabdomeres absent) and its inhibition of degeneration in *rdgB* (retinal degeneration-B). *J. Neurogenetics* **4**, 227–240.

TECHNAU, G. M. AND CAMPOS-ORTEGA, J. A. (1986). Lineage analysis of transplanted individual cells in embryos of *Drosophila melanogaster*. I. The method. *Roux's Arch. Dev. Biol.* **195**, 389–398.

TOMLINSON, A., BOWTELL, D. D. L., HAFEN, E. AND RUBIN, G. M. (1987). Localization of the *sevenless* protein, a putative receptor for positional information in the eye imaginal disc of *Drosophila. Cell* **51**, 143–150.

TOMLINSON, A., KIMMEL, B. E. AND RUBIN, G. M. (1988). *Rough*, a *Drosophila* homeobox gene required in photoreceptors R2 and R5 for inductive interactions in the developing eye. *Cell* **55**, 771–784.

TOMLINSON, A. AND READY, D. F. (1986). *Sevenless*: a cell-specific homeotic mutation of the *Drosophila* eye. *Science* **231**, 400–402.

TOMLINSON, A. AND READY, D. F. (1987). Neuronal differentiation in the *Drosophila* ommatidium. *Devl Biol.* **120**, 366–376.

WILSON, C., PEARSON, R. K., BELLEN, H. J., O'KANE, C. J., GROSSNIKLAUS, U. AND GEHRING, W. J. (1989). P-element-mediated enhancer detection: an efficient method for isolating and characterizing developmentally regulated genes in *Drosophila. Genes Dev.* **3**, 1301–1313.

J. Cell Sci. Suppl. 13, 169–189 (1990)

Printed in Great Britain © The Company of Biologists Limited 1990

The genetic control of cell proliferation in *Drosophila* imaginal discs

PETER J. BRYANT

Developmental Biology Center, University of California, Irvine, CA 92717, USA

AND OTTO SCHMIDT

Department of Microbiology, University of Stockholm, S-10691 Stockholm, Sweden

Summary

The imaginal discs of *Drosophila* provide a favorable system for the analysis of the mechanisms controlling developmental cell proliferation, because of the separation in time between cell proliferation and differentiation, and the facility with which controlling genes can be identified and characterized. Imaginal discs are established in the embryo, and grow by cell proliferation throughout the larval period. Proliferation terminates in a regular spatial pattern during the final stages of larval development and the first day of pupal development. Cell proliferation can be locally reactivated in growth-terminated imaginal discs by removing part of the disc and culturing the remaining fragment in an adult host. The pattern of proliferation in these fragments suggests that cell proliferation in imaginal discs is controlled by direct interactions between cells and their neighbors. Proliferation appears to be stimulated by positional information differences, and these differences are reduced by the addition of new cells during tissue growth. Genes involved in cell proliferation control have been identified by collecting and analyzing recessive lethal mutations which cause overgrowth of imaginal discs. In some of these mutants (*fat*, *lgd*, *c43*, *dco*) the overgrowing tissue is hyperplastic; it retains its single-layered epithelial structure and is capable of differentiating. In two of the hyperplastic mutants (*dco* and *c43*), the imaginal discs show a failure of gap-junctional cell communication, suggesting that this form of cell communication may be involved in termination of proliferation. In other mutants the overgrowing disc tissue is neoplastic: it loses its structure and ability to differentiate, becoming a tumorous growth. The two genes that give a neoplastic phenotype (*dlg* and *lgl*) have been cloned and cDNAs of one of them (*lgl*) sequenced. The *lgl* gene encodes a cell surface molecule with significant homology to calcium-dependent cell adhesion molecules (cadherins). The expression of *lgl* at the time of termination of cell proliferation suggests that there are changes in the way that cells interact with one another at these times, and that these changes may be implemented by cell adhesion molecules. Direct cell contact within the epithelium, as well as signalling through gap junctions, appears to be involved in the cell interactions needed for the termination of cell proliferation. Mutations in genes encoding the *Drosophila* homologs of growth factors, growth factor receptors and oncogenes usually show an effect on cell-fate decisions rather than cell proliferation control, but this may be because oncogenic mutations in these genes would be dominant lethals and would therefore not be identified by conventional genetic analysis.

Cell proliferation during development

The remarkably uniform size and shape of organs is the outcome of a highly controlled program of cell proliferation that occurs during embryonic and postembryonic organ development, but we know little of how these growth

Key words: imaginal discs, neoplasia, hyperplasia, cell proliferation, cell interactions.

patterns are controlled. One of the least understood aspects of the problem is the question of what causes organs to terminate growth at the appropriate final size and cell number. Our approach to this problem is to find mutations that interfere with growth termination during development of imaginal discs in larvae of the fruit fly *Drosophila*, and use them to identify the corresponding genetic functions at the molecular level. It is analogous to the strategy of identifying human growth-control genes by cloning and analyzing tumor suppressor genes (Sager, 1989).

Cell proliferation in imaginal discs

Imaginal discs are sacs of undifferentiated epithelium in the *Drosophila* larva. They have no function in the larval insect, but during metamorphosis they undergo substantial morphogenesis and differentiation to give rise to various parts of the adult body surface (see Bryant, 1978). The precursors of the adult appendages (legs, wings, antennae, mouthparts) as well as the eyes and genitalia develop from imaginal discs, whereas the abdomen develops from small nests of cells in the larval integument called histoblasts. Many internal organs develop from precursors known as imaginal nests or rings, and the central nervous system develops from precursor cell populations, known as proliferation centers, within the larval central nervous system (see Bryant and Levinson, 1985).

Imaginal discs provide an ideal experimental system in which to analyze mechanisms of cell proliferation control. They have a simple histological structure, consisting mainly of a single-layered columnar epithelium along with some muscle precursor cells basal to the epithelium. One of their advantages is that cell proliferation occurs during the larval period and is completed before differentiation begins in the pupal period. Thus the control of cell proliferation can be studied in the absence of complications due to simultaneous overt differentiation, as occurs in most tissues of vertebrates.

After the discs are established in the embryo, their cell numbers increase exponentially during the larval period, then the growth rate slows down towards the end of larval development (Bryant and Levinson, 1985). Cell proliferation in the wing disc stops in a predictable spatial pattern during the late third larval instar and early pupal period, as shown by Schubiger and Palka (1987) (Fig. 1). The first cells to stop proliferating at the end of the larval period are located in a zone about ten cells wide at the position of the presumptive dorsal/ventral margin of the wing (O'Brochta and Bryant, 1985); the presence of this zone accounts for the cell lineage restriction at this position which has been observed in mitotic recombination experiments (Bryant, 1970; Garcia-Bellido *et al.* 1976). A second, narrower band of non-proliferating cells intersects the first one at right angles and coincides in position with the presumptive third longitudinal wing vein; surprisingly, it does not coincide with the anterior/posterior lineage restriction and compartment boundary (Garcia-Bellido *et al.* 1976) which lies nearby. By 3–4 h after puparium formation, cell proliferation has terminated over the entire

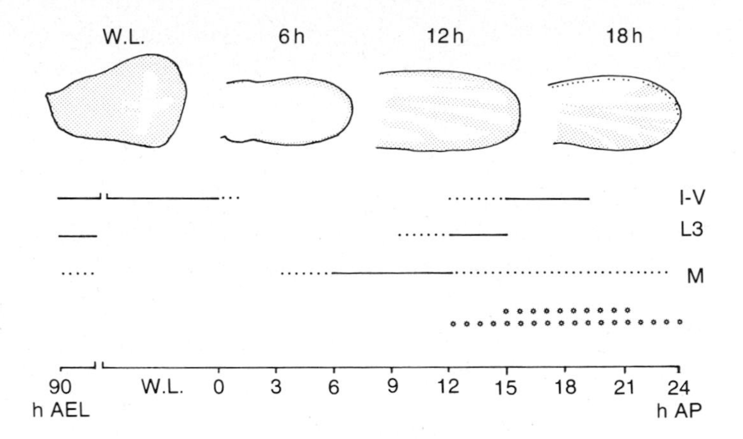

Fig. 1. Summary diagram of the spatial patterns of DNA replication in wing development. The time line at the bottom indicates the developmental time in hours after egg deposition (AEL) and after pupariation (AP). Asterisks: periods of frequent mitoses. From 15 to 21 h AP mitotic figures are especially abundant. Continuous lines: periods with abundant S-phase nuclei in the margin (M) in vein 3 (L3), and in the intervein regions (I–V). Broken lines: periods with few nuclei replicating. The figures at the top represent the spatial patterns of four selected periods (wandering larva [W.L.], 6 h, 12 h, and 18 h AP). (From Schubiger and Palka, 1987).

disc epithelium except for a row of cells along the anterior wing margin where bristles and nerve cell bodies will later develop. Measurements of DNA content indicate that the arrested cells are in the G_2 stage of the cell cycle (Fain and Stevens, 1982; Graves and Schubiger, 1982). After the epithelium has secreted the pupal cuticle there is a further cycle of mitosis between 12 and 24 h after puparium formation. This round of division occurs in a spatially ordered way, beginning in the cells along the presumptive wing veins and spreading into the inter-vein regions. After the final round of mitosis, the disc epithelium enters the stage of adult differentiation and produces the adult wing cuticle.

Since imaginal discs terminate cell proliferation at a reproducible cell number, and since the adult derivatives of imaginal discs are made up of pattern elements (e.g. bristles, sensilla, hairs) that are derived from individual cells or fixed small numbers of cells, it might be assumed that cell proliferation in imaginal discs terminates after a certain number of cell divisions. However, several studies show that growth termination in imaginal discs is not a function of a cell-division counting mechanism. The amount of proliferation within individual cell clones marked by mitotic recombination during normal development is highly variable, and clones in different individual flies occupy partly overlapping territories on the adult structure (Bryant and Schneiderman, 1969; Bryant, 1970), showing that the lineage pattern is variable. High doses of X-irradiation during larval development can kill or mitotically inhibit up to 75% of the cells in imaginal discs, without causing noticeable abnormalities of the adult structure. The surviving cells compensate and form the adult structure by undergoing additional proliferation, as shown by enlargement of mitotic recombination clones compared to controls (Haynie and Bryant, 1977). When individual cell clones are genetically altered by mitotic recombination so that they have a higher or lower proliferation rate compared to the remaining cells, they contribute more or less to the adult structure respectively, but the structure is normal (Simpson, 1976, 1979; Simpson and Morata, 1980, 1981). Thus the cell proliferation pattern is naturally variable and can be experimentally altered without noticeably altering the size or pattern of the structure produced. It therefore seems likely that proliferation is controlled by interactions between cells, independently of their lineage history. Our working hypothesis (Bryant and Simpson, 1984) is that cell proliferation is controlled by the developing map of positional information (Wolpert, 1971) in the disc and that cell proliferation stops when the map is complete.

Although the imaginal discs are undifferentiated epithelia, experimental analysis has revealed that different parts of a full-grown disc are already different from one another, so that if isolated they will construct specific parts of the adult structure (see Bryant, 1978). This covert pattern of differentiation indicates that a detailed positional information map must have become established during the growth of the disc.

Growth termination, under normal conditions, coincides with hormonal changes occurring at the onset of metamorphosis, but it also occurs in the absence of these hormonal changes. When whole discs from early larvae are implanted into the growth-permissive environment of the female adult abdomen, they grow until they reach approximately their normal final size and shape, then they stop growing (Bryant and Levinson, 1985). Thus under permissive growth conditions, the attainment of a correct size and shape is due to disc-intrinsic properties, consistent with the idea that it is controlled by the positional information map.

Growth patterns and cell communication during regeneration

Studies of transplanted disc fragments show that growth termination in regenerating imaginal disc fragments is also controlled by interactions between cells and their neighbors within the disc epithelium. When fragments of imaginal discs are cultured in adult female flies for several days before transfer to larval hosts for metamorphosis, the cuticular patterns produced after metamorphosis reveal that the tissue fragment has either regenerated or duplicated during the culture period (see Bryant, 1978). Bryant and Fraser (1988) investigated the pattern of cell proliferation in a 3/4 fragment of the wing disc, which regenerates the missing sector by cell proliferation over a period of several days. Individual cells in the starting fragment were marked by iontophoretic injection of high molecular weight, lysinated rhodamine dextran, and the positions of the marked cells were determined by fluorescence microscopy before and after various culture times. The results show that cells along the two wound edges are brought together by wound healing during the first day of the culture period. The marked cells then move apart over the next three days as new cells are added between them. The results indicate that intercalary tissue growth is restricted to the immediate region of the wound, and does not spread significantly into adjacent tissue. During the regenerative growth, the new cells become specified to make pattern elements that would normally intervene between the confronted cells, a process termed intercalation (French et *al.* 1976).

The pattern of DNA synthesis during intercalation has been monitored by analyzing the incorporation of tritiated thymidine, which is detected by autoradiography of serially sectioned disc fragments (O'Brochta and Bryant, 1987) or of the thymidine analog bromodeoxyuridine, which is detected by immunolocalization on whole mounts (Bryant and Fraser, 1988). Intense and localized DNA synthesis occurs adjacent to the healed wound, starting at 16 to 18 h after the beginning of the culture period (Fig. 2). DNA synthesis is stimulated over a distance of a few cell diameters from the wound, and localized cell proliferation continues for three to more than five days. The results suggest that DNA synthesis is stimulated by the positional information discontinuity across the healed wound, and that the stimulus is attenuated as the positional information discontinuity is reduced by the addition of cells carrying intermediate positional values.

The possible role of gap-junctional cell communication across the healed wound has been investigated by injecting small fluorescent dye molecules into cells close to the wound region, and determining the degree to which the dye passes into or across the wound area (Bryant and Fraser, 1988). In undamaged imaginal wing discs, there is a directional bias but, in contrast to a report in the literature (Weir and Lo, 1982), there are no fixed boundaries to this form of communication (Fraser and Bryant, 1985). In disc fragments, dye movement across the wound was not detectable after one day of culture. Since DNA synthesis is already underway in the growth zone at this time, we conclude that gap-junctional communication across the wound is not necessary for initiating the local cell proliferation involved in intercalary regeneration. After two days there was some dye transfer into the growth zone from cells outside it, but not as much as in other directions. After three days there was significant dye transfer into the growth zone. The time of re-establishment of gap-junctional communication therefore indicates that this form of cell communication is not involved in stimulating regenerative cell proliferation, but suggests that it could be involved in terminating it.

Imaginal disc overgrowth mutations

Genes involved in cell proliferation control in imaginal discs have been identified by selecting recessive mutations (imaginal disc overgrowth mutations) which cause the imaginal discs to continue growing by cell proliferation beyond the normal limits during an extended larval period. In some cases, the mutant animals remain in the larval stage for several weeks rather than the normal four days. The mutations are all recessive lethals, causing death of the animal either

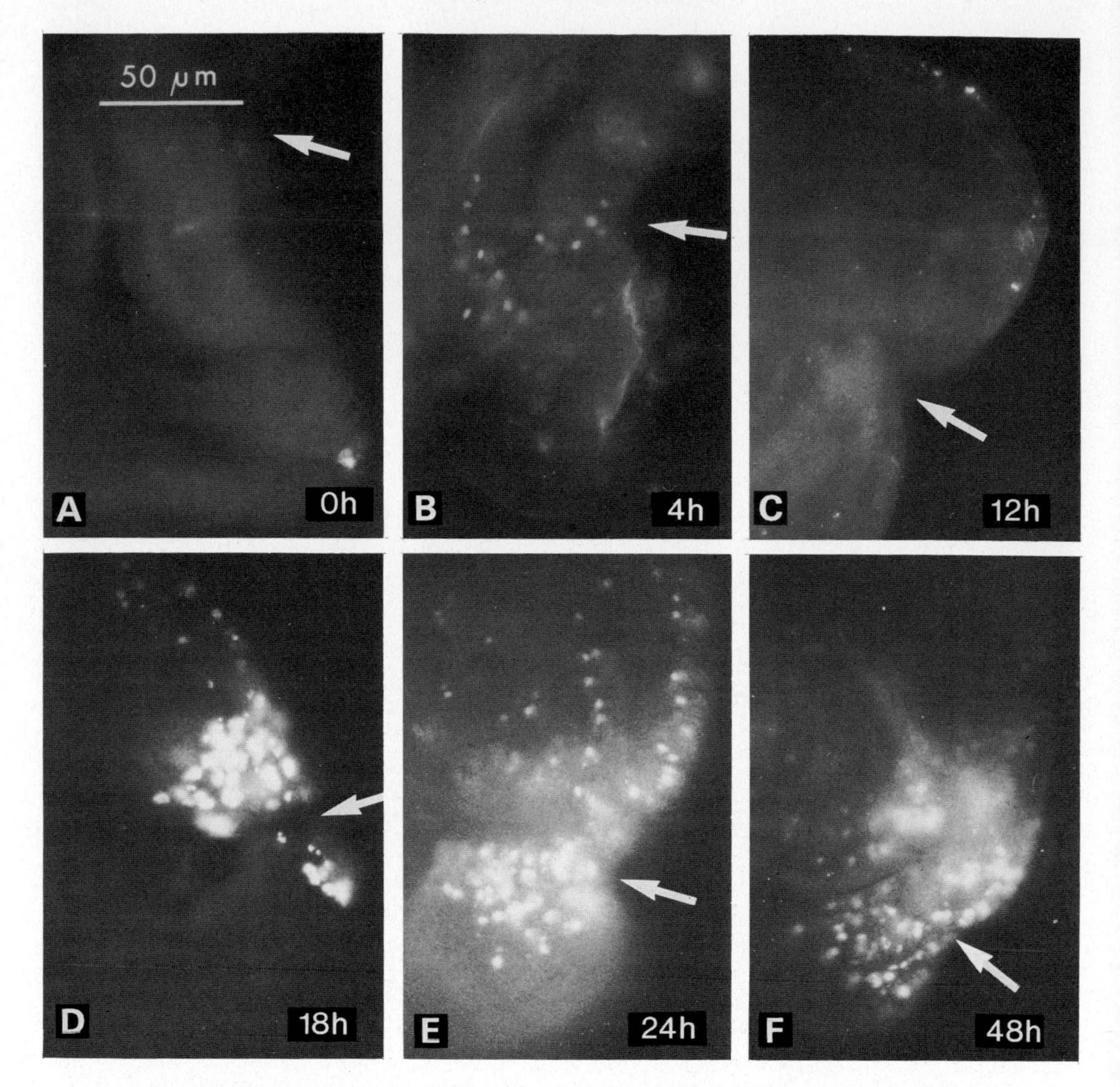

Fig. 2. Pattern of incorporation of bromodeoxyuridine into 3/4 wing disc fragments *in vitro* after various periods of culture *in vivo*, visualized by indirect immunofluorescence using a monoclonal antibody against BrdU. (A) 0 h; (B) 4 h; (C) 12 h; (D) 18 h; (E) 24 h; (F) 48 h. The position of the wound vertex (A,B) or healed wound (C–F) is indicated by an arrow. In (A) and (B), the wound has not yet healed, and there is no localized BrdU incorporation. In (C), the wound has begun to heal, but there is still no localized incorporation. In (D–F), a cluster of labeled cells can be seen adjacent to the healed wound. Trails of weakly labeled cells along the disc margin can be seen in (D) and (E). (From Bryant and Fraser, 1988).

as a late larva or as a pupa which forms after a prolonged larval period. They fall into two categories depending on the effect of the mutant lesion on the histological structure of the imaginal discs.

Table 1. *Imaginal disc overgrowth mutants of* Drosophila melanogaster

Name	Abbreviation	Chromosome	Location	Reference
Hyperplastic				
fat	*fat*	2	24D	Bryant *et al.* (1988)
giant discs	*lgd*	2	32D	Bryant and Schubiger (1971)
tumorous discs	*tud*	2	59F	Gateff and Mechler (1989)
c43	*c43*	3	85	Martin *et al.* (1977)
discs overgrown	*dco*	3	100A	Jursnich *et al.* (1990)
Neoplastic				
discs large	*dlg*	1	10B	Stewart *et al.* (1972)
giant larvae	*lgl*	2	21A	Gateff and Schneiderman (1974)

1. Hyperplastic overgrowth mutants

Five genetic loci are known in which mutations give the hyperplastic overgrowth phenotype (Table 1). In these mutants, the imaginal discs show abnormal folding patterns early in their development, then they grow to several times their normal size during the extended larval period. They maintain their single-layered epithelial structure and their ability to differentiate, but the additional tissue causes the imaginal disc to attain an abnormal morphology with additional lobes and folds as compared to normal (Fig. 3). In *lgd* and possibly in other mutants, not only the imaginal discs but also other groups of proliferating cells whose function is to produce internal adult derivatives (the imaginal rings for the foregut, hindgut, and salivary glands) show overgrowth during the extended larval period (Bryant and Levinson, 1985).

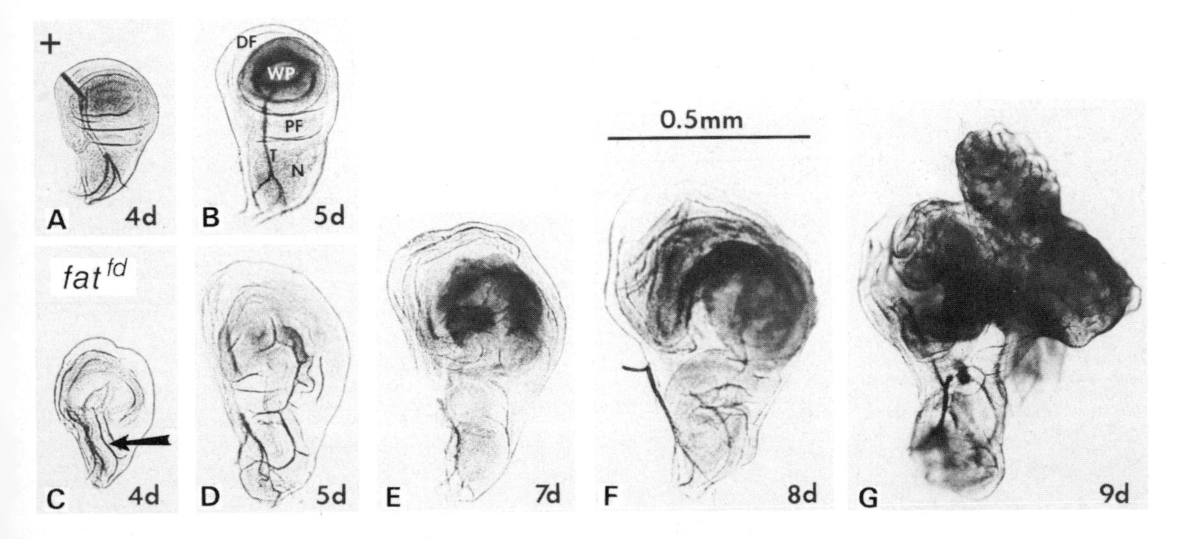

Fig. 3. Wing imaginal discs from wild type and *fat* homozygotes. (A, B) Wild type at 4 and 5 days. (C–G) fat^{fd} at 4, 5, 7, 8, and 9 days. DF, distal folds; N, presumptive notum; PF, proximal folds; T, trachea; WP, wing pouch; arrow, some of the extra folds mentioned in text. Bar, 0.5 mm. (From Bryant *et al.* 1988).

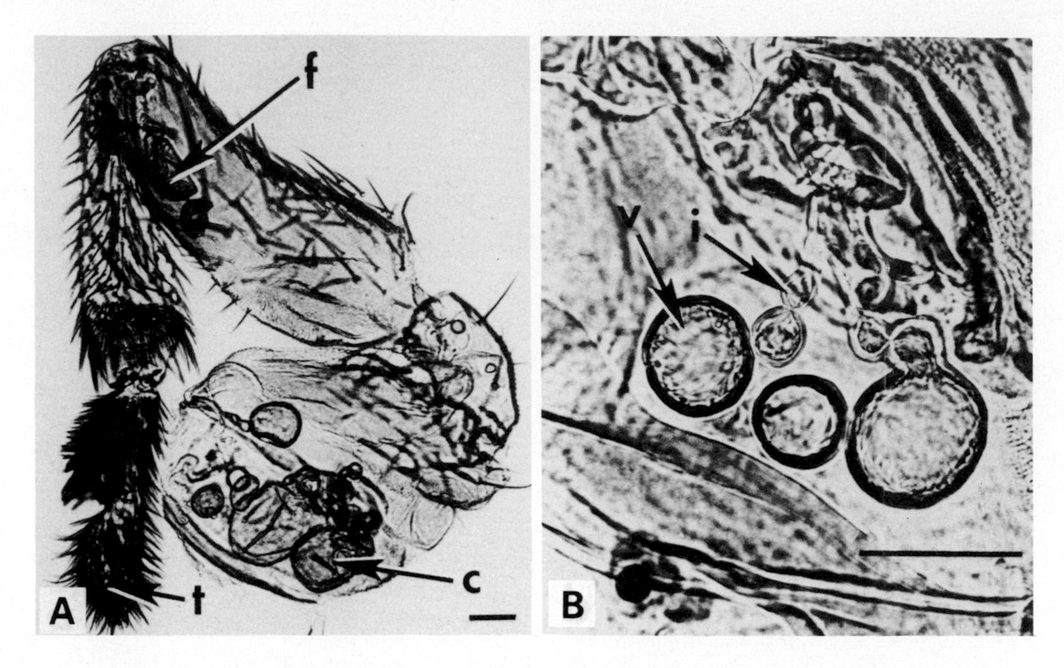

Fig. 4. Cuticular abnormalities in legs from *fat*^{fd} pharate adults. (A) Male first leg with enlarged coxa, laterally expanded and truncated tarsus (t), and vesicles with internal bristles in coxa (c) and femur (f). (B) Male first leg coxa, showing vesicles (v) apparently budding from ingrowth (i). Bars, 50 μm. (From Bryant *et al.* 1988).

In most of the hyperplastic overgrowth mutants, homozygotes are sometimes able to pupariate and to differentiate adult cuticular structures, although the resulting abnormal adult fly (pharate adult) is unable to emerge from the pupal case. The structures produced show many abnormalities specific to the genetic locus.

Pharate adults which are homozygous for lethal *fat* mutations show a characteristic syndrome of abnormalities, including invaginations and evaginations of cuticle from the body surface, completely separated cuticle vesicles (both internal and external to the body surface) and bristle polarity rosettes (Fig. 4; Bryant *et al.* 1988). These defects, especially the separated vesicles, suggest that the mutations interfere with cell adhesion within the epithelium. They show a striking resemblance to the set of abnormalities in the wing of the moth *Manduca* that are produced by transplanting pieces of pupal integument to inappropriate positions along the proximal/distal axis (Nardi and Kafatos, 1976*a*,*b*), and that have been interpreted as resulting from disruption of an adhesion gradient. It has therefore been suggested (Bryant *et al.* 1988) that the basic defect in *fat* is in cell adhesion. Defective cell adhesion could not only lead to the observed abnormalities of morphogenesis, but it could also interfere with the cell communication needed for proliferation control.

Pharate adults of *dco* show some similarities to *fat* pharate adults including separated cuticular vesicles, albeit at a much lower frequency than in *fat*

(Jursnich *et al.* 1990). Both mutants, and *c43*, show substantial widening and disruption of segmentation in the distal part of the leg, the tarsus (Bryant, 1987). *lgd* shows a different set of abnormalities, notably a drastic reduction (to about 20% of normal) in the number of bristles and other cuticular sense organs produced, subdivision of the tarsus into only two segments rather than the usual five, and large-scale mirror symmetrical pattern duplications in the leg discs when the larvae are reared under crowded conditions (Bryant and Schubiger, 1971).

Since growth termination in normal and regenerating wild-type discs appears to depend on cell interactions within the imaginal disc, there may be abnormalities in the way that cells interact in the mutant discs. Gap-junctional communication has been tested by injecting fluorescent marker dye into individual disc cells and observing passage into adjacent cells (dye coupling). Discs from wild type, *fat* and *lgd* show extensive dye coupling at all tested stages, but discs from four-day *dco* and *c43* larvae show complete lack of dye coupling (Jursnich *et al.* 1990). At later stages, when these discs are already overgrown, they start to show some dye coupling. Some *dco* genotypes, and *c43*, also show a substantial reduction in the surface density of gap junctions between imaginal disc cells when compared to wild type (Ryerse and Nagel, 1984; Jursnich *et al.* 1990). Although the defects in gap-junctional communication and in cell proliferation control may be unrelated pleiotropic effects of these mutations, another possibility is that the gap-junctional defect is responsible for the failure of *dco* and *c43* mutant discs to terminate cell proliferation. The latter possibility would be consistent with the common finding of decreased gap-junctional communication in transformed cells (Sheridan, 1989).

2. Neoplastic overgrowth mutants

In addition to causing overgrowth, these mutations cause breakdown of the single-layered epithelial structure of the imaginal discs, converting them into spongy masses of tissue as they continue to grow beyond the normal final size. The disc cells become cuboidal in shape, show irregular apical–basal polarity, and lose the ability to differentiate even after transplantation into wild-type hosts. The abnormal discs are therefore tumorous or neoplastic growths, in contrast to the hyperplastic overgrowth mutants in which the imaginal discs maintain a much more normal histological structure. According to current usage, the wild-type alleles of these genes are tumor suppressor genes.

Two genes are known in which mutations give the neoplastic overgrowth phenotype: discs large (*dlg*) (Stewart *et al.* 1972; Murphy, 1974) and giant larvae (*lgl*) (Gateff, 1978; Table 1). Mutations in both of these genes cause the imaginal discs to grow by cell proliferation beyond their normal final size, transform into solid tumors, fuse with one another and the brain, and lose their ability to differentiate (Fig. 5; Woods and Bryant, 1989). *lgl* mutations also cause overproliferation of neuroblasts in the larval brain, giving rise to ganglion mother cells which fail to differentiate into neurons (Gateff, 1978). This converts

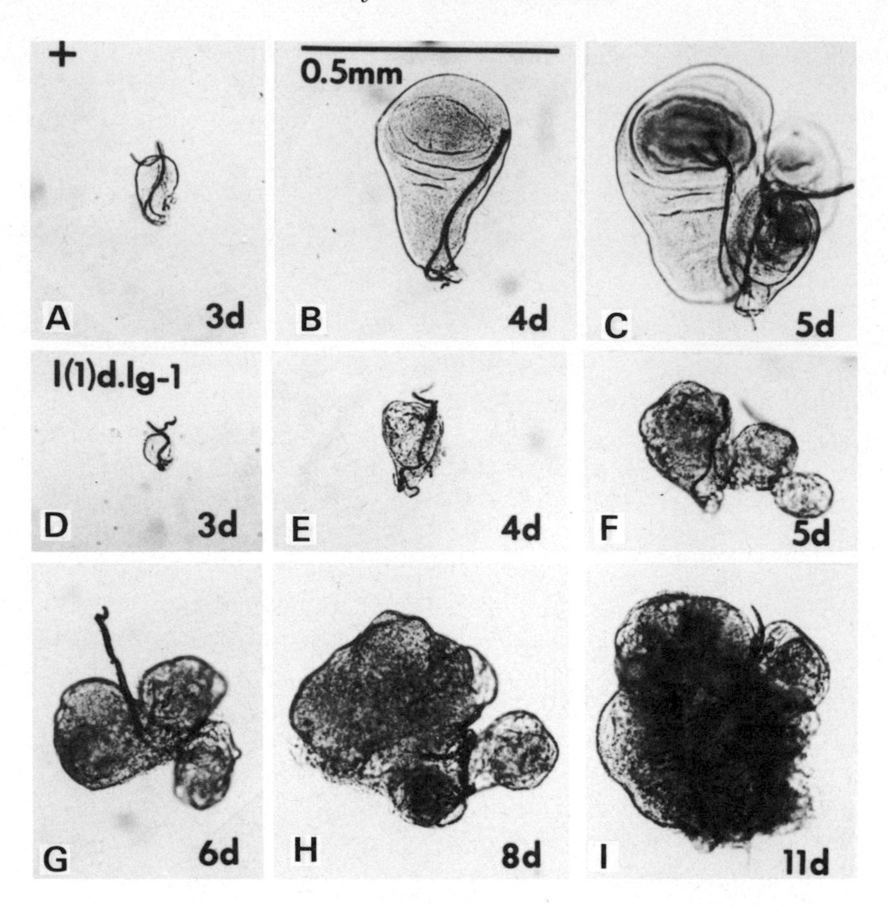

Fig. 5. Wing imaginal discs from wild type (A–C) and *dlg*$^{X1-2}$ hemizygotes (D–I); (A,B) wild type discs at 3 and 4 days; (C) wing, haltere and third leg disc at 5 days, with their connections intact; (D,E) mutant wing discs at 3 and 4 days; (F–I) mutant wing, haltere, and third leg disc complex at 5,6,8, and 11 days. Note the progressive fusion between these three discs. W, Wing disc; H, haltere disc; 3L, third leg disc. Bar, 0.5 mm. (From Woods and Bryant, 1989).

the tissue into a transplantable and invasive malignant neuroblastoma (Gateff and Schneiderman, 1974).

The *lgl* gene produces two transcripts of 5.7 and 4.3 kb in size which are detected mainly during embryogenesis and at the larval–pupal transition (Mechler *et al.* 1985). Sequencing of cDNA fragments from *lgl* revealed a conceptual reading frame with a coding capacity for a $130{\times}10^3\,M_r$ protein. Fragments of the cDNA were incorporated into bacterial expression vectors to make fusion proteins, which were then used to raise antibodies. These antibodies detect a protein of the appropriate size both during embryogenesis and at the end of larval development (Klämbt and Schmidt, 1986). The protein is found at the outer surface of dissociated cells and is attached to the cell membrane in an unknown way (Klämbt *et al.* 1989). Sequence similarity of the deduced protein to

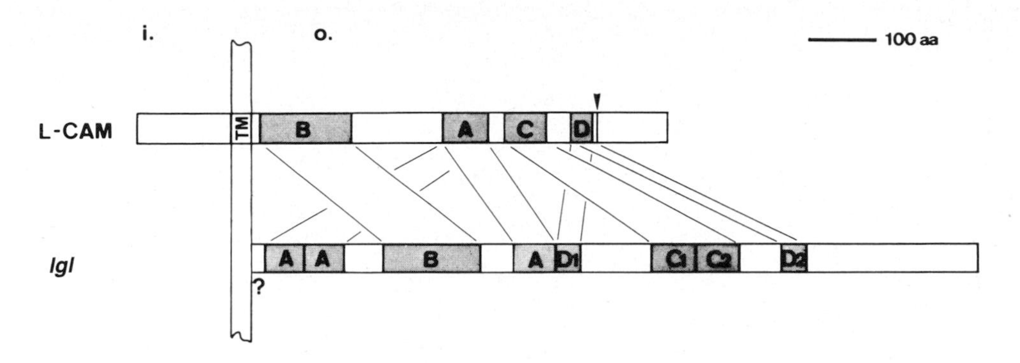

Fig. 6. *lgl* and L-CAM protein structure. Schematic drawing of the structural organization of similar protein regions in L-CAM as an example of a cadherin molecule and the *lgl* protein. The L-CAM protein (Gallin *et al.* 1987) spans the membrane (TM domain) and has a $16\times10^3 M_r$ intracellular domain (i). The *lgl* protein exhibits no transmembrane domain (Lützelschwab *et al.* 1987) and is attached to the membrane and/or the extracellular matrix by an unknown mechanism. Four regions conserved between L-CAM and *lgl* amino acid sequences are indicated. A and B are repeats containing putative N-glycosylation sites. B is a modified A-type region. C and D represent smaller regions containing negatively and positively charged amino acids, respectively. o, outside, i, inside the cell; TM, transmembrane domain (from Klämbt *et al.* 1989). The bar represents 100 amino acids.

those of cadherins (Fig. 6; Klämbt *et al.* 1989), a family of calcium-dependent cell adhesion molecules in vertebrates (Takeichi, 1988), and the presence of an RGDV motif (Lützelschwab *et al.* 1987) which could function as a binding site for cell–cell or cell–matrix interaction (Ruoslahti and Pierschbacher, 1986), suggest that the protein is involved in cell adhesion.

In sections of embryos which have been stained with antibodies against the *lgl* protein, the antigen is detected predominantly in developing larval structures (Fig. 7). The nervous system contains much less of the protein, except in areas where nerve cells are in the process of axonogenesis (Klämbt *et al.* 1989). The protein is also transiently expressed in pole cells and in the neuroblasts of the presumptive optic lobes of the brain (Klämbt and Schmidt, 1986). The protein is expressed again mainly in imaginal discs and in the central nervous system at the end of larval development.

The *lgl* protein is present in those tissues, both imaginal and larval, which are about to stop cell proliferation and begin differentiation (Klämbt and Schmidt, 1986; Klämbt *et al.* 1989). In homozygous mutant tissues, the cells show reduced contact with one another, cuboidal rather than columnar shape, and excessive proliferation. When transplanted into wild type hosts, mutant brain cells dissociate from the implant, invade host tissues and eventually kill the host (Gateff, 1978). These results indicate that the *lgl* protein functions in cell adhesion, and support the idea that cell proliferation is controlled by interaction between cells and their neighbors, with the signalling process dependent on intimate and direct cell contact.

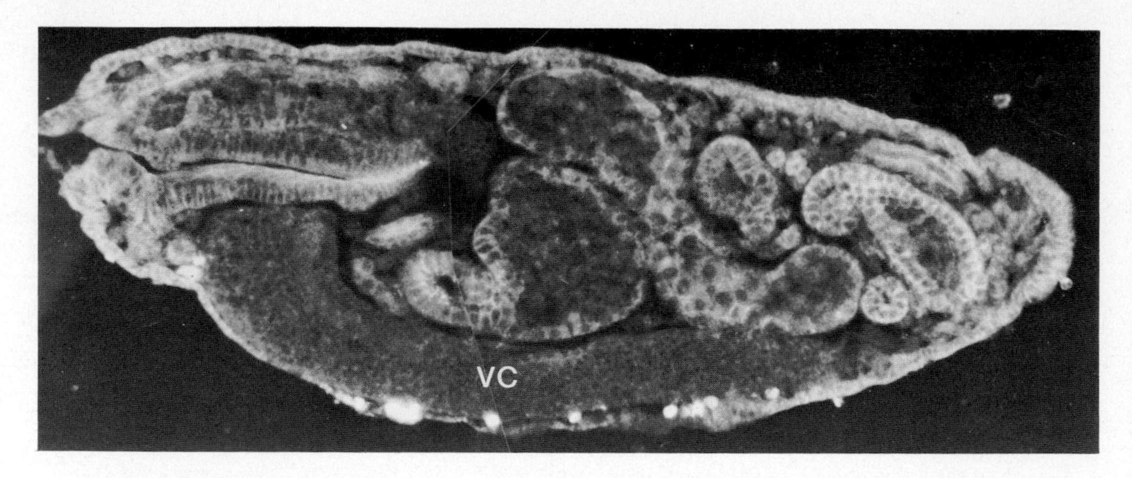

Fig. 7. Localization of *lgl* protein in a parasagittal section of a wild type *Drosophila* embryo. Binding of antibody to the *lgl* protein was visualized by indirect immunofluorescence. vc, ventral nerve cord. (O. Schmidt, unpublished).

Although both larval and imaginal primordia express the *lgl* protein in wild type embryos (Fig. 7), only imaginal primordia are phenotypically affected by *lgl* mutations (Gateff, 1978). The absence of an effect on larval tissues in homozygous embryos could be due to provision of sufficient wild type gene product by the heterozygous mother (Klämbt *et al.* 1989), or due to compensation of embryonic *lgl* functions by other gene products. The amount of rescue by either means must be insufficient to supply the imaginal disc cells, either because they have a higher requirement or because the rescuing gene products are diluted out to subthreshold levels during the extensive cell proliferation occurring in these tissues.

Tissue autonomy of overgrowth mutants

Different imaginal disc overgrowth mutants show different patterns of overgrowth, the exact pattern of extra folds and lobes being characteristic for each gene (Bryant, 1987). This suggests that these mutations exert their effects directly on the imaginal discs, rather than by affecting disc growth indirectly through an effect on some other system such as the endocrine system. This point has also been demonstrated by transplanting mutant imaginal discs into wild-type adult hosts and observing their growth properties. In all of the mutants discussed above, the growth abnormality is disc-autonomous; that is, the transplanted tissue shows overgrowth whereas wild type discs cease growth at a characteristic size and cell number (*lgl*, Gateff and Schneiderman, 1974; wild type and *lgd*, Bryant and Levinson, 1985; *c43*, Martin *et al.* 1977; *fat*, Bryant *et al.* 1988; *dco*, Jursnich *et al.* 1990; *dlg*, D. Sponaugle and D. Woods, unpublished observations). Thus, the growth-control defects are intrinsic to the imaginal discs, rather than being a result of the extended larval period or a systemic defect. The results are therefore consistent with the regeneration experiments that show the importance of disc-intrinsic growth-control mechanisms.

Developmental delay in overgrowth mutants

In all of the imaginal disc overgrowth mutants so far described, the overgrowth occurs during a greatly prolonged larval life, which is followed by death of the mutant animal either as a larva or as a pupa. In *dlg*, *fat*, *lgd*, *dco*, and *c43*, the prolonged larval life is associated with very low levels of the molting hormone 20-hydroxyecdysone during and after the time when this hormone peaks in the wild type (F. Sehnal and P. Bryant, unpublished results). Although in principle this endocrine defect may be an independent, pleiotropic effect of the mutations, we feel that it is more likely that the endocrine defect is a secondary result of the imaginal disc overgrowth. The most direct evidence for this idea is from recent work by Poodry and Woods (1990) who have shown that in *dlg* the delay in development can be prevented by eliminating imaginal discs with ionizing radiation. Not only is the delay prevented, but the irradiated animals, unlike the controls, are able to pupariate.

It has been known for many years that the presence of a regenerating imaginal disc in a larval insect can delay the onset of metamorphosis of the host, suggesting that growing imaginal disc tissue sends an inhibitory signal to the endocrine system. In the moth *Ephestia kühniella*, extirpation of a single wing disc from a mature larva (just prior to cocoon-spinning) resulted in a pupation delay of about eight days in those animals which regenerated the wing, but no delay in the animals which failed to regenerate (Pohley, 1965). If two wing discs were regenerating, the delay was about 1.5 days longer. Madhavan and Schneiderman (1969) conducted similar experiments on another moth, *Galleria mellonella*, and showed that extirpation of one wing disc delayed pupation by five days, extirpation of two discs caused an eight-day delay, and extirpation of all four wing discs caused a 14-day delay. Both Pohley (1965) and Madhavan and Schneiderman (1969) suggested that the regenerating imaginal discs somehow reduced the concentration of ecdysone in the animal.

In principle, the delay of pupation which is caused by imaginal disc extirpation could be mediated by a hormone, or it could be mediated by a signal transmitted from the imaginal discs to the endocrine glands *via* the nervous system. However, it is almost certain that the signal is an endocrine one, since a similar delay can be produced by the implantation of regenerating tissue into the hemocoel of an otherwise normal host. For example, when imaginal wing discs were implanted into last-instar *Ephestia* larvae, they underwent duplication and delayed pupation of the host by 8.6 days (forewing) or 7.7 days (hindwing) (Rahn, 1972). When wing discs which had already been cultured in one host for various periods were implanted into a second host, they delayed its pupation by an amount which decreased with the age of the regenerate. This meant that the total culture time available to the implant before pupation was practically constant. The prepupal phase was prolonged until regeneration (more properly in this case, duplication) was complete (Rahn, 1972). Similar results were obtained by Dewes (1975, 1979) using the regenerating half-genital disc of *Ephestia*. Apparently, a very important control system links developing tissues and the endocrine system in insects, and this system ensures that different parts of the insect develop in harmony. However, the nature of the signal from regenerating tissues is not known, and the question of how the endocrine glands and/or their target tissues are affected by the regenerating tissue has received little attention.

Simpson *et al.* (1980) have reported an ingenious demonstration of regeneration-induced developmental delay in *Drosophila* using several genetic techniques. Extensive regeneration in many imaginal discs of the same larva can be induced by applying a sublethal heat pulse to a larva carrying a temperature-sensitive, cell-lethal mutation (Girton and Bryant, 1980). These treated animals show delayed pupariation (Russell, 1974; Simpson and Schneiderman, 1975), and the amount of delay is correlated with the amount of extra growth that occurs during regeneration (Simpson *et al.* 1980). Furthermore, by heat-treating gynandromorphs carrying different proportions of wild type and mutant tissue, Simpson *et al.* (1980) showed that the pupariation delay was greater with larger amounts of mutant tissue. Elimination of entire imaginal discs (which, in *Drosophila*, is not followed by regeneration) did not delay pupariation, indicating that it is regeneration *per se*, rather than any other heat-induced damage, that delays pupariation. Furthermore, Simpson et al. (1980) showed that, as in the studies on other insects, the wild type (non-regenerating) tissues in the delayed animals were not enlarged. This is consistent with other results indicating that each imaginal disc has an intrinsic growth limit.

If regenerating imaginal discs cause developmental delay by an endocrine mechanism, this would suggest that imaginal tissues secrete a factor which inhibits release of 20-hydroxyecdysone from the endocrine glands. In the overgrowth mutants, the extended larval life could be a result of continuous production of this inhibitor by the imaginal discs.

Maternal effects of imaginal disc overgrowth mutations

Each zygotic lethal mutation causes death of the animal at a characteristic developmental stage, so phenotypic analysis of mutants can reveal functions of the affected gene only up to that stage. The imaginal disc overgrowth mutants die as late larvae or pupae. However, using special methods, it has been shown that the normal alleles of some of these genes also function in the adult during gametogenesis. In some cases this has been shown by testing the function of mutant germ cells transplanted into wild type hosts in the embryonic stage; in other cases homozygous mutant clones have been induced in the germ line of an otherwise heterozygous adult by X-ray induced mitotic recombination. In the case of *dlg*, homozygous germ-line clones in the ovary produce eggs in which embryonic development is disrupted because of the lack of normal gene product supplied during oogenesis. In the maternally affected embryos, an additional layer of neuroblasts is formed in the gastrula and subsequent overgrowth of the nervous system prevents completion of dorsal closure (Perrimon, 1988). The normal gene is therefore required for both cell proliferation control and epithelial integrity in the embryos, just as it is required in imaginal discs. One of the *dlg* transcripts is limited to the adult ovaries and is missing in several of the viable heteroallelic combinations (Woods and Bryant, 1989), so this may be responsible for the maternal-effect lethality shown in these genotypes.

Most of the other imaginal disc overgrowth mutations have recently been tested by germ-line transplantation and/or mitotic recombination, and all of those tested (*lgl*, *dco*, *lgd*, *fat*, *c43*) interfere with functions of the female germ line (Szabad *et al.* 1990), preventing the production of viable progeny. In one case (*lgd*) the number of cells in the egg chamber was over 40 compared to the 16 (15 nurse cells and one oocyte) found normally, suggesting an effect on cell proliferation control. In the other cases the mutant germline either failed to produce eggs or produced eggs which failed to complete their development without showing any obvious abnormalities in cell proliferation control. In imaginal discs the mutations appear to affect cell proliferation control by interfering with cell interactions controlling proliferation. We suggest that in the female germ line the mutations might also interfere with cell interactions (for example, between the oocyte and nurse cells, or between the oocyte and the surrounding follicle cells) but that these interactions control aspects of egg development other than cell proliferation.

Mutations in growth factor, growth factor receptor and oncogene homologs

Another approach to the problem of cell proliferation control in *Drosophila* is to use mutational analysis to investigate the functions of genes which show homology to various molecules thought to be involved in the control of growth and cell proliferation in vertebrates. Table 2 lists the genes showing growth factor, growth factor receptor, or oncogene homology for which a mutant phenotype has been identified in *Drosophila*. Most of these genes have recently been reviewed in detail by Hoffmann (1989). Surprisingly, for twelve of the thirteen genes listed the mutations cause a change in cell fate, whereas in only two cases is there an indication of a direct effect on growth or cell proliferation.

Mutations in four of the genes listed in Table 2 lead to cell-fate changes in the neurogenic ectoderm of the embryo. This region normally undergoes patterning to produce intermingled neuroblasts and epidermal precursors, in which neuroblasts make up about 25% of the total cells (Hartenstein and Campos-Ortega, 1984). Mutations of the *Notch*, *Delta* and *crumbs* genes (which encode molecules containing multiple EGF-like repeats) interfere with this process, leading to an abundance of neuroblasts at the expense of epidermal cells (the neurogenic phenotype; Lehmann *et al.* 1983). Mutations at other loci including *Enhancer of split*, which interacts with *Notch* and shows homology to *myc* (Knust

Table 2. Drosophila *genes showing homology to growth factors, growth factor receptors, and oncogenes, for which mutant phenotypes have been described*

Gene	Homology	Mutant Phenotype	Reference
Growth factors:			
Notch	36 EGF-like repeats	Epidermis → neuroblasts (Z)	Kidd *et al.* (1986)
		hairs → bristles (Z)	Wharton *et al.* (1985)
		transformation of retinal cell types (Z)	L. Held (personal communication)
			Cagan and Ready (1989)
Delta	9 EGF-like repeats	Epidermis → neuroblasts (Z)	Kopczynski *et al.* (1988)
			Knust *et al.* (1987*a*)
crumbs	5 EGF-like repeats	Epidermis → neuroblasts (Z)	Knust *et al.* (1987*a*)
wingless	*int-1* (secreted growth factor?)	Posterior → anterior	Baker (1987, 1988)
		embryonic half-segments (Z)	Rijsewijk *et al.* (1987)
		wing blade → notum (Z)	
decapentaplegic	TGF-beta	Dorsal → ventral	Spencer *et al.* (1982)
		embryonic hypoderm (Z)	Irish and Gelbart (1987)
		Focal degeneration in imaginal discs (Z)	Bryant (1988)
Growth factor receptors:			
faint little ball =torpedo	EGF receptor	Defective gastrulation (Z)	Price *et al.* (1989)
		Small wing, haltere and eye imaginal discs (Z)	Clifford and Schüpbach (1989)
		Dorsal → ventral egg shell and embryo (M)	Schejter and Shilo (1989)
sevenless	Receptor tyrosine kinase	Photoreceptor → cone cell in retina (Z)	Basler and Hafen (1988)
torso	Receptor tyrosine kinase	*Terminal → central embryonic segments (M)	Sprenger *et al.* (1989)
Oncogenes:			
abl	*abl* (cytoplasmic tyrosine kinase)	Disorganized retina, pupal lethal (Z)	Henkemeyer *et al.* (1987)
		Loss of embryonic cuticular structures (M)	Hoffmann (1989)
pole hole	*raf* (serine/threonine kinase)	Missing imaginal disks (Z)	Perrimon *et al.* (1983)
		Terminal → central embryonic segments (M)	Nishida *et al.* (1988)
			Ambrosio *et al.* (1989)
dorsal	*c-rel* (nuclear protein)	Ventral → dorsal embryonic hypoderm (M)	Steward (1987)
achaete-scute	*myc* (nuclear protein)	*Bristles → hairs (Z)	Villares and Cabrera (1987)
Enhancer of split	*myc* (nuclear protein)	*Epidermis → neuroblast (Z)	Knust *et al.* (1987*b*)
			Knust and Campos-Ortega (1989)

The phenotypes listed are those resulting from loss of function.
* Apparent gain-of-function mutations produce the opposite phenotype. M, maternal effect; Z, zygotic effect.

and Campos-Ortega, 1989), also produce a neurogenic phenotype (Knust *et al.* 1987*b*). *Notch* mutations also interfere with other cell-fate decisions during neurogenesis in the larval peripheral nervous system (Hartenstein and Campos-Ortega, 1986), bristle cell patterning (L. Held, personal communication) and development of the compound eye of the adult (Cagan and Ready, 1989). The neurogenic mutants do show an effect on cell proliferation in that the embryonic central nervous system is overgrown, but this is an indirect effect due to conversion of epidermal cells into neuroblasts, which are characterized by a greater proliferative potential than epidermis (Technau and Campos-Ortega, 1986).

The only genes on the list in which mutations might directly cause cell proliferation abnormalities are *raf*, in which lethal mutations cause absence or underdevelopment of imaginal discs and rings, brain lobes, lymph glands and gonads (Nishida *et al.* 1988) and *torpedo*, in which one of the zygotic effects is underdevelopment of the wing, haltere and eye imaginal discs (Clifford and Schubach, 1989). Even in these cases, it is not known whether the underdevelopment is due to inhibition of cell proliferation rather than degeneration.

The general conclusion from the *Drosophila* results is that these gene products *in vivo* usually have a role in pattern formation rather than in cell proliferation control. Although it is generally accepted that growth factors are often involved in controlling cell differentiation as well as proliferation, the scarcity of effects on proliferation is perhaps surprising. However, the phenotypes listed in Table 2 are those resulting from genetic loss of function. If mutations in these genes led to activation similar to the oncogene activation that is seen in the vertebrate oncogenes, they would show a dominant gain of function. Dominant or semi-dominant gain-of-function mutations have, in fact, been reported for *Enhancer of split* (Knust *et al.* 1987*b*), *torso* (Strecker *et al.* 1989) and *achaete-scute* (Garcia-Alonso and Garcia-Bellido, 1986), but in all cases the resultant phenotype is a cell-fate change in the opposite direction from that caused by genetic loss of function, rather than an effect on cell proliferation. A dominant gain-of-function mutation leading to excess cell proliferation would probably behave as a dominant lethal, and this is one class of mutations which *Drosophila* geneticists cannot easily study. Such alterations would not have been identified as spontaneous mutations and would not have been recovered in conventional mutagenesis screens. Possibly they could be identified if they were temperature-sensitive, but such mutants have not been reported. Therefore, the lack of clear effects on cell proliferation among the mutants listed in Table 2 could be, at least in part, due to our inability to identify and work with dominant lethal mutations.

Dr Bryant's research is supported by grants from the National Science Foundation and the Monsanto Company.

References

Ambrosio, L., Mahowald, A. P. and Perrimon, N. (1989). Requirement of the *Drosophila raf* homologue for *torso* function. *Nature* **342**, 288–291.

Baker, N. E. (1987). Molecular cloning of sequences from *wingless*, a segment polarity gene in *Drosophila*: the spatial distribution of a transcript in embryos. *EMBO J.* **6**, 1765–1773.

Baker, N. E. (1988). Transcription of the segment-polarity gene *wingless* in the imaginal discs of *Drosophila*, and the phenotype of a pupal-lethal *wg* mutation. *Development* **102**, 489–497.

Basler, K. and Hafen, E. (1988). Control of photoreceptor cell fate by the *sevenless* protein requires a functional tyrosine kinase domain. *Cell* **54**, 299–311.

Bryant, P. J. (1970). Cell lineage relationships in the imaginal wing disc of *Drosophila melanogaster*. *Devl Biol.* **22**, 389–411.

Bryant, P. J. (1978). Pattern formation in imaginal discs. In *The Genetics and Biology of Drosophila*, (ed. M. Ashburner and T. R. F. Wright), Vol. 2C, pp. 229–335. Academic Press: New York.

Bryant, P. J. (1978). Cell Interactions controlling pattern regulation and growth in epimorphic fields. In *The Molecular Basis of Cell–Cell Interaction* (ed. R. A. Lerner and D. Bergsma), Vol. XIV. Alan R. Liss, Inc.: New York.

Bryant, P. J. (1987). Experimental and genetic analysis of growth and cell proliferation in *Drosophila* imaginal discs. In *Genetic Regulation of Development* (ed. W. F. Loomis), pp. 339–372. Alan R. Liss Inc.: New York.

Bryant, P. J. (1988). Localized cell death caused by mutations in a *Drosophila* gene coding for a transforming growth factor-B homolog. *Devl Biol.* **128**, 386–395.

Bryant, P. J. and Fraser, S. E. (1988). Wound healing, cell communication and DNA synthesis during imaginal disc regeneration in *Drosophila*. *Devl Biol.* **127**, 197–208.

Bryant, P. J., Huettner, B., Held Jr, L. I., Ryerse, J. and Szidonya, J. (1988). Mutations at the *fat* locus interfere with cell proliferation control and epithelial morphogenesis in *Drosophila*. *Devl Biol.* **129**, 541–554.

Bryant, P. J. and Levinson, P. Y. (1985). Intrinsic growth control in the imaginal primordia of *Drosophila* and the autonomous action of a lethal mutation causing overgrowth. *Devl Biol.* **107**, 355–363.

Bryant, P. J. and Schneiderman, H. A. (1969). Cell lineage, growth, and determination in the imaginal leg discs of *Drosophila melanogaster*. *Devl Biol.* **20**, 263–290.

Bryant, P. J. and Schubiger, G. (1971). Giant and duplicated imaginal discs in a new lethal mutant of *Drosophila melanogaster*. *Devl Biol.* **24**, 233–263.

Bryant, P. J. and Simpson, P. (1984). Intrinsic and extrinsic control of growth in developing organs. *Q. Rev. Biol.* **59**, 387–415.

Cagan, R. L. and Ready, D. F. (1989). *Notch* is required for successive cell decisions in the developing *Drosophila* retina. *Genes Dev.* **3**, 1099–1112.

Clifford, R. J. and Schüpbach, T. (1989). Coordinately and differentially mutable activities of *Torpedo*, the *Drosophila melanogaster* homolog of the vertebrate EGF receptor gene. *Genetics* **123**, 771–787.

Dewes, E. (1975). Entwicklungsleistungen implantierter ganzer und halbierter männlicher Genitalimaginalschieben von *Ephestia kühniella* Z. und Entwicklungsdauer der Wirtstiere. *Wilhelm Roux Arch. EntwMech. Org.* **178**, 167–183.

Dewes, E. (1979). Über den Einfluß der Kulturdauer im larvalen Wirt auf die Entwicklungsleistungen implantierter männlicher Genitalimaginalscheiben von *Ephestia kuehniella* Z. *Wilhelm Roux Arch. EntwMech. Org.* **186**, 309–331.

Fain, M. J. and Stevens, B. (1982). Alterations in the cell cycle of *Drosophila* imaginal disc cells precede metamorphosis. *Devl Biol.* **92**, 247–258.

Fraser, S. E. and Bryant, P. J. (1985). Patterns of dye coupling in the imaginal wing disk of *Drosophila melanogaster*. *Nature* **317**, 533–536.

French, V., Bryant, P. J. and Bryant, S. (1976). Pattern regulation in epimorphic fields. *Science* **193**, 969–981.

Garcia-Alonso, L. and Garcia-Bellido, A. (1986). Genetic analysis of *Hairy wing* mutations. *Wilhelm Roux Arch. Devl Biol.* **195**, 259–264.

Garcia-Bellido, A., Ripoll, P. and Morata, G. (1976). Developmental compartmentalization in the dorsal mesothoracic disc of *Drosophila*. *Devl Biol.* **48**, 132–147.

Gateff, E. (1978). Malignant neoplasms of genetic origin in *Drosophila melanogaster*. *Science* **200**, 1448–1459.

Gateff, E. A. and Mechler, B. M. (1989). Tumor-suppressor genes of *Drosophila melanogaster*. *CRC Crit. Rev. Oncogen.* **1**, 221–245.

GATEFF, E. A. AND SCHNEIDERMAN, H. A. (1974). Developmental capacities of benign and malignant neoplasms of *Drosophila*. *Wilhelm Roux Arch. EntwMech. Org.* **176**, 23–65.

GIRTON, J. R. AND BRYANT, P. J. (1980). The use of cell lethal mutations in the study of *Drosophila* development. *Devl Biol.* **77**, 233–243.

GRAVES, B. J. AND SCHUBIGER, G. (1982). Cell cycle changes during growth and differentiation of imaginal leg discs in *Drosophila melanogaster*. *Devl Biol.* **93**, 104–110.

HARTENSTEIN, V. AND CAMPOS-ORTEGA, J. A. (1984). Early neurogenesis in wild type *Drosophila melanogaster*. *Wilhelm Roux Arch. Devl Biol.* **193**, 308–325.

HARTENSTEIN, V. AND CAMPOS-ORTEGA, J. A. (1986). The peripheral nervous system of mutants of early neurogenesis in *Drosophila melanogaster*. *Wilhelm Roux Arch. Devl Biol.* **195**, 210–221.

HAYNIE, J. AND BRYANT, P. J. (1977). The effects of X-rays on the proliferation dynamics of cells in the imaginal wing disc of *Drosophila melanogaster*. *Wilhelm Roux Arch. EntwMech. Org.* **183**, 85–100.

HENKEMEYER, M. J., GERTLER, F. B., GOODMAN, W. AND HOFFMANN, F. M. (1987). The *Drosophila* Abelson proto-oncogene homolog: identification of mutant alleles that have pleiotropic effects late in development. *Cell* **51**, 821–828.

HOFFMANN, F. M. (1989). Roles of *Drosophila* proto-oncogene and growth factor homologs during development of the fly. *Curr. Top. Microbiol. Immunol.* **147**, 1–29.

IRISH, V. F. AND GELBART, W. M. (1987). The *decapentaplegic* gene is required for dorsal/ventral patterning of the *Drosophila* embryo. *Genes Dev.* **1**, 868–879.

JURSNICH, V. A., HELD, L. I., JR, RYERSE, J., FRASER, S. E. AND BRYANT, P. J. (1990). Defective gap-junctional communication associated with imaginal disc overgrowth and degeneration caused by mutations of the *dco* gene in *Drosophila*. *Devl Biol.* **140**, 413–429.

KIDD, S., KELLEY, M. R. AND YOUNG, M. W. (1986). Sequence of the *Notch* locus of *Drosophila melanogaster*: relationship of the encoded protein to mammalian clotting and growth factors. *Molec. cell Biol.* **6**, 3094–3108.

KLÄMBT, C. AND SCHMIDT, O. (1986). Developmental expression and tissue distribution of the *lethal(2)giant larvae* protein of *Drosophila melanogaster*. *EMBO J.* **5**, 2955–2961.

KLÄMBT, C., MÜLLER, S., LÜTZELSCHWAB, R., ROSSA, R., TOTZKE, F. AND SCHMIDT, O. (1989). The *Drosophila melanogaster l(2)gl* gene encodes a protein homologous to the cadherin cell-adhesion molecule family. *Devl Biol.* **133**, 425–436.

KNUST, E., BREMER, K. A., VÄSSIN, H., ZEIMER, A., TEPASS, U. AND CAMPOS-ORTEGA, J. A. (1987*b*). The *Enhancer of split* locus and neurogenesis in *Drosophila melanogaster*. *Devl Biol.* **122**, 262–273.

KNUST, E. AND CAMPOS-ORTEGA, J. A. (1989). The molecular genetics of early neurogenesis in *Drosophila melanogaster*. *BioEssays* **11**, 95–100.

KNUST, E., DIETRICH, V., TEPASS, V., BREMER, K. A., WEIGEL, D., VÄSSIN, H. AND CAMPOS-ORTEGA, J. A. (1987*a*). EGF homologous sequences encoded in the genome of *Drosophila melanogaster*, and their relation to neurogenic genes. *EMBO J.* **6**, 761–766.

KOPCZYNSKI, C. C., ALTON, A. K., FECHTEL, K., KOOH, P. J. AND MUSKAVITCH, M. A. T. (1988). *Delta*, a *Drosophila* neurogenic gene, is transcriptionally complex and encodes a protein related to blood coagulation factors and epidermal growth factor of vertebrates. *Genes Dev.* **2**, 1723–1735.

LEHMANN, R., JIMENEZ, F., DIETRICH, V. AND CAMPOS-ORTEGA, J. A. (1983). On the phenotype and development of mutants of early neurogenesis in *Drosophila melanogaster*. *Wilhelm Roux Arch. Devl Biol.* **192**, 62–74.

LÜTZELSCHWAB, R., KLÄMBT, C., ROSSA, R. AND SCHMIDT, O. (1987). A protein product of the *Drosophila* recessive tumor gene, *lethal(2)giant larvae*, potentially has cell adhesion properties. *EMBO J.* **6**, 1791–1797.

MADHAVAN, K. AND SCHNEIDERMAN, H. A. (1969). Hormonal control of imaginal disc regeneration in *Galleria mellonella* (Lepidoptera). *Biol. Bull.* **137**, 321–331.

MARTIN, P., MARTIN, A. AND SHEARN, A. (1977). Studies of $l(3)c43^{hs1}$ a polyphasic, temperature-sensitive mutant of *Drosophila melanogaster* with a variety of imaginal disc defects. *Devl Biol.* **55**, 213–232.

MECHLER, B. M., MCGINNIS, W. AND GEHRING, W. J. (1985). Molecular cloning of *lethal(2)giant larvae*, a recessive oncogene of *Drosophila melanogaster*. *EMBO J.* **4**, 1551–1557.

MURPHY, C. (1974). Cell death and autonomous gene action in lethals affecting imaginal disc in *Drosophila melanogaster*. *Devl Biol.* **39**, 23–36.

NARDI, J. B. AND KAFATOS, F. C. (1976*a*). Polarity and gradients in lepidopteran wing epidermis. I. Changes in graft polarity, form, and cell density accompanying transpositions and reorientations. *J. Embryol. exp. Morph.* **36**, 469–487

NARDI, J. B. AND KAFATOS, F. C. (1976*b*). Polarity and gradients in lepidopteran wing epidermis. I. The differential adhesiveness model: Gradient of a non-diffusible cell surface parameter. *J. Embryol. exp. Morph.* **36**, 489–512.

NISHIDA, Y., HATA, M., AYAKI, T., RYO, H., YAMAGATA, M., SHIMIZU, K. AND NISHIZUKA, Y. (1988). Proliferation of both somatic and germ cells is affected in the *Drosophila* mutants of *raf* proto-oncogene. *EMBO J.* **7**, 775–781.

O'BROCHTA, D. AND BRYANT, P. J. (1985). A zone of non-proliferating cells at a lineage restriction boundary in *Drosophila*. *Nature* **313**, 138–141.

O'BROCHTA, D. AND BRYANT, P. J. (1987). Distribution of S-phase cells during the regeneration of *Drosophila* imaginal wing discs. *Devl Biol.* **119**, 137–142.

PERRIMON, N. (1988). The maternal effect of *lethal(1)discs large-1*: a recessive oncogene of *Drosophila melanogaster*. *Devl Biol.* **127**, 392–407.

PERRIMON, N., ENGSTROM, L. AND MAHOWALD, A. P. (1985). A pupal lethal mutation with a paternally influenced maternal effect on embryonic development in *Drosophila melanogaster*. *Devl Biol.* **110**, 480–491.

POHLEY, H. J. (1965). In *Regeneration in Animals* (ed. V. Kiortis and H. A. L. Trampusch), pp. 324–330. North Holland Publ. Co.: Amsterdam.

POODRY, C. A. AND WOODS D. F. (1990). Control of the Developmental Timer for *Drosophila* pupariation. *Wilhelm Roux' Arch. EntwMech. Org.* (in press).

PRICE, J. V., CLIFFORD, R. J. AND SCHÜBACH, T. (1989). The maternal ventralizing locus *torpedo* is allelic to *faint little ball*, an embryonic lethal, and encodes the *Drosophila* EGF Receptor homolog. *Cell* **56**, 1085–1092.

RAHN, P. (1972). Untersuchungen zur Entwicklung von Ganz- und Teilimplantaten der Flugelimaginalscheibe von *Ephestia kühniella* Z. *Wilhelm Roux Arch. EntwMech. Org.* **170**, 48–82.

RIJSEWIJK, F., SCHUERMANN, M., WASENAAR, E., PARREN, P., WEIGEL, D. AND NUSSE, R. (1987). The *Drosophila* homolog of the mouse mammary oncogene *int*-1 is identical to the segment polarity gene *wingless*. *Cell* **50**, 649–657.

RUOSLAHTI, E. AND PIERSCHBACHER, M. D. (1986). Arg-Gly-Asp: A versitile cell recognition signal. *Cell* **44**, 517–518.

RUSSELL, M. A. (1974). Pattern formation in the imaginal discs of a temperature-sensitive cell-lethal mutant of *Drosophila melanogaster*. *Devl Biol.* **40**, 24–39.

RYERSE, J. S. AND NAGEL, B. A. (1984). Gap junction distribution in the *Drosophila* wing disc mutants *vg*, *l(2)gd*, $l(3)c43^{hs1}$, and $l(2)gl^4$. *Devl Biol.* **105**, 396–403.

SAGER, R. (1989). Tumor Suppressor Genes: The Puzzle and the Promise. *Science* **246**, 1406–1412.

SCHEJTER, E. D. AND SHILO, B-Z. (1989). The *Drosophila* EGF receptor homolog (DER) gene is allelic to *faint little ball*, a locus essential for embryonic development. *Cell* **56**, 1093–1104.

SCHUBIGER, M. AND PALKA, J. (1987). Changing spatial patterns of DNA replication in the developing wing of *Drosophila*. *Devl Biol.* **123**, 145–153.

SHERIDAN, J. D. (1989). Cell communication and growth. In *Cell-to-Cell Communication* (ed. C. Walmor and D. Mello), pp. 187–222. Plenum Press: New York.

SIMPSON, P. (1976). Analysis of the compartments of the wing of *Drosophila melanogaster* mosaic for a temperature-sensitive mutation that reduces mitotic rate. *Devl Biol.* **54**, 100–115.

SIMPSON, P. (1979). Parameters of cell competition in the compartments of the wing disc of *Drosophila*. *Devl Biol.* **69**, 182–193.

SIMPSON, P., BERREUR, P. AND BERREUR-BONNENFANT, J. (1980). The initiation of pupariation in *Drosophila*: dependence on growth of the imaginal discs. *J. Embryol. exp. Morph.* **57**, 155–165.

SIMPSON, P. AND MORATA, G. (1980). The control of growth in the imaginal discs of *Drosophila*. In *Development and Neurobiology of* Drosophila (ed. O. Siddiqi, P. Babu, L. M. Hall and J. C. Hall), pp. 129–139. Plenum Press: New York.

Simpson, P. and Morata, G. (1981). Differential mitotic rates and patterns of growth in compartments in *Drosophila* wing. *Devl Biol.* **85**, 299–308.

Simpson, P. and Schneiderman, H. A. (1975). Isolation of temperature sensitive mutations blocking clone development in *Drosophila melanogaster*, and the effects of a temperature sensitive lethal mutation on pattern formation in imaginal discs. *Wilhelm Roux Arch. EntwMech. Org.* **178**, 247–275.

Spencer, F. A., Hoffmann, F. M. and Gelbart, W. M. (1982). *Decapentaplegic*: a gene complex affecting morphogenesis in *Drosophila melanogaster*. *Cell* **28**, 451–461.

Sprenger, F., Stevens, L. M. and Nüsslein-Volhard, C. (1989). The *Drosophila* gene *torso* encodes a putative receptor tyrosine kinase. *Nature* **338**, 478–483.

Steward, R. (1987). *Dorsal*, an embryonic polarity gene in *Drosophila*, is homologous to the vertebrate proto-oncogene, c-*rel*. *Science* **238**, 692–694.

Stewart, M., Murphy, C. and Fristrom, J. (1972). The recovery and preliminary characterization of X-chromosome mutants affecting imaginal discs of *Drosophila melanogaster*. *Devl Biol.* **27**, 71–83.

Strecker, T. R., Halsell, S. R., Fisher, W. W. and Lipshitz, H. D. (1989). Reciprocal effects of hyper- and hypoactivity mutations in the *Drosophila* pattern gene *Torso*. *Science* **243**, 1062–1066.

Szabad, J., Jursnich, V. A. and Bryant, P. J. (1990). Requirement for cell-proliferation control genes in *Drosophila* oogenesis. *Genetics* (In Press).

Takeichi, M. (1988). The cadherins: cell-cell adhesion molecules controlling animal morphogenesis. *Development* **102**, 639–655.

Technau, G. M. and Campos-Ortega, J. A. (1986). Lineage analysis of transplanted individual cells in embryos of *Drosophila melanogaster*. II. Commitment and proliferative capabilities of neural and epidermal cell progenitors. *Wilhelm Roux Arch. Devl Biol.* **195**, 445–454.

Villares, R. and Cabrera, C. V. (1987). The *achaete-scute* gene complex of *D. melanogaster*: conserved domains in a subset of genes required for neurogenesis and their homology to *myc*. *Cell* **50**, 415–424.

Weir, M. P. and Lo, C. W. (1982). Gap junctional communication compartments in the *Drosophila* wing disk. *Proc. natn. Acad. Sci. U.S.A.* **79**, 3232–3235.

Wharton, K. A., Johansen, K. M., Xu, T. and Artavanis-Tsakonas, S. (1985). Nucleotide sequence from the neurogenic locus *Notch* implies a gene product that shares homology with proteins containing EGF-like repeats. *Cell* **43**, 567–581.

Wolpert, L. (1971). Positional information and pattern formation. *Curr. Top. Devl Biol.* **6**, 183–224.

Woods, D. F. and Bryant, P. J. (1989). Molecular cloning of the *lethal(1)discs large-1* oncogene of *Drosophila*. *Devl Biol.* **134**, 222–235.

J. Cell Sci. Suppl. 13, 191–198 (1990)

Printed in Great Britain *© The Company of Biologists Limited 1990*

Retinoic acid and limb regeneration

JEREMY P. BROCKES

Ludwig Institute for Cancer Research, 91 Riding House Street, London, W1P 8BT

Summary

A key problem in the study of vertebrate development is to determine the molecular basis of positional value along a developmental axis. In amphibian regeneration, retinoic acid is able to respecify positional value in a graded fashion that is dependent on its concentration. In view of the fact that retinoic acid is a naturally occurring metabolite of vitamin A, this raises the possibility that it is deployed *in vivo* as an endogenous morphogen. Furthermore, the recent evidence that its effects are mediated by nuclear receptors of the steroid/thyroid hormone superfamily suggests the possibility of understanding the mechanism of its graded effects on morphogenesis. Such insights would be of crucial importance for our understanding of vertebrate patterning along an axis.

Introduction

In vertebrate development there are many contexts where a field of cells undergoes differentiation in a way that generates a pattern of tissues. This process is usually analysed in terms of the specification of the axes of the developing structure by the acquisition of some continually graded variable referred to as positional value (Wolpert, 1989). The validity of this general framework has been supported by the analysis of axial specification in the *Drosophila* embryo, using the methods of molecular genetics (Nusslein-Volhard *et al.* 1987). In vertebrates the molecular basis of positional value remains wholly unclear and is a major problem of current investigation. The progress in *Drosophila* has largely reflected the availability of mutants in axis formation, and this has confirmed the notion that axes are specified independently. Although such mutants have led to a detailed understanding of the genetic events that establish positional value, they have not so far led to any major insights into how such specification is implemented in terms of cell behaviour or cell–cell interactions. A major impetus in the study of pattern formation in vertebrates has come from the recognition that retinoids, and in particular retinoic acid, are able to re-specify positional value in a dose-dependent way. This has led to efforts to evaluate the possibility that these molecules are used as endogenous morphogens, as well as an interest in their mechanism of action.

The two key systems that have led to the present interest in the morphogenetic effects of retinoids are avian limb development (Tickle *et al.* 1982) and amphibian limb regeneration (Niazi and Saxena, 1978). The limb is often referred to as a secondary field, in contrast to the primary field of the whole embryo, and limb

Key words: morphogen, homeobox genes, axis specification, nuclear receptor, morphogenesis.

morphogenesis has long been the pre-eminent context for study of pattern formation in such a field. This paper focusses on limb regeneration as a context for evaluating retinoid action. There are some differences from the effects on avian limb development and I have reviewed these issues elsewhere (Brockes, 1989).

Amphibian limb regeneration

The vertebrates able to regenerate their limbs as adults are the urodele amphibians, and the principal laboratory species are the newt and the axolotl (Wallace, 1981). If the limb is amputated at any level on the proximo-distal axis (shoulder to fingertip), the wound surface is rapidly covered by epithelial cells migrating from the circumference of the amputation plane. This is followed by the appearance of the progenitor cells of the regenerate, referred to as blastemal cells. It is still a matter of some uncertainty as to how the blastemal cells arise, but they clearly are locally derived from the mesenchymal tissues within about half a millimetre from the amputation plane (Wallace, 1981). This process can be followed by using monoclonal antibodies that distinguish blastemal cells from normal limb mesenchyme (Kintner and Brockes, 1985). The blastemal cells are mesenchymal progenitors that give rise to connective tissue, cartilage and muscle, that is the cells that are the substrate for tissue patterning. They divide initially under control of the nervous system, and form a mound of undifferentiated cells called the blastema which undergoes differentiation and morphogenesis to reconstruct the regenerate. The blastema has considerable morphogenetic autonomy and can be transplanted to a location such as the anterior chamber of the eye, or the fin tunnel, where it may give rise to a limb (Stocum, 1984). This account is particularly concerned with the molecular basis of position-dependent properties of the blastema.

Positional value on the proximo-distal axis is manifest in several different properties of the blastema, most notably in the phenomenon known as distal transformation. The blastema arising at any position on the axis does not form structures proximal to its point of origin; thus a wrist blastema gives a hand while a shoulder blastema forms an arm. It is precisely this property that is altered by retinoid treatment. If a wrist blastema is treated with an appropriate dose of retinoic acid, its proliferation is arrested temporarily so that it appears retarded in comparison to a control. Nonetheless, the blastema subsequently grows out to give an entire arm arising from the wrist level (Maden, 1982). This is sometimes referred to as a serial duplication, and the extent of duplication is dependent on the dose of retinoic acid to which the blastema is exposed. Retinoids are the only class of compounds able to provoke these effects. Although the proximal effector of such morphogenetic effects is presently considered to be retinoic acid (RA), derived by oxidative metabloism from retinol (vitamin A), there is already evidence for other morphogenetically active metabolites (Thaller and Eichele, 1988).

The action of retinoids on limb morphogenesis must be considered against the background of other effects mediated by this family. The classical vitamin requirements include the role of retinol and retinal in the visual pigments and reproductive function, and that of RA in epithelial differentiation and maintenance (Pitt, 1971). It has long been recognised that high doses of vitamin A or RA provoke a well defined and wide ranging set of teratological abnormalities when administered to vertebrate embryos (Kochhar, 1977). In addition, a wide variety of cell types are affected by RA in culture. The responses show dose dependence, and include positive and negative regulation of division and differentiation. Some of these effects, for example the ability to induce differentiation of F9 teratocarcinoma cells to parietal endoderm (Strickland and Mahdavi, 1978), have been studied intensively as paradigms for understanding the molecular basis for retinoid action. Lastly, there are the morphogenetic effects which concern the specification of axes in regenerating and developing limbs. It has been difficult to analyse such a diverse set of effects in terms of a unitary mechanism, and an important step forward was the identification of hormone nuclear receptors of the steroid/thyroid superfamily which interact with RA, and mediate its effects on gene expression (Petkovich *et al.* 1987; Giguere *et al.* 1987).

Retinoic acid receptors

Most of the current information about the receptors has been reviewed elsewhere (Ragsdale and Brockes, 1990). The receptor proteins have a structure that is usually represented from the N-terminus and comprises six regions A–F. Two of these regions have the status of functionally independent domains: the C region which binds to response elements associated with retinoid modulated genes, and the E region which binds RA. Recent evidence suggests that the receptor binds to DNA in the absence of RA (de The *et al.* 1990), but is stimulated to activate transcription in its presence. The molecule acts, therefore, as a ligand-dependent transcription factor. This can be directly demonstrated in the standard co-transfection assay by introducing a reporter gene along with an expression constuct for the RA receptor. Trans-activation of the reporter occurs in the nmol l^{-1} range of RA concentration but requires much higher concentrations of retinol for any effect (Giguere *et al.* 1987; Petkovich *et al.* 1987).

At present, three retinoic acid receptor subtypes have been identified in mouse and human, and they are termed a, b and c (Brand *et al.* 1988; Krust *et al.* 1989; Zelent *et al.* 1989). The subtypes show high conservation of amino acid sequence in the C and E regions, but marked divergence in the other regions. The cross-species identity within a subtype is much greater than the identity between subtypes within a species. There is evidence for other members of the superfamily that the spectrum of genes activated is dictated, at least in part, by the A/B region at the N-terminus of the molecule (Tora *et al.* 1988; Bocquel *et al.* 1989). It seems likely that a, b and c activate distinct gene sets in response to RA, and that receptor heterogeneity and regulation may underly at least some of the diversity in the response to RA (Ragsdale and Brockes, 1990).

The urodele blastema is important for our understanding of the morphogenetical effects of RA, thus it has been of interest to identify the receptors that are expressed in this context. To date, three different receptors of the newt limb blastema have been identified in cDNA libraries. These include the newt homologue of human and mouse alpha (Ragsdale *et al.* 1989) and a partial cDNA clone that is the probable homologue of beta (Giguere *et al.* 1989). The third newt receptor, termed delta (Ragsdale *et al.* 1989), is closest to mouse and human gamma but diverges markedly in the A/B and F regions, a divergence that is particularly striking considering the close relationship of newt and human alpha. Newt gamma is expressed at higher levels in a distal than a proximal blastema, and at higher levels in the blastema compared to normal limb. It is also differentially spliced at the A/B border (Ragsdale and Brockes, 1990) and seems at present to be an attractive candidate for mediating at least some of the effects of RA on the blastema.

Although it is by no means certain that all relevant retinoid receptor subtypes have been identified, it is already clear that there is significant receptor heterogeneity and that a major challenge for the future is to identify the roles of the different subtypes. As mentioned earlier, RA has several different effects on regenerating limbs, ranging from morphogenetic effects that can be observed on the transverse and proximodistal axes, to teratogenic effects that have been widely studied in vertebrate embryos. There is a need for more basic comparative information on the receptors, for example on their affinities and regulation, and such information will help to guide hypotheses about their possible roles. Ultimately it will be necessary to modify expression of the receptors in developing and regenerating systems, and to observe the functional consequences of such manipulations. This might be done in the context of transgenic mice, or in the urodele by implantation of blastemal cells that have been genetically modified in cell culture (Ferretti and Brockes, 1988).

Regulation of gene expression by RA

In view of the complexity of the effects that RA exerts on cultured cells, it is likely that there are effects on gene expression that reflect a chain of events, for example the appearance of other transcriptional regulators. An important recent development is the identification of genes directly regulated by RA receptors. When human hepatoma cells are treated with RA, there is a rapid increase in RAR beta mRNA, whereas RAR alpha mRNA is not increased (de The *et al.* 1989). This effect occurs at the level of transcription and is mediated by a response element in the beta receptor promoter (de The *et al.* 1990). This element binds to RA receptors *in vitro* and confers RA responsiveness on a heterologous promoter. It is not clear if this is the only natural RA response element or if there are others. It will clearly be important to identify further examples of genes that are directly regulated.

A major goal of future research will be to identify the target genes that mediate the morphogenetic effects of RA. Homeobox genes show dramatic spatial and temporal regulation in vertebrate development, and they have attracted considerable attention as potential targets for RA. For example, the recent elegant analysis of the distribution of transcripts for the five contiguous members of the *Hox*-5 complex in the mouse limb bud has revealed a position-dependent pattern of activation that reflects the order of genes on the chromosome (Dolle *et al.* 1989). The importance of this order, already recognised in terms of patterns of expression, has been emphasised in a recent study of F9 cells showing that genes at the 3′ end are more responsive to RA treatment than those at the 5′ end (Papalopulu *et al.* 1990). Thus far it has not been possible to identify a homeobox gene that is regulated by RA in limb morphogenesis, although there is no shortage of examples of regulation by RA of homeobox genes in culture (Mavilio *et al.* 1988). In limb regeneration, the one example of a homeobox gene that varies on the proximo-distal axis does not show a change in expression after treatment of a distal blastema with RA (Savard *et al.* 1988). It will nevertheless be important to try to find such a gene, because it would be an excellent candidate for one that is on the 'pathway' of positional specification.

Retinoids as endogenous morphogens

Much of the current interest in RA as an endogenous morphogen has come from studies on the chick limb bud that employ HPLC analysis of solvent-extracted retinoids. RA is apparently present in the chick limb bud, and is present at somewhat higher levels in the posterior segment than in an anterior (Thaller and Eichele, 1987). These observations are consistent with the hypothesis that RA is released from the polarising region, but much remains to be done to establish if this is true. In the case of the chick limb, the grafting experiments that led to the idea of a polarising region were instrumental in providing a framework for thought about retinoid action. In the developing and regenerating urodele, such grafts do not provoke duplications in the antero-posterior axis, and the models of morphogenesis stress the importance of local cell interactions rather than the action of signalling regions.

The presence of the RARs as well as the cytoplasmic binding protein for RA (CRABP) (McCormick *et al.* 1988) in the blastema is certainly suggestive that RA plays some role in limb regeneration, but what role? Are there mechanisms that ensure a varying response along one or more axes? It seems highly unlikely that there is a gradient of retinoid in the dimensions of the adult limb, but it is possible that blastemas arising at different locations have different concentrations. It is also possible that the responsiveness of the cells is regulated, for example at the level of the receptors or the binding proteins for RA (Maden *et al.* 1988). Some evidence is needed about the concentration of RA that actually impinges on the blastemal cells, and whether this level varies with position. There are several possibilities for sources of RA. It has been suggested that the epidermis or its associated pigment cells (Baranowitz, 1989) may release retinoid precursors or

RA that are instrumental in provoking the formation of blastemal cells. Alternatively, RA may be synthesised from precursor retinol by the blastemal mesenchyme cells, as it is by the mesenchymal cells of the chick polarising region (Thaller and Eichele, 1988).

An important contribution towards solving these problems may come from the recent advances in the basic molecular biology of retinoid action. It is possible to construct plasmids in which expression of a reporter enzyme comes under control of a RA response element (de The *et al.* 1990), and hence reflects the external concentration of RA. If such plasmids can be introduced into cells *via* the transgenic troute, or by implantation of genetically modified cultured cells, then the level of the reporter may give an indication of the local concentration of RA. It is not clear if this approach will give quantitative estimates for the concentration *in vivo*, partly because of uncertainties about other cell properties that can modify the response, but the questions at present are sufficiently broad that it may give useful information.

Conclusions

It is clear that the identification of RARs has given significant impetus to studies on the morphogenetic effects of RA. In future it will clearly be important to determine functional roles for the different subtypes, and to identify the target genes that are regulated when axial specification is altered. This latter point is a particularly challenging issue, but it is one of the rare possibilities for tackling the problem of positional value in vertebrates. I have indicated how many gaps exist in our present understanding of the role of retinoids in development, and that much remains to be determined before the case for an endogenous, morphogenetic role can be considered convincing.

References

Baranowitz, S. A. (1989). Regeneration, neural crest derivatives and retinoids: a new synthesis. *J. theor. Biol.* **140**, 231–242.

Bocquel, M. T., Kumar, V., Stricker, C., Chambon, P. and Gronemeyer, H. (1989). The contribution of the N- and C-terminal regions of steroid receptors to activation of transcription is both receptor- and cell-specific. *Nucl. Acids Res.* **17**, 2581–2595.

Brand, N. J., Petkovich, M., Krust, A., Chambon, P., de The, H., Marchio, A., Tiollais, P. and Dejean, A. (1988). Identification of a second human retinoic acid receptor. *Nature* **332**, 850–853.

Brockes, J. P. (1989). Retinoids, homeobox genes, and limb morphogenesis. *Neuron* **2**, 1285–1294.

de The, H., del Mar Vivanco-Ruiz, M., Tiollais, P., Stunnenberg, H. and Dejean, A. (1990). Identification of a retinoic acid-responsive element in the retinoic acid receptor B gene. *Nature* **343**, 177–180.

de The, H., Marchio, A., Tiollais, P. and Dejean, A. (1989). Differential expression and ligand regulation of the retinoic acid receptor alpha and beta genes. *EMBO J.* **8**, 429–433.

Dolle, P., Izposua-Belmonte, J-C., Falkenstein, H., Renucci, A. and Duboule, D. (1989). Coordinate expression of the murine Hox-5 complex homeobox-containing genes during limb pattern formation. *Nature* **342**, 767–772.

Ferretti, P. and Brockes, J. P. (1988). Culture of newt cells from different tissues and their expression of a regeneration-associated antigen. *J. exp. Zool.* **247**, 77–91.

Giguere, V., Ong, E. S., Evans, R. M. and Tabin, C. J. (1989). Spatial and temporal expression of the retinoic acid receptor in the regenerating amphibian limb. *Nature* **337**, 566–569.

Giguere, V., Ong, E. S., Segui, P. and Evans, R. M. (1987). Identification of a receptor for the morphogen retinoic acid. *Nature* **330**, 624–629.

Kintner, C. R. and Brockes, J. P. (1985). Monoclonal antibodies to the cells of a regenerating limb. *J. Embryol. exp. Morph.* **89**, 37–55.

Kochhar, D. M. (1977). Cellular basis of congenital limb deformity induced in mice by vitamin A. In *Morphogenesis and Malformation of the Limb*, (ed. D. Bergsma and W. Lenz), pp. 111–154. New York: Alan R. Liss.

Krust, A., Kastner, P., Petkovich, M., Zelent, A. and Chambon, P. (1989). A third human retinoic acid receptor, hRAR-gamma. *Proc. natn. Acad. Sci. U.S.A.* **86**, 5310–5314.

Maden, M. (1982). Vitamin A and pattern formation in the regenerating limb. *Nature* **295**, 672–675.

Maden, M., Ong, D. E., Summerbell, D. and Chytil, F. (1988). Spatial distribution of cellular protein binding to retinoic acid in the chick limb bud. *Nature* **335**, 733–735.

Mavilio, F., Simeone, A., Boncinelli, E. and Andrews, P. (1988). Activation of four homeobox gene clusters in human embryonal carcinoma cells induced to differentiate by retinoic acid. *Differentiation* **37**, 73–79.

McCormick, A. M., Shubeita, H. E. and Stocum, D. L. (1988). Cellular retinoic acid binding protein: detection and quantitation in regenerating axolotl limbs. *J. exp. Zool.* **245**, 270–276.

Niazi, I. A. and Saxena, S. (1978). Abnormal hind limb regeneration in tadpoles of the toad, *Bufo andersoni*, exposed to excess vitamin A. *Folia biol., Praha* **26**, 3–11.

Nusslein-Volhard, C., Frohnhofer, H. G. and Lehmann, R. (1987). Determination of anteroposterior polarity in the *Drosophila* embryo. *Science* **238**, 1675–1681.

Papalopulu, N., Hunt, P., Wilkinson, D., Graham, A. and Krumlauf, R. (1990). Hox-2 homeobox genes and retinoic acid: potential roles in patterning the vertebrate nervous system. In *Advances in Neural Regeneration Research*, (ed. F. Seil). Alan R. Liss Inc: New York.

Petkovich, M., Brand, N. J., Krust, A. and Chambon, P. (1987). A human retinoic acid receptor which belongs to the family of nuclear receptors. *Nature* **330**, 444–450.

Pitt, G. A. J. (1971). Vitamin A. In *Carotenoids* (ed. O. Isler), pp. 717–742. Birkhauser Verlag: Basel und Stuttgart.

Ragsdale, C. W. and Brockes, J. P. (1990). Retinoic acid receptors and vertebrate limb morphogenesis. In *Structure and function of hormone nuclear receptors*, (ed. M. G. Parker). London: Academic Press. (In press.)

Ragsdale, C. W., Petkovich, M., Gates, P. B., Chambon, P. and Brockes, J. P. (1989). Identification of a novel retinoic acid receptor in regenerative tissues of the newt. *Nature* **341**, 654–657.

Savard, P., Gates, P. B. and Brockes, J. P. (1988). Position dependent expression of a homeobox gene transcript in relation to amphibian limb regeneration. *EMBO J.* **7**, 4275–4282.

Stocum, D. L. (1984). The urodele limb regeneration blastema. Determination and organization of the morphogenetic field. *Differentiation* **27**, 13–28.

Strickland, S. and Mahdavi, V. (1978). The induction of differentiation in teratocarcinoma stem cells by retinoic acid. *Cell* **15**, 393–403.

Thaller, C. and Eichele, G. (1987). Identification and spatial distribution of retinoids in the developing chick limb bud. *Nature* **327**, 625–628.

Thaller, C. and Eichele, G. (1988). Characterisation of retinoid metabolism in the developing chick limb bud. *Development* **103**, 473–483.

Tickle, C., Alberts, B., Wolpert, L. and Lee, J. (1982). Local application of retinoic acid to the limb bud mimics the action of the polarising region. *Nature* **296**, 564–565.

Tora, L., Gronemeyer, H., Turcotte, B., Gaub, M.-P. and Chambon, P. (1988). The N-terminal region of the chicken progesterone receptor specifies target gene activation. *Nature* **333**, 185–188.

Wallace, H. (1981). *Vertebrate Limb Regeneration*. J. Wiley: Chichester.
Wolpert, L. (1989). Positional information revisited. *Development (Suppl.)* **107**, 3–12.
Zelent, A., Krust, A., Petkovich, M., Kastner, P. and Chambon, P. (1989). Cloning of murine retinoic acid receptor alpha and beta cDNAs and of a novel third receptor gamma predominantly expressed in skin. *Nature* **339**, 714–717.

J. Cell Sci. Suppl. 13, 199–208 (1990)

Printed in Great Britain © The Company of Biologists Limited 1990

Signals in limb development: STOP, GO, STAY and POSITION

LEWIS WOLPERT

Department of Anatomy and Developmental Biology, University College and Middlesex School of Medicine, London, W1P 6DB, UK

Summary

Cell-to-cell interactions in early limb development are considered within the framework of the extracellular signals STOP, GO, STAY and POSITION, a classification which emphasises that the signals are elective rather than instructive, and that complexity arises from cells' response. Patterning in the limb is analysed in terms of signals that specify positional values along the anteroposterior axes, and retinoic acid is thought to be a positional morphogen. There is however, evidence for patterning which does not depend on a positional signal. In the early bud the mesenchyme gives POSITION signals to the apical ridge, which in turn provides a STAY signal to the mesenchyme in the progress zone. Non-ridge ectoderm produces a STOP signal with respect to cartilage differentiation. The pattern of cartilage differentiation is specified well before cartilage condensation. Growth factors affect both cartilage and muscle differentiation in culture. Pigment patterns result from feather germs providing STOP or GO signals to the melanoblasts which enter all feather germs. The pathways for the cell-to-cell signals are not known but may involve gap junctions.

Introduction

Limb development is both important in its own right and as a model system for studying pattern formation in development. Early development of the limb involves outgrowth of a bud from the flank, the loose mesenchyme of the bud giving rise to all the connective tissues of the limb except muscle. Muscle has a different origin, presumptive muscle cells migrating from the somites into the limb at a very early stage. Outgrowth of the limb bud is dependent on a distal thickening in the ectoderm, the apical ectodermal ridge. Beneath this ridge lies the progress zone containing mutliplying cells; as the cells leave the progress zone they begin to differentiate into cartilage and other connective tissues. The forelimb elements are laid down in a proximo-distal sequence – first the humerus, then the radius and ulna, and finally the wrist followed by the digits.

A number of cell-to-cell interactions have been identified in limb development and it may be instructive to consider them in terms of the effect they have on the cells. The extracellular signals controlling development may be classified in terms of their effect on cell development: STOP, GO, STAY and POSITION. This classification emphasizes that the factors operating on cells have at any one time a limited capacity for changing cell behaviour, and that the complexity of development lies within the cell and not in intercellular signalling. The complex tranducing mechanisms and second messenger systems offer further support for

Key words: limb, development, signals.

this view. For a factor to be classified in terms of one of these activities, its concentration must change either spatially or temporally and be correlated with the observed effect. Thus a distinction is drawn between the role of those factors which are necessary for development but which do not change and those which do. For example, if TGF-β is present throughout limb development at a constant concentration then it does not qualify as a control factor even if *in vitro* it stimulates chondrogenesis in a dose-dependent manner.

The four kinds of signal are defined with respect to their effect on cell development. However, it may not always be easy to assign unequivocally a signal to one of them. STOP is effectively an inhibitory signal and may be linked to a GO signal, inhibiting a cell proceeding along one pathway and directing it along another. GO signals may direct a cell along a maturation pathway or along one of several pathways and so it might best be represented by GO (x) where x represents the different pathways. At any one time the number of pathways, x, open to the cell is quite small, often being only two. STAY signals maintain the cells in their particular programme and the removal of such a signal results in the cell changing programmes. POSITION signals assign cells a positional value and this helps to distinguish them from GO signals.

Patterning

We have proposed a model to account for the spatial pattern of cell differentiation during limb development that is based on the concept of interpretation of positional information (Wolpert, 1981). Positional value, a cell parameter that is related to the cell's position, is thought to be specified in the progress zone. For the antero-posterior axis, position may be specified by a signal from the polarizing region at the posterior margin of the limb. When an additional polarizing region is grafted to the anterior margin of the limb bud a mirror image limb develops; the pattern of digits, for example, is now 432 234 compared to the 234 of a normal limb. From such experiments it has been concluded that the polarizing region provides a POSITION signal to the cells in the progress zone (Table 1,1). The signal from the polarizing region can be mimicked by a localized source of retinoic acid (Tickle *et al.* 1985) and retinoic acid is present in the bud with a higher concentration in the posterior half (Thaller and Eichele, 1987). Retinoic acid can alter positional values and is thus a very strong candidate for being the positional signal, but further evidence is required to establish that the cells respond to retinoic acid and not some other signal evoked by local application of retinoic acid. Even so the presence of both nuclear retinoic acid receptor (Dolle *et al.* 1989) and cytoplasmic binding protein (Maden *et al.* 1989) in the progress zone is very encouraging.

There are quite good reasons to think that these signals, other than those from the polarizing region, are involved in patterning along the anteroposterior axis. The only telling experiments involve disaggregation of the mesodermal cells, followed by reaggregation and placement of the aggregate in an ectodermal

Table 1. *Cell-to-cell interactions in early limb development*

Source	Target	Signal
1. Polarizing region	Mesenchyme in progress zone	POSITION
2. Mesenchyme progress zone	Mesenchyme progress zone	STOP/GO
3. Mesenchyme	Apical ridge	STAY/POSITION
4. Apical ectoderm	Mesenchyme in progress zone	STAY
5. Cartilage	Cartilage	GO
6. Ectoderm	Mesenchyme	STOP
7. Muscle connective tissue	Presumptive muscle cells	GO
8. Feather follicle	Melanoblast	GO/STOP
9. Mesenchyme	Ectoderm	POSITION
10. Feather germ	Feather germ	STOP

jacket. Without a discrete polarizing region, moderately good digits form (e.g., see Patou, 1973), and this suggests that another mechanism, such as one generating an isomorphic prepattern, is involved. Such mechanisms have, in fact, been put forward (see Wolpert and Stein, 1984).

The two types of mechanism make rather specific predictions with respect to the relationship between the number of elements along the anteroposterior axis and the width of the limb bud. We have tested these predictions with respect to the development of the humerus and found that some of the results do not conform with the predictions from a positional signal and suggest, rather, that a prepattern mechanism may be involved (Wolpert and Hornbruch, 1987). The nature of the prepattern mechanism is unknown but would involve interactions between mesenchyme cells. If this were based on a reaction diffusion mechanism (Murray, 1989) then it is not unreasonable to think of both inhibitory STOP signals and positive GO signals (Table 1,2).

The early bud

One of the earliest interactions in limb development is that involving the specification and maintenance of the apical ridge. The ridge is maintained by a signal from the underlying mesenchyme which provides a STAY signal. The polarizing region also plays a role in this process since both the length of the ridge and its position on the bud is specified by the polarizing region *via* the POSITION signal on the mesenchyme (Tickle *et al.* 1989). The polarizing region signal also specifies the position of that mesenchyme which specifies the position of the ridge (Table 1,3).

The effect of the apical ridge on the underlying mesenchyme can also be thought of as a STAY signal because it maintains the underlying mesenchyme at a high rate of proliferation and also prevents it differentiating into cartilage (see below) (Table 1,4). Indeed the high rate of proliferation in the early bud has its origin in the reduction of proliferation in the adjacent flank rather than an increase in the bud region. In this sense, thinking of the ridge providing a STAY signal rather than promoting proliferation may be the most appropriate way of describing the factor's action.

When an additional polarizing region is grafted to the anterior margin the limb bud always widens and this requires additional cell proliferation. Aono and Ide (1988) have some rather indirect evidence that this may involve an FGF-like factor produced by the polarizing region which brings about the additional proliferation.

Cartilage differentiation

In the previous section emphasis was placed on POSITION with STOP/GO signals patterning the cartilage. There are however, additional signals involved because cartilage will not develop in the progress zone nor adjacent to the ectoderm. Moreover it seems that all the cells in the progress zone will differentiate into cartilage when all signals are removed. Thus one aspect of cartilage patterning is a STOP signal.

If progress zone cells are cultured in micromass culture, all the cells differentiate as cartilage giving rise to a sheet of cartilage (Cottrill *et al.* 1987*a*). The inference is that within the progress zone the cells are subject to a STAY signal from the apical ridge that keeps them proliferating and prevents them from differentiating (Table 1,4). When the cells begin to form cartilage it seems that they produce the GO factor which promotes cartilage differentiation. Evidence for such a GO factor comes from studies in micromass which suggest that mesenchymal cell aggregates must reach a threshold size before chondrogenesis can proceed (Cottrill *et al.* 1987*b*) (Table 1,5).

The ectoderm may provide an important component in patterning cartilage by providing a STOP signal (Table 1,6) which prevents cartilage differentiation. When ectoderm is placed on the surface of a micromass culture, cartilage is inhibited in the region both below and adjacent to the ectoderm (Solursh *et al.* 1981). We (Gregg *et al.* 1989) have found that the ectoderm completely inhibited the accumulation of cartilage specific type II collagen transcripts in the mesenchyme cells, and that the inhibition was not mediated by a change of cell shape as suggested by Zanetti and Solursh (1986). The ectoderm might also produce a GO signal directing the cell to a non-cartilage pathway.

Several authors have suggested that the cell contacts present at condensation and the associated cell-to-cell interactions are an important feature of cartilage differentiation. In one particular model of cartilage element patterning it is suggested that the condensation process itself and the associated mechanical changes are responsible for the patterning of the cartilaginous elements (Oster *et al.* 1985). A different interpretation of condensation is offered here which has important implications for interpreting the action of factors which promote or block cartilage differentiation: condensation is seen as an early manifestation of cartilage differentiation and not a cause. If cells from wing buds at stage 22–23 are placed as single cells in a collagen gel, a significant number of the cells will form cartilage (Solursh *et al.* 1982). This shows that single cells can form cartilage without the necessity for any of the cell-to-cell interactions associated with condensation. Furthermore, when double anterior limbs are constructed before condensation occurs, two humeri or a thickened humerus develop in a substantial number of cases (Wolpert and Hornbruch, 1990). This shows that the humerus rudiment is specified well before condensation occurs, a result which is not consistent with the physico-mechanical model.

Taken together these results argue for an early specification of both cartilage and the spatial pattern of its differentiation. In these terms condensation should be seen as an early manifestation of cartilage differentiation, reflecting, most likely, a change in the nature of the extracellular matrix, causing the cells to come close together. More specifically, hyaluronic acid may be removed from the matrix and replaced by cartilage matrix (Singley and Solursh, 1981).

TGF-β is a potent stimulation of cartilage differentiation in micromass cultures (Kulyk *et al.* 1989) and we have recently shown that bFGF also stimulates chondrogenesis. Together their effects are additive. Of great interest is the suggestion by Lyons *et al.* (1989) that the TGF-β super family is required to control the progression of cell types through their differentiation pathway – a GO signal. Thus bone morphogenetic protein may stimulate early cartilage and TGF-β-2 would promote chondrogenesis; and at later stages similar molecules may be involved in maturation and hypertrophy.

Direct investigation of the ability of retinoic acid to control the differentiation of cartilage cells has been carried out in culture. It is well known that retinoic acid can inhibit chondrogenesis in culture and this is related to its teratogenic effect which is to inhibit cartilage differentiation. There is no reason to believe that this is in any way related to its effect on patterning. In serum-free culture, Paulsen *et al.* (1988) found that retinoic acid stimulates chondrogenesis in a dose-related manner at about $5\,\mathrm{mg\,ml^{-1}}$. The overall morphology of the cultures was unchanged. Ide and Aono (1988) found that retinoic acid at similar low concentrations promoted both proliferation and chondrogenesis of distal cells, while cells from proximal regions were unresponsive. It is far from clear whether these effects are related to the ability of retinoic acid to alter positional values in the limb.

There is a widespread view that molecules of the extracellular matrix play a controlling role in morphogenesis. The changes in these molecules in limb development have been catalogued by Solursh (1990), but there is little evidence for their role as signals. Newman (1988) has proposed a role for fibronectin, which is stimulated by TGF-β, in promoting aggregation which he considers important for cartilage differentiation. The role of such factors should be treated with caution since prevention of proteoglycan accumulation may inhibit chondrogenesis. A number of matrix components could act at this late stage of differentiation.

Muscle differentiation

In the limb, muscle cells have a lineage quite distinct from that of the connective tissue cells, since they are all derived from a small population of presumptive muscle cells that migrate into the limb at an early stage of development (Chevallier, 1979). The patterning of muscle results from the migrating muscle cells accumulating in specific regions. This patterning of muscle cells probably involves specific adhesive interactions with muscle connective tissue which may also provide a GO signal for muscle differentiation (Table 1,7).

Seed and Hauschka (1988) looked at the muscle colony-forming (MCF) potential of stage 23 chick myoblasts and concluded that two subclasses of MCF cells exist. The FGF-independent subclass showed a delay in differentiation in the presence of FGF, whilst the FGF-dependent subclass required FGF for terminal differentiation. In micromass culture we have found that TGF-β on its own has little effect on muscle cell differentiation while bFGF has a striking inhibitory action which has a clear dosage dependence. However, when TGF-β1 is added together with bFGF the inhibitory effect of bFGF is blocked (Schofield and Wolpert, 1990). Both the number and the morphology of differentiated muscle cells is restored to that seen in cultures grown in the absence of bFGF. Thus both TGF-β and bFGF could act together in the developing chick limb to control cell differentiation. bFGF is probably dispersed throughout the limb although not necessarily with a uniform distribution, whilst TGF-β is produced in the precartilaginous cellular condensation. We therefore hypothesise that TGF-β is able to affect both cells of the precartilaginous and the adjacent premuscle masses. If TGF-β and bFGF behave *in vivo* as they do in our culture system, their interaction could provide a possible mechanism for both enhancing cartilage differentiation and ending the inhibitory action of bFGF on myoblasts, thus allowing myogenic differentiation to precede.

Pigment patterns in the limb

The pigment patterns of birds are as beautiful as they are varied. How many factors might be required to specify such patterns? Only some of the patterns are based on the differentiation of pigment cells that migrate into the feather germs from the neural crest. Other patterns are structural and are due to properties of the feathers themselves.

We have investigated the cellular basis of local pigment patterns in the wings of the quail embryo. A local pigment pattern refers to a set of feathers which have a different pigmentation to those around them. There are also variations of pigment patterns within a feather and that is a separate issue. The now almost classical view of pigment patterning was put forward by Rawles (1948). On the basis of grafting together of melanoblasts between closely related strains she concluded that the pattern corresponded to that of the melanoblast donor. On this basis I proposed that the melanoblast cells must read the local positional values and interpret them according to its genetic constitution. However, almost all of

Rawles' experiments focussed on patterns within feathers rather than local patterns and our recent experiments lead to a quite different conclusion.

The dorsal feathers of the quail wing are all pigmented whereas there is a local pattern on the ventral surface, in which several rows of feathers are unpigmented. The unpigmented feathers lack pigmentation because of an inhibitory STOP signal exerted within the feather germ. The evidence for this is that if unpigmented germs are cultured and the melanoblasts allowed to migrate away from their ectoderm they differentiate and become pigmented (Richardson *et al.* 1990). Thus the pattern is determined by the character of the individual feather germs rather than the pigment cells. Consistent with such a mechanism, guinea fowl melanoblasts give a quail pattern on the ventral surface of the wing even though the guinea fowl wing is fully pigmented both ventrally and dorsally.

Local factors acting as STOP signals thus control pigment pattern formation. In addition there are GO signals that determine when pigment differentiation occurs, and thus control temporal expression (Table 1,8). The GO factors are clearly seen in an operation in which a quail wing bud is grafted onto a guinea fowl host before the neural crest cells have entered the bud. The bud now becomes populated with guinea fowl neural crest cells and pigment cell differentiation is visible in the quail wing many hours before differentiation occurs in the normal contralateral guinea fowl wing (Richardson *et al.* unpublished data).

There remains the key question of what specifies which feathers will inhibit melanocyte differentiation and which will not: that is, the specification of the basic pattern. The simplest view is that it reflects the basic pattern of positional information within the mesenchyme specifying POSITION in the ectoderm (Table 1, 9). In a mirror image duplication brought about by a graft of a polarizing region, the pigment pattern follows that of the underlying structures (Richardson *et al.* 1990).

While positional information may specify the character of the feather germs it is only indirectly involved in the patterning of the feather germs themselves. The feather germs are arranged in a highly ordered hexagonal pattern, and this spacing pattern is not directly linked to positional information, but probably involves an inhibitory STOP signal (Table 1, 10) during spacing of the feather germs. Positional information specifies which regions will have feather germs but the actual spacing of the feather germs within such regions is due to another mechanism. For example, if growth of the limb bud is reduced at the time when the feather spacing is being specified then fewer feathers are specified but their spacing remains the same as normal (McLachlan, 1986).

Discussion

The presence of growth factors in the early amphibian embryo and the evidence for their role in specifying the pattern in the mesoderm (Smith, J. C. *et al.* 1989; Slack, 1990) has given rise to the hope that they may be playing a similar role in limb morphogenesis, and be the crucial signals in cell-to-cell interactions. Thus far only FGF has been shown to be present at the crucial early stages of pattern formation (Seed *et al.* 1988) while production of TGF-β is only detected in the limb itself at later stages (Heine *et al.* 1987). IGF-II mRNA has also been detected in the limb bud (Engstrom *et al.* 1987), and Ralphs *et al.* (1990) using antibody staining, have shown IGF to be present in the tip of the limb while it is absent in regions undergoing chondrogenesis. At later stages the peptides are present in chondrocytes. In more general terms we remain ignorant about the identity of the signals involved in early limb morphogenesis as listed in Table 1. Retinoic acid is a most promising candidate for specifying position.

Finally, we must consider the pathway the signals may take. Do they pass directly *via* cell-to-cell contact, *via* the matrix, through the extracellular space, or through gap junctions? The flow of blood through the limb suggests that extracellular diffusion is unlikely (Georgiello and Caplan, 1983). A lipid soluble molecule like retinoic acid could diffuse within cell membranes and pass from cell to cell at points of cell contact. Allen *et al.* (1990) have investigated the role of gap junctions in positional signalling by the polarizing region and have concluded that gap junctions may play a role in enabling the polarizing region cells to communicate with anterior mesenchyme cells.

References

ALLEN, F., TICKLE, C. AND WARNER, A. (1990). The role of gap junctions in patterning of the chick limb bud. *Development* **108**, 623–634.

AONO, H. AND IDE, H. (1988). A gradient of responsiveness to the growth-promoting activity of ZPA (Zone of polarizing activity) in the chick limb bud. *Devl Biol.* **128**, 136–141.

CHEVALLIER, A. (1979). Role of somitic mesoderm in the development of the thorax in bird embryos. II. Origin of the thoracic and appendicular musculature. *J. Embryol. exp. Morph.* **49**, 73–88.

COTTRILL, C. P., ARCHER, C. W., HORNBRUCH, A. AND WOLPERT, L. (1987*a*). The differentiation of normal and muscle-free distal limb bud mesenchyme in micromass culture. *Devl Biol.* **119**, 143–151.

COTTRILL, C. P., ARCHER, C. W., HORNBRUCH, A. AND WOLPERT, L. (1987*b*). Cell sorting and chondrogenic aggregate formation in micromass culture. *Devl Biol.* **122**, 503–515.

DOLLE, P., RUBERTE, E., KASTNER, P., PETKOVICH, M., STONER, C. M., GUDAS, L. J. AND CHAMBON, P. (1989). Differential expression of retinoic acid receptor α, β, τ and CRAMP genes in the developing limb of the mouse. *Nature* **342**, 702–704.

ENGSTROM, W., BELL, K. M. AND SCHOFIELD, P. N. (1987). Expression of the insulin-like growth factor II gene in the developing chick limb. *Cell biol. int. Rep.* **11**, 415–421.

GEORGIELLO, D. M. AND CAPLAN, A. I. (1983). The fluid flow dynamics in the developing chick wing. In *Limb Development and Regeneration* (ed. J. F. Fallon and A. I. Caplan), pp. 143–154. A. R. Liss Inc: New York.

GREGG, B. C., ROWE, A., BRICKELL, P., DEVLIN, C. J. AND WOLPERT, L. (1989). Ectodermal inhibition of cartilage differentiation in micromass culture of chick limb bud mesenchyme in relation to gene expression and cell shape. *Development* **105**, 769–778.

HEINE, U. L., MURRAY, E. F., FLANDERS, K. C., ELLINGWORTH, L. R., LAM, H., THOMPSON, N. L., ROBERTS, A. B. AND SPORN, M. B. (1987). Role of transforming growth factor-β in the development of the mouse embryo. *J. Cell Biol.* **105**, 2861–2876.

IDE, H. AND AONO, H. (1988). Retinoic acid promotes proliferation and chondrogenesis in the distal mesoderm cells of chick limb bud. *Devl Biol.* **130**, 767–773.

KULYK, W. M., RODGERS, B. J., GREER, K. AND KOSHER, R. (1989). Promotion of embryonic limb cartilage differentiation by transforming growth factor β. *Devl Biol.* **135**, 424–430.

LEE, J. AND TICKLE, C. (1985). Retinoic acid and pattern formation in the developing chick wing: SEM and quantitative studies of the early effects on the apical ridge and bud outgrowth. *J. Embryol. exp. Morph.* **90**, 851–860.

LYONS, K. M., PELTON, R. W. AND HOGAN, B. L. M. (1989). Patterns of expression of murine VGr-1 and BMP-Za RNA suggest that transforming growth factor-β-like genes coordinately regulate aspects of embryonic development. *Genes Dev.* **3**, 1657–1668.

MCLACHLAN, J. C. (1986). The effect of 6-aminonicotinamide on limb development. *J. Embryol. exp. Morph.* **55**, 307–318.

MADEN, M., ONG, O. F., SUMMERBELL, D. AND CHYTIL, F. (1989). The role of retinoid-binding proteins in the generation of pattern in the developing limb and the nervous system. *Development* **107** (Supplement), 109–119.

MURRAY, J. D. (1989). *Mathematical Biology*, pp. 372–430. Springer: Berlin.

NEWMAN, S. A. (1988). Lineage and pattern in the developing vertebrate limb. *Trends Genet.* **4**, 329–332.

OSTER, G. F., MURRAY, J. D. AND MAINI, P. (1985). A model for chondrogenic condensations in the developing limb: the role of extracellular matrix and cell tractions. *J. Embryol. exp. Morph.* **89**, 92–112.

PATOU, M. P. (1973). Analyse de la morphogenese due pied des Oiseaux a laide de melange cellulaires interspecifique. *J. Embryol. exp. Morph.* **89**, 93–112.

PAULSEN, D. F., LARGILLE, R. M., DRESS, V. AND SOLURSH, M. (1988). Selective stimulation of *in vitro* limb-bud chondrogenesis by retinoic acid. *Differentiation* **39**, 123–130.

RALPHS, J. R., WYLIE, L. AND HILL, D. J. (1990). Distribution of insulin-like growth factor peptides in the developing chick embryo. *Development* **109**, 51–58.

RAWLES, M. E. (1948). Origin of melanophores and their role in the development of color patterns in vertebrates. *Physiol. rev.* **28**, 383–408.

RICHARDSON, M. K., HORNBRUCH, A. AND WOLPERT, L. (1990). Mechanisms of pigment pattern formation in the quail embryo. *Development* **109**, 81–89.

SCHOFIELD, J. N. AND WOLPERT, L. (1990). Effect of TGF-β1, TGF-β2, and bFGF on chick cartilage and muscle differentiation. *Expl Cell Res.* (in press).

SEARLS, R. L. AND JANNERS, M. Y. (1971). The initiation of limb bud outgrowth in the embryonic chick. *Devl Biol.* **24**, 198–213.

SEED, J., COLWIN, B. B. AND HAUSCHKA, S. D. (1988). Fibroblast growth factor levels in the whole embryo limb bud during chick development. *Devl Biol.* **128**, 50–57.

SEED, J. AND HAUSCHKA, S. D. (1988). Clonal analysis of vertebrate myogenesis. VIII. Fibroblast growth factor (FGF)-dependent and FGF-independent muscle colony types during chick wing development. *Devl Biol.* **128**, 40–49.

SINGLEY, C. T. AND SOLURSH, M. (1981). The spatial distribution of hyaluronic acid and mesenchymal condensation in the embryonic wing. *Devl Biol.* **84**, 102–120.

SLACK, J. M. W. (1990). Growth factors as inducing agents in early *Xenopus* development. *J. Cell Sci.* (Suppl.) **13**, 119–130.

SMITH, J. C., COOKE, J., GREEN, J. B. A., HAWES, G. AND SYMES, K. (1989). Inducing factors and the control of mesodermal pattern in *Xenopus laevis*. *Development* **107**, 149–160.

SMITH, S. M., PANG, K., SUNDIN, O., WEDDEN, S. E., THALLER, C. AND EICHELE, G. (1989). Molecular approaches to vertebrate limb morphogenesis. *Development* **107** (Suppl.) 121–132.

SOLURSH, M. (1990). The role of extracellular matrix molecules in limb development. *Seminars in Devl Biol.* **1**, 45–54.

SOLURSH, M., LINSENMAYER, T. F. AND JENSON, K. L. (1982). Chondrogenesis from simple limb mesenchyme cells. *Devl Biol.* **94**, 259–264.

SOLURSH, M., SINGLEY, C. T. AND REITER, R. A. (1981). The influence of epithelia on cartilage and loose connective tissue formation by limb mesenchyme cultures. *Devl Biol.* **86**, 471–482.

THALLER, C. AND EICHELE, G. (1985). Identification and spatial distribution of retinoids in the developing chick limb bud. *Nature* **327**, 625–628.

TICKLE, C., CRAWLEY, A. AND FARRAR, J. (1989). Retinoic acid application to chick wing buds leads to a dose-dependent reorganization of the apical ectodermal ridge that is mediated by the mesenchyme. *Development* **106**, 691–705.

TICKLE, C., LEE, J. AND EICHELE, G. (1985). A quantitative analysis of the effect of all-*trans*-retinoic acid on the pattern of limb development. *Devl Biol.* **109**, 82–95.

WOLPERT, L. (1981). Positional information and pattern formation. *Phil. Trans. R. Soc.* B **295**, 441–450.

WOLPERT, L. AND HORNBRUCH, A. (1987). Positional signalling and the development of the humerus in the chick limb bud. *Development* **100**, 333–338.

WOLPERT, L. AND HORNBRUCH, A. (1990). Double anterior chick limb buds and models for cartilage rudiment specification. *Development* **109**, 961–966.

WOLPERT, L. AND STEIN, W. D. (1984). Positional information and pattern formation. In *Pattern Formation* (ed. G. M. Malacinski and S. V. Bryant), pp. 2–21. MacMillan: New York.

ZANETTI, N. C. AND SOLURSH, M. (1986) Epithelial effects on limb morphogenesis involve extracellular matrix and cell shape. *Devl Biol.* **113**, 110–118.

INDEX

Activins, mesoderm inducing factors, 123–124
Alzheimer's disease, chromaffin cell grafting, 108
Amphibians, limb regeneration, 191–196
Antiestrogens, chemoprevention and therapy of epithelial malignancy, 145–147
Apical ridge, effects on mesenchyme, 201
Astrocytes, effect of bFGF, 103

Betaglycan, soluble/insoluble form, 136
Binding proteoglycans, 131–137
Blastema,
  newt limb blastema, 194
  retinoic acid receptors, 196
Blastula, mesoderm induction, 119–129
Bombesin,
  action on Swiss 3T3 cells, 45–46
  mitogenic stimulation, 1–3
Bombesin receptor,
  coupling to G protein, 48–52
  modulation by guanine nucleotides, 47–48
  stabilization by Mg^{2+}, 47
Bone morphogenetic proteins,
  1–7, schematic representation, 151
  identification and molecular cloning of growth factors, 150–151
  in vivo activity of BMP-2, 154–155
  influencing bone development, 149–155
  interrelationships and homologies with other TGF-β superfamily members, 151–153
  mRNAs, 153–154
Bradykinin, mitogenic stimulation, 1–3
Brain tumors, basic fibroblast growth factor, 108
Brain-derived neurotrophic factor (BDNF), sequence homology to nerve growth factor, 97

Ca^{2+},
  Ca^{2+}-dependent Ca^{2+} release, 1–3
  internal stores, 1–3
  oscillation mechanisms, 1–3
Cadherins, and *lgl* encoded surface molecule homology, 179
Carcinogenesis, and transforming growth factors, 139–147
Cartilage, differentiation, 202–203
Cell activation, cytosolic ions and messengers, imaging and manipulation, 1–3
Cell development, developmental signals, 199–206
Cell–cell interactions, limb development, 201
Central nervous system lesions, 106–107
Cerebrospinal fluid, basic fibroblast growth factor, 108
Chick limb, retinoids, 195–196
Chimeras, mapping and dissociating properties of PDGF A and B, 31–40
Chromaffin cells, basic fibroblast growth factor, 104–105
COS-1 cells, *int*-2 transcription, 91–94
Cytokines *see* Interleukins; Tumour necrosis factor
Cytosolic free Ca^{2+}, 1–3
Cytosolic ions and messengers, imaging and manipulation during cell activation, 1–3

Developmental signals, limb development, 199–206
DIA/LIF *see* Differentiation inhibiting activity/leukemia inhibitory factor
Diacylglycerol, cytosolic messenger, 1–3
Differentiation inhibiting activity/leukemia inhibitory factor, 75–83
  alternative forms, 80
  G protein, 78
  PIN-type, 78
  receptor, 77–82
  tyrosine kinase receptor, 78
Differentiation inhibiting activity/leukemia inhibitory factor receptor,
  control of stem cell function, 77–78
  molecular and biological characteristics, 78–80
  multiple functional forms, 80–81
  stem cell derived signals, 82
Drosophila,
  genes,
    boss gene, 159
    ectopic expression of *sevenless* and *rough*, 157–167
    lethal *fat* mutations, pharate adults, 176
    and mutant phenotypes, 184
    showing homology to growth factor receptors, 184
    ultrabithorax gene, expression, 162
  genetic control of cell proliferation in imaginal discs, 169–185
  imaginal disc neoplastic overgrowth mutations, 173–180
    developmental delay, 181–182
    maternal effects, 182–183
    tissue autonomy, 180–181
  imaginal disc neoplastic overgrowth phenotype, *dlg* and *lgl*, 177
  positional signalling in the developing eye, 157–167
  wing development,
    growth patterns and cell regeneration, 172
    spatial patterns, 171

EGF *see* Epidermal growth factor
Embryonal carcinoma cells, lines, 90
Embryonic stem cells, growth and differentiation factors, 75–83
Eosinophilic progenitor cells, promotion by, interleukin 5, 67
Epidermal growth factor,
  structure-function relationships, 5–9

variants, mitogenic activity, 5–9
Epidermal growth factor receptor, *Drosophila*, homology to various genes, 184
Epidermal growth factor-like growth factor, *Drosophila*, homology to various genes, 184
Epithelial malignancy, chemoprevention and therapy, 145–147
Eye development,
insect compound eye, ectopic expression of *sevenless* and *rough*, 157–167
positional signalling in *Drosophila*, 157–167

F9 embryonal carcinoma cells, 90
FDCP-mix cells, myeloid growth factor-stimulated development, 69–70
Feather patterns, quail, 204–205
b-Fibroblast growth factor,
action on neuroectodermal cells *in vivo*, 105–107
basic and acidic, 87–88, 97–98
cells and tissues expressing bFGF, 99
chromaffin cells, 104–105
functions outside the nervous system, 99–100
mesoderm induction, 119–129
molecular properties, 98–99, 100–101
muscle differentiation, 204
in neurons and glial cells, 101–104
pattern formation, 206
possible clinical applications, 107–108
receptors, 105–106
structure, 99
Fibroblast growth factor,
human homologues, amino acid sequences, 88
int-2 gene, 87–95
K-FGF, 82, 95
Fibroblast growth factor receptor, in *Xenopus*, 127
K-Fibroblast growth factors, 82

G proteins,
DIA/LIF, 78
pertussis toxin-sensitive, 45–46
role in bombesin signal transduction pathways, 45
Gap junctions, polarizing region communications, 206
Gastrin releasing peptide, mitogenic activity, Swiss 3T3 cells, 44
Gestodene, binding site in malignant mammary tissue, 145–146
Glial cells, effect of bFGF, 103
Granulocyte colony-stimulating factor, 66
Granulocyte-macrophage colony-stimulating factor, 65–66
Growth factor receptors, *Drosophila*, showing homology to genes, 184
Growth factors,
Drosophila, showing homology to genes, 184
influencing bone development, 149–155
mesoderm patterns, 205–206
Growth inhibition, and transforming growth factors, 139–147
Guanine nucleotide regulatory proteins *see* G proteins
Guanosine-5′-*O*-(3-thiotriphosphate) (GTPγS), 1–3

Haemopoiesis,
basic mechanisms, 60
cell growth factors, biochemistry and biology, 57–70
CFU-S assay, 59
HPP-CFC assay, 59
model systems, 67–68
Heat shock protein 70, *Drosophila*, *rough* minigene, 161
Heparin, binding to fibroblast growth factors, 123
HL-60 cell line, 68–70
Homeobox genes,
potential targets for retinoic acid regulation, 195
rough, lethal indiscriminate expression, 161

IGFI *see* Insulin-like growth factor
Imaginal discs,
genetic control of cell proliferation in *Drosophila*, 169–185
model of cell proliferation, 170
neoplastic overgrowth mutations, 173–180
X-irradiation, 171
Indicator dyes, fluorescence, 1–3
Inositol 1,4,5-trisphosphate,
cytosolic messenger, 1–3
mitogenic activity of PMT, 52
Insect compound eye, positional signalling in *Drosophila*, 157–167
Insulin receptor, 5–9
Insulin-like growth factor, structure, simulations of molecular dynamics, 5–9
Insulin-like growth factor receptor, types 1 and 2, 5–9
Int-2 gene,
embryonic expression, 89–90
fibroblast growth factor family, 87–95
structure, 90
timing of transcription, 90
Int-2 protein,
alternative initiation codons, 93–95
functions, 95
synthesis and processing, 90–93
Interleukin 1,
and 5-fluorouracil, 62
activity, 62
Interleukin 3, activity on myeloid progenitor cells, 61–62
Interleukin 4, differentiation of multipotent cells, 62–63
Interleukin 5, promotion of eosinophilic progenitor cells, 67
Interleukin 6,
and action of DIA/LIF, 78
effects on multipotent cells, 63–64
Interleukins, human interleukin for DA cells (HILDA), bioregulatory activities, 79

Kallikreins, serine proteases, 19–29
Keratinocyte growth factor, amino acid sequence, 88–89

Leukemia inhibitory factor, 75–83
Lgl encoded surface molecule, homology to cadherins, 169–185
Limb development,
pigment patterns, 204–205
progress zone, 200
signals, 199–206
Limb regeneration, and retinoic acid, 191–196
Lymphotoxin, structure, compared with tumour necrosis factor, 14–16

M1 cell line, 69–70
Macrophage colony-stimulating factor, 67
Macrophage inflammatory protein 1 *see* Stem cell inhibitor
Mammary tumours,
binding site for gestodene, 145–146
int-2 gene as proto-oncogene, 89
Mesoderm inducing factors,
bFGF, 128–129
biological activity, 124–125
candidates, 122–124
FGF family, 97–108
Xenopus, 119–129
Mesoderm patterns, growth factors, 205–206
Messengers, cytosolic, 1–3
Mink lung epithelial cells *see* Mv1Lu epithelial cells
Mitogenic signalling, G protein and bombesin receptor, 43–53
Mitogenic stimulation, 1–3
Mitogens, *Pasteurella multocida* toxin, 52
Morphogens, retinoic acid, 191–196
Mouse mammary tumour virus (MMTV), *int*-2 gene, 89
Muscle,
activin A induced, 124
differentiation, 204
Mv1Lu epithelial cells, lack of response to TGFs, 134–136
Myc, *Drosophila*, homology to various genes, 184
Myeloid growth factor-stimulated development, cell lines, 68–70
Myeloid haemopoietic cell growth factors,
biochemistry and biology, 57–70
list, 58

Na^+,
influx, 1–3
intracellular $[Na^+]$, 1–3
Nerve growth factor,
murine,
7S NGF precursor, 19–29
crystal forms, 26
electron micrograph, 27
UV spectra, 28
X-ray crystallography, 19–29
sequence homology to brain-derived neurotrophic factor, 97–98
various species, sequence alignment, 20
β-Nerve growth factor,
amino acid sequence, 23
multiple isomorphous replacement, 21–24
protein–protein interaction with α and γ-NGF, 28
purification and crystallisation, 21
γ-Nerve growth factor, as an esteropeptidase, 19
7S-Nerve growth factor precursor, crystal forms, 26
Neuroectodermal tumors, basic fibroblast growth factor, 108
Neuron death, prevention by neurotrophic factors, 97, 106
Neurotrophic factors,
defined, 97
prevention of neuron death, 97, 106
X-ray crystallography, 19–29
Newt limb blastema, 194
NIH/3T3 transfectants, 89
activation of α and β-PDGF receptors, 37
Nuclear localization, *int*-2 transcription in COS-1 cells, 91–94
Nuclear Magnetic Resonance, high resolution, 5–9
Nuclear Overhauser effects, on TGF and hEGF, 5–9

Ommatidia, insect compound eye, 157–167
Oncogenes,
Drosophila, homology to various genes, 184
v-*sis*, encoding PDGF B chain, 31
Oscillation mechanisms, Ca^{2+}, 1–3
Osteoblasts, osteoclasts, regulation *see* Bone morphogenetic proteins
Overhauser effects, on TGF and hEGF, 5–9

Parietal endoderm, early site of *int*-2 gene expression, 90
Parkinson's disease, chromaffin cell grafting, 108
Pasteurella multocida toxin, as mitogen for Swiss 3T3 cells, 52
Pattern formation, limb regeneration in vertebrates, 191–196
PCC4 embryonal carcinoma cells, 90
PDGF *see* Platelet-derived growth factor
Peripheral nervous system lesions, 107
Pertussis toxin-sensitive G proteins, 45–46
Phosphoinositide, breakdown of, 1–3
Pigment patterns, limb development, 204–205
Platelet-derived growth factor A and B,
construction of chimeric molecules, 32–34
site-directed mutagenesis, 39
transforming and secretory properties, 31–40
Platelet-derived growth factor chimeric proteins, 31–40
Platelet-derived growth factor receptor,
compartmentalization, 36
tyrosine phosphorylation, 37
PMT *see Pasteurella multocida* toxin
Polarizing region,
cell–cell interactions in early limb development, 201
gap junction communications, 206
grafted, 202

Polyphosphoinositides, enzymic breakdown
coupled to bombesin receptor, 45–46
Positional information, limb development,
199–200
Progestin, synthetic, 145–146
Progestin receptor, binding site for gestodene,
145–146
Progress zone, limb development, 200
Protein kinase C,
activation, specific marker, 46
activation by bFGF, 99
Protein phosphorylation, permeabilized cells,
46–47
Proteoglycans, binding, and transforming
growth factor receptors, 131–137

Quail, feather patterns, 204–205

Receptor binding, 5–9, 11–17
tumour necrosis factor, 17
Retinal precursor cells, insect compound eye
ommatidia, 163–165
Retinoic acid,
cytoplasmic binding protein, blastema, 195
and cytoplasmic binding protein, progress
zone, 200
and limb regeneration, 191–196
receptors, 193–194, 196
progress zone, limb development, 200
regulation of gene expression, 194–195
Retinoids,
morphogenetic effects, 191–196
and steroids, regulation of transforming
growth factor-beta, 139–147
Rhabdomeres, insect compound eye ommatidia,
163
Rough homeobox genes, *Drosophila*, 161

Serine proteases, kallikrein family, 19–29
Serine/threocine kinase, *Drosophila*, homology
to various genes, 184
Sevenless gene, *Drosophila*, heat shock-
inducible, 162
Signal transduction, bFGF, 99
Signals, limb development, 199–206
Site-specific mutagenesis, 5–9
Solution structure, 5–9
Stem cells,
embryonic, growth and differentiation factors,
75–83
inhibitor, action on CSU-S, 64–65
Steroid hormone superfamily, regulation of
transforming growth factor-beta, 139–147
Swiss 3T3 cells,
model for small cell lung cancer, 44
protein phosphorylation, 46

TGF *see* Transforming growth factor
Thrombin, mitogenic stimulation, 1–3
Transformation mechanism, PDGF, 32–34
Transforming growth factor receptors,
and binding proteoglycans, 131–137
type I and type II binding proteins, 131–137
Transforming growth factors,
binding to betaglycan, 135–136
regulation of morphogenesis and
differentiation, 131–137
superfamily, 132
Transforming growth factor-alpha,
structure–function relationships, 5–9
variants, mitogenic activity, 5–9
Transforming growth factor-beta,
activation by proteases, 131–137
biologically latent form, 143–145
blocking effect of bFGF, 204
distribution in various cell types, 132–133
effects on haemopoietic cells, 64
in limb development, 206
regulation by steroid hormone superfamily,
139–147
subtypes, post-transcriptional regulation,
141–142
Transforming growth factor-beta-1,
active/latent forms, plasma half-lives, 144
mRNA, potential stem–loop structures, 142
Transforming growth factor-beta-2, mesoderm
inducing factors, 123–126
Transmembrane signalling, bombesin receptor,
43–53
Tumorigenesis,
mammary cancer, 87–95, 145
and transforming growth factors, 139–147
Tumour necrosis factor,
receptor binding, 17
structure and function, 11–17
Tyrosine kinase receptor, *Drosophila*, homology
to various genes, 184
Tyrosine phosphorylation, platelet-derived
growth factor receptor, 37

Ultrabithorax gene, *Drosophila*, expression, 162
Urodele blastema, 194

V-*sis* oncogene, encoding human PDGF B, 31
Vasopressin, mitogenic stimulation, 1–3

WEHI-3B cell line, 69–70

X-ray crystallography,
growth factors, 5–9
murine nerve growth factor, 19–29
tumour necrosis factor, 11–17
Xenopus,
growth factors as inducing agents, 119–129
mesoderm induction, 119–129

The Company of Biologists Limited is a non-profit-making organization whose directors are active professional biologists. The Company, which was founded in 1925, is the owner and publisher of this and *The Journal of Experimental Biology* and *Development* (formerly *Journal of Embryology and Experimental Morphology*).

Journal of Cell Science is devoted to the study of cell organization. Papers will be published dealing with the structure and function of plant and animal cells and their extracellular products, and with such topics as cell growth and division, cell movements and interactions, and cell genetics. Accounts of advances in the relevant techniques will also be published. Contributions concerned with morphogenesis at the cellular and sub-cellular level will be acceptable, as will studies of micro-organisms and viruses, in so far as they are relevant to an understanding of cell organization. Theoretical articles and occasional review articles will be published.

Subscriptions

Journal of Cell Science will be published 13 times in 1990 in the form of 3 volumes, each of 4 parts, and 1 Supplement. The subscription price of volumes 95, 96, 97 plus Supplement 13 is £455 (USA and Canada, US $820; Japan, £500) post free. Supplements may be purchased individually – prices on application to the Portland Press Ltd. Orders for 1990 may be sent to any bookseller or subscription agent, or to Portland Press Ltd, PO Box 32, Commerce Way, Colchester CO2 8HP, UK. Copies of the journal for subscribers in the USA and Canada are sent by air to New Jersey for delivery with the minimum delay.

Back numbers of the *Journal of Cell Science* may be ordered through Portland Press Ltd. This journal is the successor to the *Quarterly Journal of Microscopical Science*, back numbers of which are obtainable from Messrs William Dawson & Sons, Cannon House, Park Farm Road, Folkestone, Kent CT19 5EE, UK.

Copyright and reproduction

1. Authors may make copies of their own papers in this journal without seeking permission from The Company of Biologists Limited, *provided that such copies are for free distribution only: they must not be sold.*
2. Authors may re-use their own illustrations in other publications appearing under their own name, without seeking permission.
3. Specific permission will *not* be required for photocopying copyright material in the following circumstances.
   (a) For private study, provided the copying is done by the person requiring its use, or by an employee of the institution to which he/she belongs, without charge beyond the actual cost of copying.
   (b) For the production of multiple copies of such material, to be used for bona fide educational purposes, provided this is done by a member of the staff of the university, school or other comparable institution, for distribution without profit to student members of that institution, provided the copy is made from the original journal.
4. For all other matters relating to the reproduction of copyright material written application must be made to Dr R. J. Skaer, Company Secretary, The Company of Biologists Limited, Department of Zoology, Downing Street, Cambridge CB2 3EJ, UK.

© The Company of Biologists Limited 1990

Journal of Cell Science Supplements

No.	Title	£	U.S.$	Year
No. 1	**Higher Order Structure in the Nucleus** Edited by P. R. Cook and R. A. Laskey ISBN: 0 9508709 4 3 234 pp. Proceedings of 1st BSCB–Company of Biologists (COB) Symposium	£12.00	U.S.$23.00	1984
No. 2	**The Cell Surface in Plant Growth and Development** Edited by K. Roberts, A. W. B. Johnston, C. W. Lloyd, P. Shaw and H. W. Woolhouse ISBN: 0 9508709 7 8 350 pp. The 6th John Innes Symposium	£15.00	U.S.$30.00	1985
No. 3	**Growth Factors: Structure and Function** Edited by C. R. Hopkins and R. C. Hughes ISBN: 0 9508709 9 4 242 pp. BSCB–COB Symposium	£15.00	U.S.$30.00	1985
No. 4	**Prospects in Cell Biology** Edited by A. V. Grimstone, Henry Harris and R. T. Johnson ISBN: 0 948601 01 9 458 pp. An essay volume to mark the journal's 20th anniversary	**SOLD OUT** ~~£15.00~~	~~U.S.$30.00~~	1986
No. 5	**The Cytoskeleton: Cell Function and Organization** Edited by C. W. Lloyd, J. S. Hyams and R. M. Warn ISBN: 0 948601 04 3 360 pp. BSCB–COB Symposium	**SOLD OUT** ~~£15.00~~	~~U.S.$30.00~~	1986
No. 6	**The Molecular Biology of DNA Repair** Edited by A. R. S. Collins, R. T. Johnson and J. M. Boyle ISBN: 0 948601 06 X 353 pp.	£40.00	U.S.$70.00	1987
No. 7	**Virus Replication and Genome Interactions** Edited by J. W. Davies *et al.* ISBN: 0 948601 10 8 350 pp. The 7th John Innes Symposium	£40.00	U.S.$70.00	1987
No. 8	**Cell Behaviour: Shape, Adhesion and Motility** Edited by J. Heaysman, A. Middleton and F. Watts ISBN: 0 948601 12 4 449 pp. BSCB–COB Symposium	£35.00	U.S.$60.00	1987
No. 9	**Macrophage Plasma Membrane Receptors: Structure and Function** Edited by S. Gordon ISBN: 0 948601 13 2 200 pp.	£29.00	U.S.$50.00	1988
No. 10	**Stem Cells** Edited by Brian I. Lord and T. Michael Dexter ISBN: 0 948601 16 7 280 pp.	£35.00	U.S.$65.00	1988
No. 11	**Protein Targeting** Edited by K. F. Chater, N. J. Brewin, R. Casey, K. Roberts, T. M. A. Wilson and R. B. Flavell ISBN: 0 948601 21 3 270 pp. The 8th John Innes Symposium	£35.00	U.S.$65.00	1989
No. 12	**The Cell Cycle** Edited by Robert Brooks, Peter Fantes, Tim Hunt and Denys Wheatley ISBN: 0 948601 23 X 300 pp. BSCB–Journal of Cell Science Symposium	£40.00	U.S.$60.00	1989
No. 13	**Growth Factors in Cell and Developmental Biology** Edited by M. D. Waterfield ISBN: 0 948601 27 2 210 pp. BSCB–Journal of Cell Science Symposium	£30.00	U.S.$55.00	1990

This series of supplementary casebound volumes deals with topics of outstanding interest to cell and molecular biologists

These are provided free to subscribers to *Journal of Cell Science*. They may be purchased separately from:

Portland Press Ltd, PO Box 32, Commerce Way, Colchester CO2 8HP, UK